AF334181

Springer Tracts in Modern Physics

Volume 269

Springer Tracts in Modern Physics provides comprehensive and critical reviews of topics of current interest in physics. The following fields are emphasized:

- Elementary Particle Physics
- Condensed Matter Physics
- Light Matter Interaction
- Atomic and Molecular Physics
- Complex Systems
- Fundamental Astrophysics

Suitable reviews of other fields can also be accepted. The Editors encourage prospective authors to correspond with them in advance of submitting a manuscript. For reviews of topics belonging to the above mentioned fields, they should address the responsible Editor as listed in "Contact the Editors".

More information about this series at http://www.springer.com/series/426

Youichi Murakami · Sumio Ishihara
Editors

Resonant X-Ray Scattering in Correlated Systems

 Springer

Editors
Youichi Murakami
Institute of Materials Structure Science,
    Condensed Matter Research Center
High Energy Accelerator Research
    Organization
Tsukuba, Ibaraki
Japan

Sumio Ishihara
Department of Physics
Tohoku University
Sendai, Miyagi
Japan

ISSN 0081-3869   ISSN 1615-0430   (electronic)
Springer Tracts in Modern Physics
ISBN 978-3-662-53225-6   ISBN 978-3-662-53227-0   (eBook)
DOI 10.1007/978-3-662-53227-0

Library of Congress Control Number: 2016948268

Printed on acid-free paper

This Springer imprint is published by Springer Nature
The registered company is Springer-Verlag GmbH Germany
The registered company address is: Heidelberger Platz 3, 14197 Berlin, Germany

Preface

Electronic and magnetic properties in condensed matters are dominated by the electronic structure as well as the crystal structure. Synchrotron X-ray diffraction is one of the most important methods to determine the crystal structure including the electron density distribution, while spectroscopic experiments using the synchrotron radiation have become indispensable to study the electronic structure. The resonant X-ray scattering (RXS), which has the combined characters of diffraction and spectroscopy, is a powerful tool to investigate the physical properties of solids and thin films. The RXS can provide valuable information on the valence electrons, especially a variety of ordered states for the electron degrees of freedom. In the d-electron systems, we call the electronic ordered states as charge, spin, and orbital orders, while in the f-electron systems, they are referred to as multipole orders. The RXS can give us the element- and site-specific information when the beam energy is tuned to the corresponding absorption energy. We can also elucidate the details of the ordered states using the polarization property of synchrotron X-rays. When we change sample environments such as temperature, pressure, magnetic field or electric field, we sometimes observe a phase transition accompanied by a drastic change of electronic states, which are directly detected by the RXS. Some examples of these in-situ observations are shown in this book. We also have learned over the last two decades that the resonant inelastic X-ray scattering (RIXS) can be effective to detect electronic excitations such as charge, spin and orbital waves in strongly correlated electron systems. In this way, the RXS and RIXS have played increasingly significant roles in many scientific fields as well as condensed matter physics.

This book is written with the intention of presenting systematic descriptions of the RXS and RIXS techniques and the applications. In Chapter 1, the RXS and RIXS are reviewed from theoretical point of view. The Chapters 2 and 3 contain RXS studies of $3d$ and $4f$ electron systems, respectively. Magnetism study using RXS is described in Chapter 4. The RXS using soft X-ray is introduced in Chapter 5. Chapter 6 is dedicated to the description of the RIXS technique for the study of electronic excitations. This book has been written by six authors who are experts at RXS and RIXS studies in condensed matter physics. The readers will find the original points of view of these authors about physics in strongly correlated electron systems. We hope this book will be a useful textbook for researchers and students to study a variety of science using RXS and RIXS.

It is a pleasure to thank Prof. Atsushi Fujimori for his invaluable help in editing this book. The cooperation with Dr. Ute Heuser and Dr. Thorsten Schneider of Springer has been extremely helpful.

Tsukuba, Japan   Youichi Murakami
Sendai, Japan   Sumio Ishihara

Contents

Resonant X-ray Scattering and Orbital Degree of Freedom in Correlated Electron Systems

Sumio Ishihara

1 Introduction

The concept of the electronic orbital emerges in a wide variety of phenomena in condensed matter physics. The $2l + 1$-fold degeneracy of the electronic orbitals in an isolated atom is lifted in crystalline solids, in which the spherical symmetry in a potential is reduced. In some classes of solids with high crystalline symmetries, the orbital (quasi) degeneracy partially remains. In transition-metal compounds, electrons fill partially the outer-most degenerated orbitals. There is a degree of freedom what orbital is occupied by electrons. This orbital degree of freedom comes out in several magnetic, optical, and structural properties in crystalline materials. The two famous examples are known; the Cu $3d_{x^2-y^2}$ orbital characters of doped holes are crucially important for the high transition-temperature superconductivity in cuprates, and the multiple Fe $3d$ orbital bands govern the electronic structure at vicinity of the Fermi level in the iron based superconductors.

Instead of the great progresses of the orbital concept in correlated electron system, the direct experimental observations have been left behind for a long time. A reduction of the lattice symmetry is usually supposed as a consequence of the spontaneous lifting of the degenerated electronic orbitals. The direct observations of the anisotropic electronic wave functions and charge clouds have been recognized as tough experimental works for a prolonged period of time. In 1998, the resonant X-ray scattering (RXS) technique was applied to the layered manganites, and observed directly the long-range orbital order [1, 2]. After this first observation, this technique has been developed widely and deeply, and has successfully observed a number of the orbital related phenomena.

S. Ishihara (✉)
Department of Physics, Tohoku University, Sendai 980-8578, Japan
e-mail: ishihara@cmpt.phys.tohoku.ac.jp

© Springer-Verlag Berlin Heidelberg 2017
Y. Murakami and S. Ishihara (eds.), *Resonant X-Ray Scattering in Correlated Systems*, Springer Tracts in Modern Physics 269,
DOI 10.1007/978-3-662-53227-0_1

In this chapter, we review RXS and resonant inelastic X-ray scatteiring (RIXS) in the correlated electron systems with the orbital degrees of freedom from the fundamental view points. In Sect. 2, we review RXS and the orbital orders. The interacting orbital models and several characteristics in the orbital orders are introduced in Sect. 2.1. A historical review of observations of the orbital order and RXS are presented in Sect. 2.2. The scattering cross section of RXS and RIXS are formulated in Sect. 2.3. As a case study of RXS applied to the correlated electron systems with the orbital degree of freedom, theoretical and experimental studies in impurity effects on the orbital order are introduced in Sect. 2.4. In Sect. 3, we review RIXS and the orbital excitations. The theoretical studies in the collective orbital and vibronic excitations are introduced in Sect. 3.1. The non-resonant inelastic X-ray scattering (NIXS) is another experimental tool to detect the orbital excitation. We briefly review NIXS from the orbital excitations in Sect. 3.2. We introduce RIXS from the orbital excitations in Sect. 3.3, where the polarization effects of the X-rays are focused on. Section 4 is devoted to summary and future perspective of RXS and RIXS. Also see other reviews for RXS and RIXS in correlated electron systems with the orbital degree of freedom [3–6].

2   RXS and Orbital Order

2.1   *Interacting Orbital Model and Orbital Order*

The RXS and RIXS techniques are widely recognized as the powerful experimental tools to detect the long-ranged orbital orders and the orbital excitations, respectively, which originate from the interactions between the electronic orbital degree of freedom in crystalline solids. In this subsection, before explaining the RXS and RIXS for the orbital degree of freedom, we introduce the interacting orbital models and several characteristics in the orbital orders.

The orbital degree of freedom is identified as the electronic multipole degrees of freedom in solids. In a crystal with the inversion symmetry, the electronic charge distributions are classified by the 2^l-poles degree of freedom with a positive even integer l. This is mathematically represented by the spherical harmonic function $Y_l^m(\theta, \phi)$ where $-l \leq m \leq l$. In particular, the charge quadrupole ($l = 2$) plays a leading role in the d and f electron systems with strong electron correlation. In this chapter, we mainly focus on the $3d$ orbital degrees of freedom in the transition-metal compounds. The $3d$ orbital wave functions in a transition-metal ion are approximately represented by the wave functions in a hydrogen-like isolated atom under the centrosymmetric potential due to the effective positive charge. The wave functions are given as $\psi_{nlm}(\mathbf{r}) = R_{nl}(\mathbf{r})Y_l^m(\theta, \phi)$ with $n = 3$, $l = 2$ and $-2 \leq m \leq 2$. In a cubic crystalline field, the five-fold degeneracy is lifted to the two- and three-fold degenerated orbitals termed the $d_{e_g} = \{d_{3z^2-y^2}, d_{x^2-y^2}\}$ and $d_{t_{2g}} = \{d_{yz}, d_{zx}, d_{xy}\}$ orbitals, respectively. These are explicitly given as

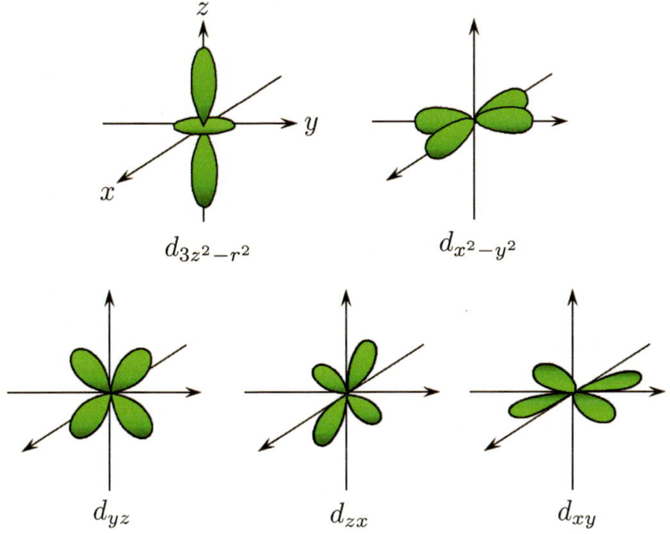

Fig. 1 Schematic spatial distributions of the $3d$ orbitals

$$\psi_{3z^2-r^2} = \psi_{3d0}, \tag{1}$$

$$\psi_{x^2-y^2} = \frac{1}{\sqrt{2}}(\psi_{3d2} + \psi_{3d-2}), \tag{2}$$

$$\psi_{yz} = \frac{i}{\sqrt{2}}(\psi_{3d1} + \psi_{3d-1}), \tag{3}$$

$$\psi_{zx} = -\frac{1}{\sqrt{2}}(\psi_{3d1} - \psi_{3d-1}), \tag{4}$$

$$\psi_{xy} = -\frac{i}{\sqrt{2}}(\psi_{3d2} - \psi_{3d-2}), \tag{5}$$

and the schematic spacial distributions are shown in Fig. 1.

Several kinds of the interactions between the local orbital degrees of freedom are known. (i) The most popular interactions are the kinetic-exchange type interactions in Mott insulators, which are described theoretically by the so-called Kugel–Khomskii type Hamiltonian [7–11]. (ii) The long-range Coulomb interaction brings about the interactions between the anisotropic charge distributions [12]. (iii) The cooperative Jahn–Teller (JT) effect in insulating solids induces the interaction between the electronic orbitals mediated by the lattice distortions or phonons [13–15]. (iv) The Ruderman–Kittel–Kasuya–Yoshida type interaction mediated by the itinerant electrons exists in rare-earth compounds [16]. The orbital interactions are expressed in a unified fashion as

$$\mathscr{H} = \sum_{ij} \sum_{\alpha\beta} J_{ij}^{\alpha\beta} \, O_i^{\alpha} O_j^{\beta}, \tag{6}$$

where $J_{ij}^{\alpha\beta}$ is the interaction between the orbitals at sites i and j. The operator O_i^{α} represents the orbital degree of freedom at site i where α describes the component of the operator. This is often written using the projection operator as $O_i^{\alpha} = |i\gamma_1\rangle\langle i\gamma_2|$ where $|i\gamma\rangle$ denotes a state in which an electron occupies the orbital γ at site i.

One of the simple and well known orbital models is the interacting e_g orbital model in a Mott insulator where the kinetic-exchange type interactions are considered between the e_g orbitals. The doubly-degenerate e_g orbitals are introduced at each site in a cubic lattice. When an average electron number per site is set to be one, the system is identified as a Mott insulator under the strong on-site Coulomb interactions. This situation is applied to $KCuF_3$ in a perovskite-type crystal structure. The nominal valence of a Cu ion is $2+$ where the electron configuration of the $3d$ orbitals is $(t_{2g})^6(e_g)^3$ owing to the crystalline-field effect in the cubic lattice. The electronic structure is well described by the Hubbard model with the orbital degeneracy. This is explicitly given as

$$\mathscr{H} = - \sum_{\langle ij\rangle \gamma\gamma's} t_{ij}^{\gamma\gamma'} \left(c_{i\gamma s}^{\dagger} c_{j\gamma's} + \text{H.c.} \right) + U \sum_{i\gamma} n_{i\gamma\uparrow} n_{i\gamma\downarrow} + U' \sum_{i\gamma>\gamma'} n_{i\gamma} n_{i\gamma'}$$
$$+ J \sum_{iss'\gamma>\gamma'} c_{i\gamma s}^{\dagger} c_{i\gamma's'}^{\dagger} c_{i\gamma s'} c_{i\gamma's} + I \sum_{i\gamma\neq\gamma'} c_{i\gamma\uparrow}^{\dagger} c_{i\gamma\downarrow}^{\dagger} c_{i\gamma'\downarrow} c_{i\gamma'\uparrow}, \tag{7}$$

where $c_{i\gamma s}$ is the electron annihilation operator with orbital $\gamma\,(=3z^2-r^2, x^2-y^2)$ and spin $s\,(=\uparrow,\downarrow)$ at site i, and $n_{i\gamma s} = c_{i\gamma s}^{\dagger} c_{i\gamma s}$ is the electron density operator. The first term represents the electron transfer between the site i with orbital γ and a NN site j with orbital γ'. The 2nd and 3rd terms are the intra-orbital and inter-orbital Coulomb interactions, respectively. The 4th and 5th terms represent the Hund coupling and the pair-hopping interaction, respectively. It is crucially important that the electron-hopping integrals $(t_{ij}^{\gamma\gamma'})$ explicitly depend on the bond directions and the occupied orbitals. These are factorized by the Slater–Koster parameters as $t_{ij}^{\gamma\gamma'} = C_{ij}^{\gamma\gamma'} t$ with numerical constants $C_{ij}^{\gamma\gamma'}$ and the orbital-independent hopping integral t. In the interaction terms of the Hamiltonian, the two relations, $U = U' + 2J$ and $I = J$, exist in the atomic d orbitals. These are not satisfied in a system, where the d orbitals are hybridized with the p orbitals in the ligand ions.

Since one electron is localized at each site in a Mott insulator, it is convenient to introduce the pseudo-spin (PS) operator with amplitude of $1/2$ in order to describe the local orbital degree of freedom. This is defined by

$$\mathbf{T}_i = \frac{1}{2} \sum_{\gamma\gamma's} c_{i\gamma s}^{\dagger} \sigma_{\gamma\gamma'} c_{i\gamma's}, \tag{8}$$

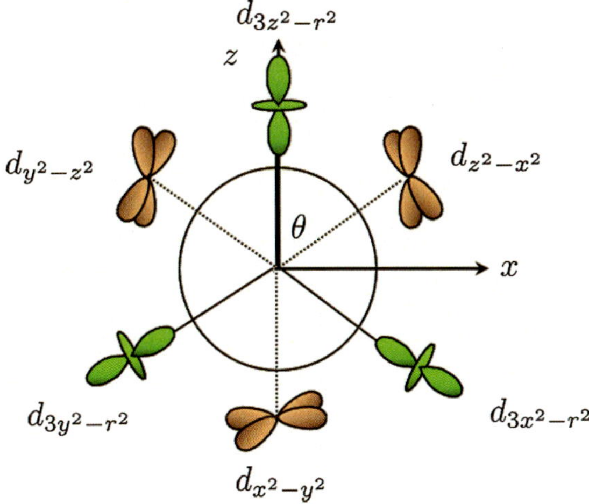

Fig. 2 The orbital pseudo spin and the corresponding $3d\,e_g$ orbitals

where $\boldsymbol{\sigma}$ are the Pauli matrices. The polar and azimuthal angles of PS characterize the orbital wave functions as

$$|\theta, \phi\rangle = \cos\left(\frac{\theta}{2}\right)|d_{3z^2-r^2}\rangle + e^{i\phi}\sin\left(\frac{\theta}{2}\right)|d_{x^2-y^2}\rangle. \tag{9}$$

In the real wave function at $\phi = 0$, relations between the PS angles and the wave functions are shown in Fig. 2.

From the two-orbital Hubbard model introduced above, the low-energy orbital structure is well described by the effective orbital model derived by the canonical transformation. This is justified in the case that the on-site Coulomb interactions are much larger than the hopping integrals. Through the perturbational calculation with respect to the electron hopping integrals, the interacting orbital model is obtained as [8, 17–19]

$$\mathcal{H} = -2J_1 \sum_{\langle ij \rangle} \left(\frac{3}{4} + \mathbf{S}_i \cdot \mathbf{S}_j\right)\left(\frac{1}{4} - \tau_i^l \tau_j^l\right)$$
$$- 2J_2 \sum_{\langle ij \rangle} \left(\frac{1}{4} - \mathbf{S}_i \cdot \mathbf{S}_j\right)\left[\left(\frac{1}{4} - \tau_i^l \tau_j^l\right) + \left(\frac{1}{2} + \tau_i^l\right)\left(\frac{1}{2} + \tau_j^l\right)\right]$$
$$- 2J_3 \sum_{\langle ij \rangle} \left(\frac{1}{4} + \mathbf{S}_i \cdot \mathbf{S}_j\right)\left(\frac{1}{2} + \tau_i^l\right)\left(\frac{1}{2} + \tau_j^l\right), \tag{10}$$

where $\mathbf{S}_i$ is the spin operator with amplitude of $1/2$ defined as $\mathbf{S}_i = \frac{1}{2}\sum_{\gamma ss'} c_{i\gamma s}^\dagger \boldsymbol{\sigma}_{ss'}$ $c_{i\gamma s'}$. The PS operators are rearranged as

$$\tau_i^l = \cos\left(\frac{2\pi n_l}{3}\right) T_i^z + \sin\left(\frac{2\pi n_l}{3}\right) T_i^x, \tag{11}$$

where $l(=x, y, z)$ denotes a direction of the bond connecting sites i and j in a cubic lattice, and n_l is defined as $(n_x, n_y, n_z) = (1, 2, 3)$. The exchange constants are given by $J_1 = t^2/(U' - J)$, $J_2 = t^2/(U - I)$ and $J_3 = t^2/(U + I)$. The magnitude relations, $J_1 > J_2 > J_3$, are satisfied.

Several characteristics in the interacting e_g orbital model introduced in Eq. (10) are summarized. (i) The first term in Eq. (10) is the dominant term which provides the ferromagnetic interaction and the staggered-type orbital interaction. This is attributed to the Hund coupling in the intermediate states in the perturbational processes. This interaction is believed to be origins of the magnetic and orbital structures in LaMnO$_3$ and KCuF$_3$. (ii) Each term is represented by a product of the spin and orbital parts. It is worth noting that $(\frac{3}{4} + \mathbf{S}_i \cdot \mathbf{S}_j)$ and $(\frac{1}{4} - \mathbf{S}_i \cdot \mathbf{S}_j)$ are the projection operators for the spin-triplet and spin-singlet states, respectively. (iii) Any continuous symmetries do not exist in the orbital parts of the Hamiltonian. (iv) The orbital interactions explicitly depend on the bond directions.

Several simplified orbital models are known. The following two approximations are introduced in Eq. (10). The hopping integrals are diagonal as $t_{ij}^{\gamma\gamma'} = t\delta_{\gamma\gamma'}$, and a relation for the exchange interactions is imposed as $J_2 = J_3$. Equation (10) is simplified as

$$\begin{aligned}
\mathcal{H} = &- 2J_1 \sum_{\langle ij \rangle} \left(\frac{3}{4} + \mathbf{S}_i \cdot \mathbf{S}_j\right)\left(\frac{1}{4} - \mathbf{T}_i \cdot \mathbf{T}_j\right) \\
&- 2J_2 \sum_{\langle ij \rangle} \left(\frac{1}{4} - \mathbf{S}_i \cdot \mathbf{S}_j\right)\left(\frac{3}{4} + \mathbf{T}_i \cdot \mathbf{T}_j\right),
\end{aligned} \tag{12}$$

which is termed the SU(2) $\times$ SU(2) spin-orbital model [20, 21]. The first (second) term in Eq. (12) favors the spin-triplet and orbital-singlet states (the spin-singlet and orbital-triplet states). The more simplified model is proposed as

$$\mathcal{H} = 2J \sum_{\langle ij \rangle} \left(\frac{1}{4} + \mathbf{S}_i \cdot \mathbf{S}_j\right)\left(\frac{1}{4} + \mathbf{T}_i \cdot \mathbf{T}_j\right), \tag{13}$$

termed the SU(4) spin-orbital model [22]. This is rewritten as $\mathcal{H} = J \sum_{\langle ij \rangle} \sum_{l=1}^{15} X_i^l X_j^l$ where $X^l = \{2S^\alpha, 2T^\beta, 4S^\alpha T^\beta\}$ are the generators in the SU(4) group.

Another type of the simplified orbital model has been studied. This is derived by neglecting the spin degrees of freedom in Eq. (10), and is often termed the orbital only model defined by

$$\mathcal{H} = J \sum_{\langle ij \rangle} \tau_i^l \tau_j^l, \tag{14}$$

where a relation $\sum_l \tau_i^l = 0$ is used. The same form of the Hamiltonian is derived from the model where the electronic orbitals couple with the lattice distortion as follows [14, 23–25]. The doubly degenerate e_g orbitals are located at each site in a three-dimensional perovskite lattice, where the ligand anions, such as the oxygens or chlorines, are introduced at a middle of the bonds connecting NN cations. The orbital-lattice Hamiltonian is given as

$$\mathcal{H}_{cJT} = g \sum_i \left(T_i^z Q_{iu} + T_i^x Q_{iv} \right) + \frac{1}{2} K \sum_{i,m(=u,v)} Q_{im}^2$$
$$+ g_0 \sum_i \left(T_i^z u_u + T_i^x u_v \right) + c_0 \sum_{m(=u,v)} u_m^2, \tag{15}$$

where Q_{iu} and Q_{iv} represent the two vibrational modes in an anion octahedra centered at site i with the $d_{3z^2-r^2}$ and $d_{x^2-y^2}$ symmetries, respectively, and u_u and u_v are the uniform strains. The 1st and 3rd terms represent the JT coupling between the electronic orbitals and the vibraitonal modes, and that between the orbitals and the uniform strains, respectively. The 2nd and 4th terms are for the elastic energies. By integrating out the lattice degrees of freedom using the canonical transformation, the effective orbital interaction is derived. When the spring constant between the NN metal and ligand ions is considered, the interaction between the PS operators is given as the same form with that in Eq. (14). The interaction constant in this case is given by $J = g^2/K$, which is induced by the virtual exchange of the JT phonons between the electronic orbitals.

We focus on the characteristics in the orbital only model in Eq. (14). The interactions along the three axes in a cubic lattice are explicitly written as

$$\mathcal{H} = J \sum_{\langle ij \rangle_x} \left(-\frac{1}{2} T_i^z + \frac{\sqrt{3}}{2} T_i^x \right) \left(-\frac{1}{2} T_i^z + \frac{\sqrt{3}}{2} T_i^x \right)$$
$$+ J \sum_{\langle ij \rangle_y} \left(-\frac{1}{2} T_i^z - \frac{\sqrt{3}}{2} T_i^x \right) \left(-\frac{1}{2} T_i^z - \frac{\sqrt{3}}{2} T_i^x \right)$$
$$+ J \sum_{\langle ij \rangle_z} T_i^z T_j^z, \tag{16}$$

where $\langle ij \rangle_l$ implies the NN pair connecting sites i and j along the direction l. It is shown that the interactions explicitly depend on the bond direction. For example, in a NN bond along the x axis, a staggered alignment of the $d_{3x^2-r^2}$ and $d_{y^2-z^2}$ orbitals is favored. The all bond energies cannot be minimized simultaneously within the naïve orbital configurations. A similar situation is often seen in magnets on geometrical frustrated lattices. The present case is sometime termed "the orbital frustration" [26–28]. It is convenient to introduce the Fourier transformation of the interaction between the orbitals as [17]

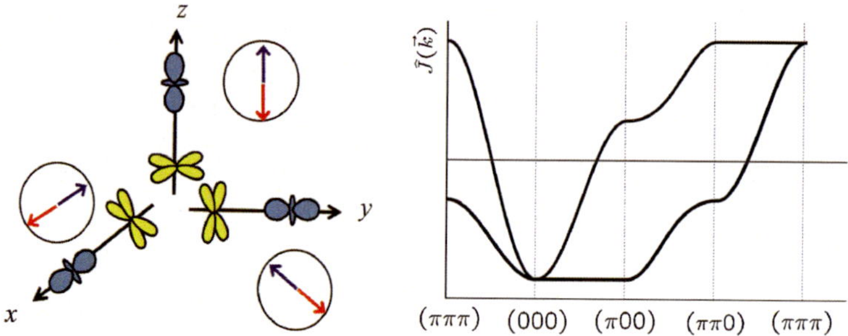

Fig. 3 (*Left*) A concept of the orbital frustration in a cubic lattice. *Arrows* represent directions of the orbital PSs. (*Right*) Momentum dependences of the effective orbital interactions in Eq. (18) in the orbital only model [17]

$$\mathscr{H} = J \sum_{\mathbf{k}} \hat{\mathbf{T}}^{t}(\mathbf{k}) \cdot \hat{J}(\mathbf{k}) \cdot \hat{\mathbf{T}}(\mathbf{k}), \tag{17}$$

where $\hat{\mathbf{T}}(\mathbf{k}) = [T^{z}(\mathbf{k}), T^{x}(\mathbf{k})]$ are the Fourier transformations of the PS operators, and the $\hat{J}(\mathbf{k})$ is the 2×2 matrix. By diagonalizing the interaction matrix, the eigen values of $\hat{J}(\mathbf{k})$ are obtained as

$$J_{\pm}(\mathbf{k}) = -\sum_{l} \cos ak_{l} \pm \sqrt{\sum_{l} \cos^{2} ak_{l} - \sum_{lmn} \frac{1}{2}\varepsilon_{lmn} \cos ak_{m} \cos ak_{n}}, \tag{18}$$

where ε_{lmn} is the Levi-Civita antisymmetric tensor, and a is a lattice constant. The dispersion relations are plotted in Fig. 3 which take their maxima (minima) along the M–R directions (Γ–X directions) and other equivalent directions in the Brillouin zone for a cubic lattice. This result implies that all of the orbital configurations characterized by the momenta along the Γ–X axis are the possible candidates for the stable orbital configurations in the ground state. Continuous degeneracies exist, since the two eigen interactions are merged at the points Γ and R. This is highly in contrast to the Heisenberg model with the NN exchange interaction where $J(\mathbf{k})$ takes its minimum at the Γ or R points. These degeneracies in the orbital configurations are classified into the following two types [29]: (i) The rotational-type degeneracy. Let us consider a staggered-type orbital order where the PS angles at the two sublattices are given as $(\theta_{A}, \theta_{B}) = (\theta_{0}, \theta_{0} + \pi)$ with any value for θ_{0}. We define the PS angle as $\theta = \tan^{-1}((\langle T^{x}\rangle)/(\langle T^{z}\rangle))$. (ii) The stacking-type degeneracy. Suppose a staggered type orbital configuration denoted as $(\theta_{0}, \theta_{0} + \pi)$. When all PSs in any xy planes are transformed by $T^{x} \rightarrow -T^{x}$, the energy is not changed. This is owing to the fact that the orbital interaction along the z axis does not depend on T^{x}.

These degenerated orbital configurations are lifted by taking into account the thermal/quantum fluctuations. This is the so-called "order by fluctuation" mechanism well known in the frustrated magnets. We introduce the calculated results of the orbital order by thermal fluctuation. In Fig. 4, the orbital order parameters calculated

Fig. 4 **a** Temperature
dependences of the orbital
order parameter, and **b** those
of the orbital angle
correlation in the orbital only
model. The classical Monte
Carlo method in finite size
clusters of the L^3 cubic
lattice is used. Data are
excerpt from Ref. [30]

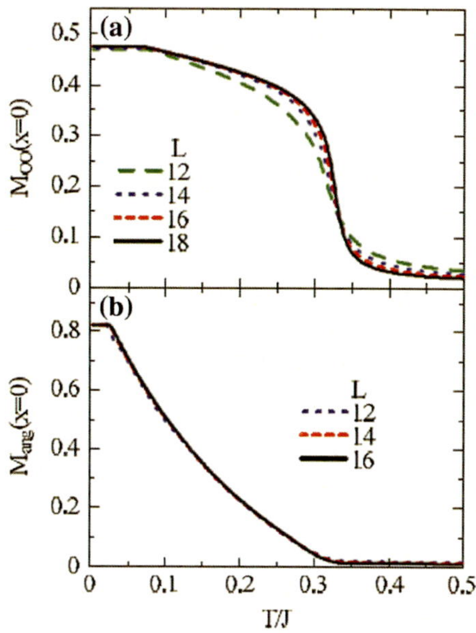

by the classical Monte Carlo method are presented [30]. The orbital order parameter
at the R point is defined as $M_{OO} = N^{-1} \langle (\sum_i e^{i\mathbf{Q}\cdot\mathbf{r}_i} \mathbf{T}_i)^2 \rangle$ with $\mathbf{Q} = (\pi, \pi, \pi)$. This
increases at about $T = 0.34 J$ corresponding to the orbital ordering temperature. The
numerical results for the PS angle are also shown in Fig. 4, where the correlation of the
PS angles is defined as $M_{\text{angle}} = N^{-1} \langle (\sum_i e^{i\mathbf{Q}\cdot\mathbf{r}_i} \cos 3\theta_i)^2 \rangle$. The two results imply
that the both types of the degeneracies explained above are lifted by the thermal
fluctuation, and the orbital order at the momentum (π, π, π) with the PS angles
$(\theta_A, \theta_B) = (2n\pi/3, 2n\pi/3 + \pi)$ (n: integer) is realized.

In the real transition-metal compounds with the orbital degree of freedom, the
non-linear JT coupling and the anharmonic lattice potentials provide important contributions to the orbital order. For example, the long range orbital order accompanied with the JT-type lattice distortions set in at about $800\,\text{K}$ in KCuF$_3$. The
hole-orbital wave functions are characterized as $(d_{y^2-z^2}, d_{z^2-x^2})$ with the momentum at (π, π, π) as shown in Fig. 5. Another type of the orbital alignment with
$(\pi, \pi, 0)$ is also reported. These orbitals are represented by the orbital angles
$(\pi/3, -\pi/3)$, which are different from $(2n\pi/3, 2n\pi/3 + \pi)$ suggested by the orbital
only model explained above. This discrepancy may be attributable to the anhar-
monic lattice potential $\mathscr{H}_{anh} = B \sum_i (Q_{ui}^3 - 3 Q_{vi}^2 Q_{ui})$, and the higher-order JT
coupling $\mathscr{H}_{h-JT} = C \sum_i [(Q_{ui}^2 - Q_{vi}^2) T_i^z - 2 Q_{vi} Q_{ui} T_i^x]$, which give the potential minima/maxima at $\theta = 2n\pi/3$ with an integer n in the adiabatic potential [14].
More detailed studies of the role of the JT distortion in KCuF$_3$ is given in Refs. [31–
33].

Fig. 5 A schematic orbital order in KCuF$_3$

2.2 An Overview of Observations of Orbital Order and RXS

Experimental observations of the long-ranged orbital order have been recognized as a crucially important issue in the condensed matter physics for a long time. Since the orbital orders in the transition-metal compounds are often accompanied with the lattice distortion due to the cooperative JT type interaction, the observations of the lattice distortions by the X-ray and neutron diffractions have been accepted as the evidences of the orbital order. A softening of the elastic constants observed by the ultrasonic measurements has been also useful to identify a precursor of the orbital ordering, although this is limited to the information of the zero momentum.

One of the breakthrough for the direct observation of the orbital order was performed by the polarized-neutron diffraction experiments in K$_2$CuF$_4$ [34]. This method is valid in the case where the electrons participated in the orbital order are responsible for the magnetic order. In the neutron diffraction experiments, the scattering intensity with the scattering vector $\mathbf{K}$ is given as $I(\mathbf{K}) \propto |F(\mathbf{K})|^2$ where $F(\mathbf{K})$ is the structure factor and is divided into the nucleus part and the magnetic part as $F(\mathbf{K}) = F_N(\mathbf{K}) + F_M(\mathbf{K})$. The structure factor is represented by the atomic form factors as $F(\mathbf{K}) = \sum_i e^{i\mathbf{K}\cdot\mathbf{r}_i} f_i(\mathbf{K})$ with $f_i(\mathbf{K}) = f_i^N(\mathbf{K}) + f_i^M(\mathbf{K})$. The nucleus and magnetic parts of the atomic form factors are given by $f_i^N = \langle\psi|e^{i\mathbf{K}\cdot\mathbf{r}_i}\rho_N(\mathbf{r})|\psi\rangle$ and $f_i^M = \langle\psi|e^{i\mathbf{K}\cdot\mathbf{r}_i}\mathbf{S}_\perp(\mathbf{r})|\psi\rangle$, respectively, where $\rho_N(\mathbf{r})$ and $\mathbf{S}_\perp(\mathbf{r})$ are the nucleus density and the spin density which is a perpendicular component of $\mathbf{S}(\mathbf{r})$ to $\mathbf{K}$. By using the polarized neutrons, the nucleus and magnetic scatterings can be observed separately. Since the magnetic scattering provides information of the orbital order in the magnetic ordered phase, shapes of the electronic orbitals are determined by analyzing the magnetic part of the atomic form factor. The polarized-neutron diffraction experiments were performed in KCu$_2$F$_4$ at 4.2 K [34]. This material shows the ferromagnetic transition at $T_c = 6.25$ K. The staggered-type orbital order was believed to be realized from the observations of the lattice distortions of the F$_6$ octahedron. The ratios of the magnetic part of the atomic form factor to the nucleus one, i.e. $F_M(\mathbf{K})/F_N(\mathbf{K})$, were measured experimentally, and were compared with the calculations, in which the alternate order of the $d_{y^2-z^2}$ and $d_{z^2-x^2}$ orbital is assumed. The model calculations show good agreements with the experimental results. This techniques were also applied to observations of the orbital orders in $YTiO_3$ and $Lu_2V_2O_7$ [35, 36].

Direct observations of the orbital order are also possible through the measurements of the anisotropic electronic charge density in the outermost orbital. This is usually tough measurement utilizing the X-ray diffraction techniques, since the diffraction intensity is proportional to the square of the charge densities, and the amount of the anisotropic charge density is less than one. The charge density distribution was investigated by the synchrotron X-ray powder diffraction in the double-layered manganite $Nd_2SrMn_2O_7$ [37]. The layer-type antiferromagnetic order, term the A-type antiferromagnetic order, starts to realize at $T_N = 150\,\mathrm{K}$ below which the $d_{x^2-y^2}$ orbital components were expected to be increased. The electron density distributions above and below T_N were determined by a combination method of the Rietveld refinement and the maximum-entropy method. It was shown that the charge density distributions of Mn ions are more anisotropic below T_N. This is reasonable to the magnetic structure, since the magnetic energy gains due to the double-exchange and superexchange interactions are favored in this orbital order. Direct observations of the anisotropic charge distribution are developed recently using the accurate X-ray diffractions in single crystals where the multiple scatterings are avoided as possible. The methods were applied to observations of the several orbital ordered compounds [38, 39].

The RXS phenomena, often termed the anomalous X-ray scattering (AXS), is the X-ray diffraction in which the X-ray energy is set up around the absorption edge in an element included in a crystal. In addition to the conventional atomic form factor $(f_0(\mathbf{K}))$, the anomalous term $(f'(\mathbf{K}) + if''(\mathbf{K}))$ participates in the X-ray scattering. The AXS phenomena have been long known since 1930s and have been applied to the studies of the crystallographic polarity, the alloy structures, as well as the structural studies of the glass and non-crystalline materials and so on [40–42]. The wide and genuine applications of AXS have started, since the modern synchrotron-radiation technique have become popular. The dichroism in the X-ray scattering, which is crucial in observation of the orbital order, was first observed in Refs. [43, 44]. The double refraction, i.e. the birefringence, in the X-ray regime was measured in $VO^{2+}\cdot2(C_3H_7O_2)^-$ near the vanadium K-edge. This observation was attributed to the anisotropic character in the atomic scattering factor near the absorption edge. The X-ray dichroism was also observed in $Rb\cdot UO_2\cdot2NO_3$ near the uranium L_1- and L_3-edges. From the polarization dependences of the absorption coefficients, the anisotropic components in the real and imaginary parts of the atomic scattering factors were deduced. A possibility of the X-ray forbidden reflections due to the anomalous X-ray scattering was predicted theoretically in Ref. [45]. It was proposed that the tensor character in the anomalous part of the atomic scattering factors, reflecting the local symmetry, leads to the finite forbidden reflections. When f_0 is only taken into account, the scattering intensity vanishes at these reflections. This was termed the anisotropy of the tensor of susceptibility reflection.

The first direct observation of the orbital order utilizing RXS was performed by Murakami and coworkers in perovskite manganites, $LaSrMnO_4$ and $LaMnO_3$ [1, 2].

Here, the experimental results of RXS in LaMnO$_3$ are introduced. It was known that the structural phase transition from the cubic to the orthohombic structures occurs at around 800 K, and the A-type antiferromagnetic order occurs at $T_N = 141$ K. A nominal valence of a Mn ion in LaMnO$_3$ is 3+. The local electron configuration is $(t_{2g})^3(e_g)^1$ where the e_g orbital degeneracy exists in a cubic crystalline field, in the same way in KCuF$_3$. The orbital order, where the $d_{3x^2-r^2}$ and $d_{3y^2-r^2}$ orbitals are aligned alternately in the xy plane, was expected from the lattice distortions in the O$_6$ octahedron. The X-ray diffraction measurements were performed at (300) in the orthorhombic notation, which corresponds to the superlattice reflection point for the orbital ordered state. The X-ray scattering factor is proportional to the difference of the atomic form factors in the orbital sublattices A and B as

$$F(\mathbf{K}) = f_A(\mathbf{K}) - f_B(\mathbf{K})$$
$$= \int d\mathbf{r} e^{i\mathbf{r}\cdot\mathbf{K}} \{\rho_A(\mathbf{r}) - \rho_B(\mathbf{r})\}, \tag{19}$$

where $\rho_{A(B)}(\mathbf{r})$ is the charge density at a Mn ion in the sublattice A (B). Since the charge distributions have a relation $\rho_A(x, y, z) = \rho_B(y, x, z)$, $F(\mathbf{K}) = 0$ at $\mathbf{K} = (hh0)$ in the cubic notation. This corresponds to the forbidden reflection. The experimental observation in the X-ray diffraction intensity is presented in Fig. 6 as a function of the X-ray energy. It is shown that the intensity is remarkably enhanced around the Mn K absorption edge of around 6.55–6.56 KeV. That is, the forbidden reflection occurs near the Mn K-edge. The temperature dependences of the RXS intensity at the orbital order reflection are shown in Fig. 7. With decreasing temperature, the RXS intensity rapidly increases at about 780 K, which corresponds to the structural phase transition.

The polarization dependence of the X-ray diffraction intensity is crucial to identify the scattering in which the anomalous term of the atomic form factor is concerned. The azimuthal angle (Ψ) dependence of the scattering intensity is presented in Fig. 6, in which the scattering intensity was measured as a function of the rotation angle with respect to the scattering vector. The experimental set up is shown in the inset of the lower panel in Fig. 6. The scattering intensity as a function of the azimuthal angle and the analyzer angle (ϕ) is given by [46]

$$I(\psi, \phi) = |F_A|^2 |D_{\sigma\sigma} \cos\phi - D_{\pi\sigma} \sin\phi|^2, \tag{20}$$

with

$$D_{\sigma\sigma} = F_{zz}(\mathbf{K}) \cos^2\Psi + \frac{1}{2}\{F_{xx}(\mathbf{K}) + F_{yy}(\mathbf{K})\} \sin^2\psi, \tag{21}$$

$$D_{\pi\sigma} = \frac{1}{2}\{F_{xx}(\mathbf{K}) + F_{yy}(\mathbf{K}) - 2F_{zz}(\mathbf{K})\} \sin\psi \cos\psi \sin\theta$$
$$+ \frac{1}{2}\{-F_{xx}(\mathbf{K}) + F_{yy}(\mathbf{K})\} \sin\psi \cos\theta, \tag{22}$$

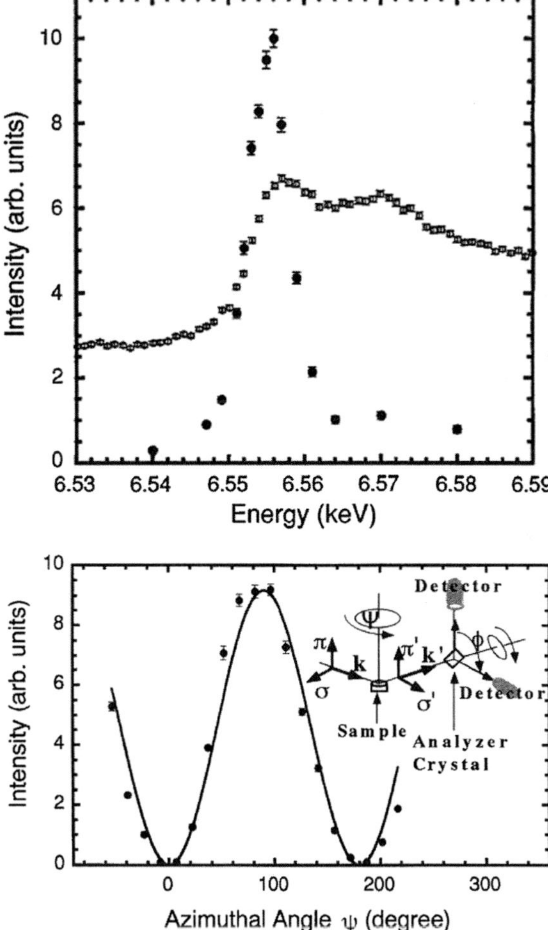

Fig. 6 (*Upper*) *Filled circles* show the energy dependence of the integrated RXS intensity in LaMnO₃ at (300). *Open squares* show the fluorescence. (*Lower*) Azimuthal-angle dependence of the RXS intensity of the orbital order reflection (300). *Inset* shows a schematic view of the experimental configuration and definition of the polarization directions. Excerpt from Ref. [2]

where θ is the scattering angle. The scattering angle for the analyzer crystal is set to be $\pi/4$. The tensor elements of the scattering factor are represented as $F_{\alpha\beta}(\mathbf{K})$. In the orbital ordered state in LaMnO₃, the atomic form factors in the two orbital sublattices are given phenomenologically by the symmetry considerations as

$$f_A(\mathbf{K}) = \begin{pmatrix} f_l(\mathbf{K}) & & \\ & f_s(\mathbf{K}) & \\ & & f_s(\mathbf{K}) \end{pmatrix}, \qquad f_B(\mathbf{K}) = \begin{pmatrix} f_s(\mathbf{K}) & & \\ & f_l(\mathbf{K}) & \\ & & f_s(\mathbf{K}) \end{pmatrix}, \tag{23}$$

when the Cartesian coordinates are taken as the principal axes. Then, we have $I(\psi, \phi = 0) = 0$ and

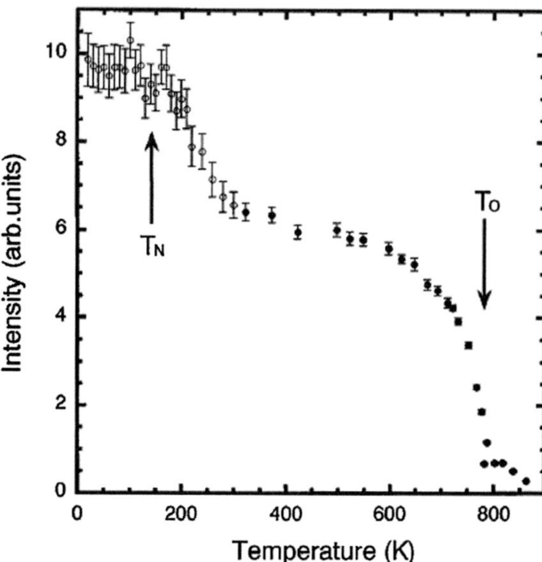

Fig. 7 Temperature dependence of the RXS intensity at (3 0 0) in LaMnO$_3$. Excerpt from Ref. [2]

$$I\left(\psi, \phi = \frac{\pi}{2}\right) = \left|\frac{1}{2}\left(f_l(\mathbf{K}) - f_s(\mathbf{K})\right)\sin\psi\cos\theta\right|^2, \tag{24}$$

which is proportional to $|f_l(\mathbf{K}) - f_s(\mathbf{K})|^2 \sin^2\Psi$. This phenomenological analyse reproduces well the experimental observations presented in Fig. 6, showing a direct evidence of the anisotropic tensor characters in the atomic scattering factors.

The RXS study was performed for the observation of the orbital order in V$_2$O$_3$ [47, 48], independently of the study in manganites in Refs. [1, 2]. The forbidden reflection peaks were observed at (1 1 1) which is enhanced at the pre-edge region of the V K absorption edge. Through the comparison with the theoretical calculation as well as the azimuthal angle analyses, these peaks were identified as the orbital order reflection peaks. After the first observation of RXS in the orbital ordered manganites, this method was applied immediately to a number of the correlated electron systems with the orbital degree of freedom. Some topical issues are summarized.

- The RXS has clarified the orbital orders in the several 3d transition-metal compounds, e.g. manganites [49–61, 104], titanates [62–66], vanadates [67–69], cobaltites [70, 71], nickelates [72, 73], cuprates [74–76] and magnetites [77, 78]. The theoretical researches in RXS as a probe to detect the 3d orbital order have been also developed [46, 48, 79–89]. A part of these studies is introduced in Chap. 2.
- The quadrupole and higher-order multipole orders in the rare-earth compounds have been observed by the RXS measurements [90–97] under the supports by the theoretical calculations [98–100]. A detailed review is presented in Chap. 3.

- The electronic orbitals in the surface and interfaces in transition-metal compounds, as well as the artificial super lattices have been studied by RXS [101–103]. A part of Chap. 2 is devoted to this issue.
- As explained in Sect. 2.3, the hard X-ray RXS does not access directly to the $3d$ orbitals but the $4p$ ones in the $3d$ transition-metal compounds. Direct accesses to the electronic orbitals, concerned in the orbital order, have been performed by the soft X-ray RXS in the $3d$ transition-metal compounds [104–108], and the hard-X-ray RXS in the $5d$ ones, where the spin and orbital are strongly entangled with each other [109–111]. These are reviewed in Chaps. 4 and 5.
- The RIXS spectroscopy has been applied to a wide variety of correlated electron systems with the orbital degree of freedom in order to search the orbital excitations [112–119]. These are introduced in Sect. 3.3 and Chap. 6.
- Recently, RXS and RIXS methods are applied to study of the transient and non-equilibrium dynamics of the orbital degree of freedom [120–123]. This new direction is introduced in Chap. 5.

2.3 Scattering Cross Section of RXS

The scattering cross section for RXS and RIXS is introduced in this subsection. Also see other reviews and textbooks [124–126]. We start from the Hamiltonian for the electrons interacting with the electro-magnetic field given by

$$\mathcal{H} = \mathcal{H}_{PA} + \mathcal{H}_Z + \mathcal{H}_{LS}. \tag{25}$$

The first term represents the coupling between the electron momentum ($\mathbf{P}_i$) and the vector potential of the X-ray [$\mathbf{A}(\mathbf{r})$] given as

$$\mathcal{H}_{PA} = \sum_i \left[\frac{1}{2m}\mathbf{P}_i - \frac{e}{c}\mathbf{A}_\perp(\mathbf{r}) \right]^2, \tag{26}$$

where c, e and m are the light velocity, the unit charge, and the electron mass, respectively. The 2nd and 3rd terms in the Hamiltonian represent the Zeeman effect and the relativistic spin-orbit interaction given by

$$\mathcal{H}_Z = -\frac{e}{mc} \sum_i \mathbf{S}_i \cdot [\nabla \times \mathbf{A}(\mathbf{r})], \tag{27}$$

and

$$\mathcal{H}_{LS} = \frac{1}{2m^2c^2}\frac{1}{r}\frac{dV(|\mathbf{r}|)}{dr} \sum_i \mathbf{S}_i \cdot \mathbf{L}_i, \tag{28}$$

respectively, with the electronic spin $\mathbf{S}_i = (\hbar/2)\boldsymbol{\sigma}_i$, the orbital angular momentum $\mathbf{L}_i = \mathbf{r}_i \times \mathbf{P}_i$, and the atomic potential $V(|\mathbf{r}|)$. The vector potential is divided into the transverse and longitudinal components as $\mathbf{A}(\mathbf{r}) = \mathbf{A}_\perp(\mathbf{r}) + \mathbf{A}_\parallel(\mathbf{r})$, where the conditions $\nabla\mathbf{A}_\perp(\mathbf{r}) = 0$ and $\nabla \times \mathbf{A}_\parallel(\mathbf{r}) = 0$ are satisfied. By introducing the gauge transformation $\mathbf{P}_i \rightarrow \mathbf{P}_i - (e/c)\mathbf{A}(\mathbf{r})$, $\mathscr{H}_{LS}$ is rewritten as

$$\mathscr{H}_{LS} = -\frac{e\hbar}{2m^2c^2} \sum_i \mathbf{E} \times \mathbf{S}_i - \frac{e^2\hbar}{2m^2c^4} \sum_i \mathbf{S}_i \cdot (\dot{\mathbf{A}}(\mathbf{r}) \times \mathbf{A}(\mathbf{r})), \tag{29}$$

where the equations $\mathbf{E} = -(\mathbf{r}/r)(dV/dr) = -\partial\mathbf{A}/\partial t$ are used.

By introducing the quantum-field theoretical formalisms for electron and photon, we reformulate the interaction part of the Hamiltonian. In the the first order terms of $\mathbf{A}(\mathbf{r})$ originating from $\mathscr{H}_{PA}$, the exponential factor in an integral is expanded with respect to a power of $\mathbf{k} \cdot \mathbf{r}$ as

$$\int d\mathbf{r}e^{\pm i\mathbf{k}\cdot\mathbf{r}}\phi_\gamma^*(\mathbf{r})\nabla\phi_{\gamma'}(\mathbf{r}) \sim \int d\mathbf{r} \left[1 \pm i\mathbf{k}\cdot\mathbf{r} - (\mathbf{k}\cdot\mathbf{r})^2 + \cdots\right]\phi_\gamma^*(\mathbf{r})\nabla\phi_{\gamma'}(\mathbf{r}), \tag{30}$$

where the first and second terms describe the dipole and quadrupole transitions, respectively. The results are summarized as

$$\mathscr{H} = \mathscr{H}_{dip} + \mathscr{H}_{quad} + \mathscr{H}_\rho + \mathscr{H}_{LS} + \mathscr{H}_Z, \tag{31}$$

where the first three terms originate from $\mathscr{H}_{PA}$ in Eq. (26). The first term describes the dipole transition given by

$$\mathscr{H}_{dip} = i\frac{e\hbar}{mc} \sum_{\mathbf{k}\lambda} C_{\mathbf{k}} \left(a_{\mathbf{k}\lambda}e^{-i\omega_{\mathbf{k}}t} + a_{\mathbf{k}\lambda}e^{i\omega_{\mathbf{k}}t}\right) \sum_{i\gamma\gamma's} \varepsilon_{\mathbf{k}\lambda} \cdot \mathbf{D}_{i\gamma\gamma'}c_{i\gamma s}^\dagger c_{i\gamma's}, \tag{32}$$

where $a_{\mathbf{k}\lambda}^\dagger$ ($a_{\mathbf{k}\lambda}$) is the creation (annihilation) operator of a photon with momentum $\mathbf{k}$ and polarization $\lambda(=1, 2)$, and $c_{i\gamma s}^\dagger$ ($c_{i\gamma s}$) is the operator for an electron with orbital γ and spin s at site i. We define a constant $C_{\mathbf{k}} = \sqrt{\hbar c^2/(2V\omega_{\mathbf{k}})}$, the polarization vector of X-ray $\varepsilon_{\mathbf{k}\lambda}$, and the photon energy $\omega_{\mathbf{k}} = c|\mathbf{k}|$. The dipole-transition matrix in Eq. (32) is written using the electronic wave functions as

$$\mathbf{D}_{i\gamma\gamma'} = \int d\mathbf{r}\phi_\gamma^*(\mathbf{r} - \mathbf{r}_i)\nabla\phi_{\gamma'}(\mathbf{r} - \mathbf{r}_i) = \langle\gamma|\nabla|\gamma'\rangle, \tag{33}$$

where $\phi_\gamma(\mathbf{r})$ is the electronic wave function with orbital γ. The quadrupole transition term and the Zeeman term are combined as

$$\mathcal{H}_Z + \mathcal{H}_{quad} = -i\frac{e\hbar}{2mc}\sum_{\mathbf{k}\lambda} C_{\mathbf{k}}a_{\mathbf{k}\lambda}e^{-i\omega_{\mathbf{k}}t}\sum_{i\gamma\gamma'ss'}c^{\dagger}_{i\gamma s}c_{i\gamma's'}$$

$$\times\left[\frac{i}{\hbar}(\mathbf{k}\times\boldsymbol{\varepsilon}_{\mathbf{k}\lambda})\cdot\left(\mathbf{L}_{\gamma\gamma'}\delta_{ss'}+\boldsymbol{\sigma}_i f_{\gamma\gamma'}(\mathbf{k})\right)+\sum_{\alpha\beta}\mathbf{k}_{\alpha}T^{\alpha\beta}_{\gamma\gamma'}\varepsilon^{\beta}_{\mathbf{k}\lambda}\delta_{ss'}\right]+H.c.. \quad (34)$$

We introduce the orbital angular momentum

$$\mathbf{L}_{\gamma\gamma'} = \langle\gamma|(\mathbf{r}\times\boldsymbol{\nabla})|\gamma'\rangle, \quad (35)$$

and the second-rank tensor for the quadrupole transition

$$T^{\alpha\beta}_{\gamma\gamma'} = \langle\gamma|(r_{\alpha}\nabla_{\beta}+\nabla_{\alpha}r_{\beta})|\gamma'\rangle, \quad (36)$$

where the following identity is utilized

$$2(k_{\alpha}r_{\alpha})(\varepsilon^{\beta}\nabla_{\beta}) = \left[(k_{\alpha}r_{\alpha})(\varepsilon^{\beta}\nabla_{\beta})+(k_{\alpha}\nabla_{\alpha})(\varepsilon^{\beta}r_{\beta})\right]$$
$$+ \left[(k_{\alpha}r_{\alpha})(\varepsilon^{\beta}\nabla_{\beta})-(k_{\alpha}\nabla_{\alpha})(\varepsilon^{\beta}r_{\beta})\right], \quad (37)$$

as a calculation technique [126]. The remaining terms in Eq. (31), $\mathcal{H}_{LZ}$ and $\mathcal{H}_{\rho}$, are given as

$$\mathcal{H}_{LS} = -\frac{e^2\hbar}{2m^2c^4}\sum_{\mathbf{k}\mathbf{k}'\lambda\lambda'}C_{\mathbf{k}}C_{\mathbf{k}'}2i\omega_{\mathbf{K}}a^{\dagger}_{\mathbf{k}\lambda}a_{\mathbf{k}'\lambda'}(\boldsymbol{\varepsilon}_{\mathbf{k}\lambda}\times\boldsymbol{\varepsilon}_{\mathbf{k}'\lambda'})\cdot\sum_{i}\mathbf{S}_i(\mathbf{K})e^{i\mathbf{K}\cdot\mathbf{r}_i}e^{-i\omega_{\mathbf{K}}t}, \quad (38)$$

and

$$\mathcal{H}_{\rho} = \frac{e^2}{2mc^2}\sum_{\mathbf{k}\mathbf{k}'\lambda\lambda'}C_{\mathbf{k}}C_{\mathbf{k}'}\boldsymbol{\varepsilon}_{\mathbf{k}\lambda}\cdot\boldsymbol{\varepsilon}_{\mathbf{k}'\lambda'}2a^{\dagger}_{\mathbf{k}\lambda}a_{\mathbf{k}'\lambda'}\sum_{i}\rho_i(\mathbf{K})e^{i\mathbf{K}\cdot\mathbf{r}_i}e^{-i\omega_{\mathbf{K}}t}, \quad (39)$$

respectively, with the momentum transfer $\mathbf{K} = \mathbf{k} - \mathbf{k}'$. We introduce the spin density

$$\mathbf{S}_i(\mathbf{K}) = \frac{1}{2}\sum_{\gamma\gamma'}\sum_{ss'}f_{\gamma\gamma'}(\mathbf{K})c^{\dagger}_{i\gamma s}\boldsymbol{\sigma}_{ss'}c_{i\gamma's'}, \quad (40)$$

and the charge density

$$\rho_i(\mathbf{K}) = \sum_{\gamma\gamma'}\sum_{s}f_{\gamma\gamma'}(\mathbf{K})c^{\dagger}_{i\gamma s}c_{i\gamma's}, \quad (41)$$

where the atomic scattering factor is introduced as

$$f_{\gamma\gamma'}(\mathbf{k}) = \langle\gamma|e^{i\mathbf{k}\cdot\mathbf{r}}|\gamma'\rangle. \quad (42)$$

In Eq. (31), $\mathcal{H}_{dip}$, $\mathcal{H}_Z$, and $\mathcal{H}_{quad}$ are the first order of $\mathbf{A}(\mathbf{r})$, and $\mathcal{H}_{LS}$ and $\mathcal{H}_\rho$ are the second order of $\mathbf{A}(\mathbf{r})$. These are summarized in a unified fashion as

$$\mathcal{H}_1 = \sum_{\mathbf{k}\lambda} \sum_\eta \sum_i \left[C_\mathbf{k} \mathbf{J}_i^\eta(\mathbf{k}) \varepsilon_{\mathbf{k}\lambda}^\eta a_\mathbf{k} e^{i\mathbf{k}\cdot\mathbf{r}_i} e^{-i\omega_\mathbf{k}t} + \text{H.c.} \right], \tag{43}$$

and

$$\mathcal{H}_2 = \sum_{\mathbf{k}\mathbf{k}'\lambda\lambda'} C_\mathbf{k} C_{\mathbf{k}'} a_{\mathbf{k}\lambda}^\dagger a_{\mathbf{k}'\lambda'} \sum_{\alpha\beta} \varepsilon_{\mathbf{k}\lambda}^\alpha \varepsilon_{\mathbf{k}'\lambda'}^\beta \sum_i R_i^{\alpha\beta}(\mathbf{K}) e^{i\mathbf{K}\cdot\mathbf{r}_i} e^{-i\omega_\mathbf{K}t}, \tag{44}$$

where we introduce the generalized current operator

$$\mathbf{J}_i^\eta(\mathbf{k}) = i\frac{e\hbar}{mc} \left[\sum_{\gamma\gamma's} \mathbf{D}_{i\gamma\gamma'}^\eta c_{i\gamma s}^\dagger c_{i\gamma's} + \sum_{\gamma\gamma's} \frac{1}{2} \sum_\alpha k_\alpha T_{\gamma\gamma'}^{\alpha\eta} c_{i\gamma s}^\dagger c_{i\gamma's} \right.$$
$$\left. - \frac{i}{2\hbar} \sum_{\alpha\beta} \varepsilon_{\alpha\beta\eta} \left\{ \mathbf{L}_i^\alpha + 2\mathbf{S}_i^\alpha(\mathbf{k}) \right\} k_\beta \right], \tag{45}$$

and the charge and spin density operators

$$R_i^{\alpha\beta}(\mathbf{K}) = \frac{e^2}{mc^2} \left\{ \delta_{\alpha\beta}\rho_i(\mathbf{K}) - i\left(\frac{\hbar\omega_\mathbf{K}}{mc^2}\right) \sum_\gamma \varepsilon_{\alpha\beta\gamma} \mathbf{S}_i^\gamma(\mathbf{K}) \right\}. \tag{46}$$

We define $\mathbf{L}_i = \sum_{\gamma\gamma's} c_{i\gamma s}^\dagger c_{i\gamma's} \mathbf{L}_{\gamma\gamma'}$. The X-ray scattering occurs by the second order of $\mathcal{H}_1$ and the first order of $\mathcal{H}_2$.

Now we consider the X-ray scattering from the incident X-ray with the wave vector $\mathbf{k}_i$, the frequency $\omega_{\mathbf{k}_i}$ and the polarization λ_i to the scattered X-ray with $\mathbf{k}_f$, $\omega_{\mathbf{k}_f}$, λ_f, as shown in Fig. 8. The electronic states are changed by the scattering from $|i\rangle$ with energy ε_i to $|f\rangle$ with ε_f. The differential scattering-cross section is obtained by the perturbational calculation termed the Kramers–Heisenberg formula. The Hamiltonian in the electron-photon coupled system is divided into the unperturbed and perturbed terms are $\mathcal{H} = \mathcal{H}_0 + \mathcal{H}'$. The scattering matrix is given as

$$\begin{aligned} S_{fi} &= \langle \Phi_f | U(+\infty, -\infty) | \Phi_i \rangle \\ &= \delta_{fi} - 2\pi i \delta\left(E_f - E_i\right) \langle \Phi_f | \mathcal{H}' | \Psi_i \rangle, \end{aligned} \tag{47}$$

Fig. 8 Scattering processes in RXS and RIXS. *Straight* and *wavy lines* represent the electronic and photon states, respectively

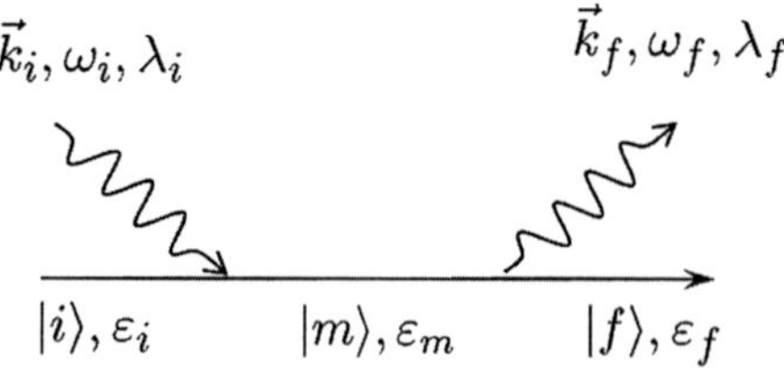

where $|\Phi_{i(f)}\rangle$ is the egen state of $\mathcal{H}_0$ with energy $E_{i(f)}$, and $U(t, t')$ is the time-evolution operator. The wave function $|\Psi_i\rangle$ is given by the Lippman–Schwinger equation as

$$\begin{aligned}
|\Psi_i\rangle &= U(0, -\infty)|\Phi_i\rangle \\
&= |\Phi_i\rangle + \frac{1}{E_i - \mathcal{H} + i\gamma}\mathcal{H}'|\Phi_i\rangle,
\end{aligned} \tag{48}$$

where γ is an infinitesimal constant. Up to the second order of $\mathcal{H}'$, the scattering cross section is obtained as

$$\frac{d^2\sigma}{d\Omega d\omega_i} = \sigma_T \sum_f |S|^2 \delta(\varepsilon_f - \varepsilon_i + \omega_f - \omega_i), \tag{49}$$

where $\sigma_T = [e^2/(mc^2)]^2$ is the scattering cross section in the Thomson scattering. The scattering matrix is classified by $S = S_1 + S_2$ where the first and second terms originates from $\mathcal{H}_1$ and $\mathcal{H}_2$, respectively, and are defined by $S_1 = \langle f|\mathcal{H}_1(\varepsilon_i - \mathcal{H}_0 + i\Gamma)^{-1}\mathcal{H}_1|i\rangle$ and $S_2 = \langle f|\mathcal{H}_2|i\rangle$ where $\mathcal{H}_0$ is the electronic part of the Hamiltonian, and Γ is the inverse of the core-hole lifetime. The explicit forms for the scattering cross sections are given as

$$S_1 = \left(\frac{m}{e^2}\right) \sum_{\alpha\beta} \varepsilon^\alpha_{\mathbf{k}_f \lambda_f} \sum_m \left[\frac{\langle f|J^\alpha_{-\mathbf{k}_f}|m\rangle\langle m|J^\beta_{\mathbf{k}_i}|i\rangle}{\varepsilon_i - \varepsilon_m - \omega_f} + \frac{\langle f|J^\beta_{\mathbf{k}_f}|m\rangle\langle m|J^\alpha_{-\mathbf{k}_i}|i\rangle}{\varepsilon_f - \varepsilon_m + \omega_i + i\Gamma} \right] \varepsilon^\beta_{\mathbf{k}_i \lambda_i}, \tag{50}$$

and

$$S_2 = \sum_{\alpha\beta} \varepsilon^\alpha_{\mathbf{k}_f \lambda_f} \langle f|R^{\alpha\beta}_{\mathbf{K}}|i\rangle \varepsilon^\beta_{\mathbf{k}_i \lambda_i}, \tag{51}$$

where we define $\mathbf{J}_\mathbf{k} = \sum_i \mathbf{J}_i(\mathbf{k})e^{i\mathbf{k}\cdot\mathbf{r}_i}$, and $R^{\alpha\beta}_\mathbf{K} = \sum_i R^{\alpha\beta}_i(\mathbf{K})e^{i\mathbf{K}\cdot\mathbf{r}_i}$. In Eq. (50), the intermediate electronic state in the second order perturbation is denoted by $|m\rangle$ with energy ε_m. When the X-ray energy is far from the absorption edge, S_2 and the parts of S_1 obtained in the limit of $\omega_i, \omega_f \gg \varepsilon_m - \varepsilon_i, \varepsilon_m - \varepsilon_f$ are the dominant contributions in the scattering cross section. On the other hand, when the X-ray energy is tuned around the absorption edge, the second term in Eq. (51) is resonantly enhanced, and often overcomes the contributions from S_2.

When RXS is utilized as a probe to detect the orbital order and the orbital excitation in the $3d$ transition-metal compounds, the dipole transition term in Eq. (45) is dominant. This is given explicitly as

$$\mathbf{J}^\eta_i(\mathbf{k}) = \frac{ie\hbar}{mc} \sum_{\gamma\gamma's} c^\dagger_{i\gamma s} c_{i\gamma's} \mathbf{D}^\eta_{i\gamma\gamma'}. \tag{52}$$

As shown in Eq. (50), the atomic scattering factor for RXS is the second rank tensor with respect to the X-ray polarization. When the X-ray energy is tuned around the $3d$ transition-metal K-edge, the indices γ and γ' in the dipole transition matrix in Eq. (33) are chosen to be the $1s$ orbital and one of the $4p$ orbitals given as

$$\mathbf{D}^{\eta}_{4p_\alpha 1s} = \delta_{\eta\alpha} \langle 4p_\alpha | \nabla^\alpha | 1s \rangle, \tag{53}$$

where α takes the Cartesian coordinate. When the X-ray energy is tuned around the L-edge of the $3d$ transition-metal ions, the $2p$ electrons are excited to the $3d$ orbitals, and S_1 directly reflects the symmetry of the $3d$ orbitals. Because of the strong spin-orbit interaction in the $2p$ orbitals, the dipole transition matrix is given as

$$\mathbf{D}^{\eta}_{3d_{\gamma\sigma} 2p(JJ^z)} = \langle 3d_{\gamma\sigma} | \nabla^\eta | JJ^z \rangle. \tag{54}$$

The wave functions for the $2p$ orbitals are represented explicitly by

$$|3/2, +3/2\rangle = \frac{1}{\sqrt{2}} \left(|p_{x\uparrow}\rangle + i |p_{y\uparrow}\rangle \right), \tag{55}$$

$$|3/2, +1/2\rangle = \frac{1}{\sqrt{6}} \left[2 |p_{z\uparrow}\rangle - \left(|p_{x\downarrow}\rangle + i |p_{y\downarrow}\rangle \right) \right], \tag{56}$$

$$|3/2, -1/2\rangle = \frac{1}{\sqrt{6}} \left[2 |p_{z\downarrow}\rangle + \left(|p_{x\uparrow}\rangle - i |p_{y\uparrow}\rangle \right) \right], \tag{57}$$

$$|3/2, -3/2\rangle = \frac{1}{\sqrt{2}} \left(|p_{x\downarrow}\rangle - i |p_{y\downarrow}\rangle \right), \tag{58}$$

for the L_3-edge case, and

$$|1/2, +1/2\rangle = \frac{1}{\sqrt{3}} \left[|p_{z\uparrow}\rangle + \left(|p_{x\downarrow}\rangle + i |p_{y\downarrow}\rangle \right) \right], \tag{59}$$

$$|1/2, -1/2\rangle = \frac{1}{\sqrt{3}} \left[|p_{z\downarrow}\rangle - \left(|p_{x\uparrow}\rangle - i |p_{y\uparrow}\rangle \right) \right], \tag{60}$$

for the L_2-edge case, where $|p_{\alpha s}\rangle$ represents the $2p$ wave function with orbital α and spin s.

In the RXS at the $3d$ transition-metal K-edge, the dipole transition occurs from the $1s$ to $4p$ orbitals in the intermediate states of RXS. Since the $3d$ orbitals are responsible mainly for the orbital order, the interactions between the $4p$ and $3d$ orbitals bring about the anisotropic tensor component in S_1. The orbital-dependent Coulomb interactions between the $3d$ and $4p$ electrons bring about the anisotropy in the $4p$ orbital states. The tensor elements in the scattering factor based on this mechanism was examined in the calculations by the exact diagonalization methods and the projection operator methods [46, 79, 83]. Another type of the mechanisms based on the JT-type lattice distortion was proposed for manganites; the anisotropic lattice distortion due to the $3d$ orbital order lifts the orbital degeneracy in the three

$4p$ orbitals. This was studied by using the first-principle band structure calculation, and the multiple-scattering theory [80–82]. In the two mechanisms, the relative level schemes of the $4p$ orbitals are opposite, and a magnitude correlation of $f_l(\mathbf{K})$ and $f_s(\mathbf{K})$ in Eq. (23) is different with each other. However, it is usually difficult to distinguish the two mechanisms through the conventional measurement of the RXS intensity at the orbital order peak. This is because the scattering intensity is proportional to $|f_l(\mathbf{K}) - f_s(\mathbf{K})|^2$ as shown in Eq. (24). Experimental studies to resolve the RXS mechanisms from the orbital order are introduced in a part of Chap. 2.

2.4 A Case Study of RXS: Impurity Effect in Orbital Order in KCuF$_3$

In this subsection, we introduce the theory and experimental researches for the impurity effects in the orbital ordered state as a case study of RXS applied to the correlated electron system with the orbital degree of freedom.

Impurity effects on the long range ordered state have been studied in a number of the correlated electron system. The high transition-temperature cuprate superconductors, and the colossal magnetoresistive manganites are the famous examples. Nonmagnetic impurity effect in the low-dimensional quantum spin systems is another example. A few percent doping of the magnetic ions, such as Mg and Zn, into the two-leg ladder cuprate $SrCu_2O_3$ [127, 128] and the spin-Peierls system $CuGeO_3$ [129, 130] brings about the antiferromagnetic long range orders.

Impurity effect in the long-range orbital ordered state was investigated in KCuF$_3$ utilizing RXS [131, 132]. The orbital order, in which the $d_{x^2-z^2}$ and $d_{y^2-z^2}$ orbitals for holes are alternately aligned with the momentum at (π, π, π), is confirmed by the RXS experiment. The Cu ions are substituted by the Zn ions where the electron configuration is d^{10} without the orbital degree of freedom. The RXS experiments where the X-ray energy is tuned around the Cu K-edge were performed at the (3/2 3/2 3/2) reflection point corresponding to the orbital superlattice reflection. The energy dependences of the RXS intensity in $KCu_xZn_{1-x}O_3$ for several values of x are shown in the inset of Fig. 9. The intensity is resonantly enhanced around the Cu K edge, and no large changes in the energy dependence are seen by the orbital dilution. However, the scattering intensity is rapidly reduced by doping. The intensity as a function of the Zn concentration is plotted in Fig. 9. The results indicate that the long-ranged orbital order disappears around $x = 0.6$, which almost coincides to the concentration where the cubic to tetragonal structural phase transition occurs. It is also suggested experimentally that the peak profiles are factorized as a form expected from the fractal pattern as $I(q) \sim 1 + Cq^{-(d-D_f)}$ with $d = 3$, a constant C, and the fractal dimension D_f. Value of D_f was deduced to be 3.2. Another type of the orbital dilution effect, and the impurity induced structural and magnetic phase transitions were examined in $Fe_{1-x}Mn_xCr_2O_4$, where the orbital and lattice structures were measured by the high-resolution X-ray diffraction [133].

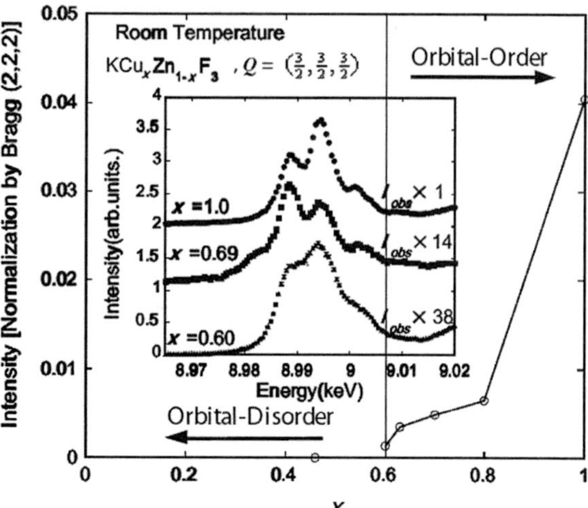

Fig. 9 The RXS intensities at (3/2 3/2 3/2) reflection in KCu$_{1-x}$Zn$_x$F$_3$. The *inset* shows the energy dependences of the RXS intensities for several Zn concentrations. Excerpt from Ref. [131]

In dilute magnets, such as, KMn$_{1-x}$Mg$_x$F$_3$, the x dependence of the magnetic ordering temperature as well as the critical concentration where the magnetic order vanishes are well explained by the percolation theory [134, 135]. On the contrary, the critical concentration in KCu$_x$Zn$_{1-x}$F$_3$, where the orbital order disappears, is much larger than the percolation threshold in a simple-cubic lattice, $x_p = 0.31$. Thus, the experimental observations introduced above imply that the dilution effects on the orbital order are not understood within the conventional percolation theory. The impurity effects on the orbital order were examined theoretically [30, 136, 137]. The orbital only model introduced in Eq. (14) was modified to analyze the quenched impurity effects as

$$\mathcal{H} = J \sum_{\langle ij \rangle} \tau_i^l \tau_j^l \varepsilon_i \varepsilon_j, \tag{61}$$

where ε_i takes 1 or 0 for an ion with and without the orbital degree of freedom at site i. The calculated orbital ordering temperatures are presented in Fig. 10, where we note that the impurity concentration is represented by x. As comparisons, the magnetic ordering temperatures in the XY, Ising and Heisenberg models are plotted. It is shown that the reduction of the orbital ordering temperature is much stronger than that of the magnetic ordering temperatures. The real space configurations of the orbital PSs are shown in Fig. 10 for $x = 0.1$ and 0.3. Near the impurity sites, remarkable tiltings of the PSs from the ordered state are confirmed. This disturbing the PS configuration becomes remarkable at $x = 0.3$. This is caused by the local-symmetry breaking by the orbital dilution. Let us focus on an orbital PS at site i and an impurity located at $\mathbf{r}_i + \hat{x}a$ where $\hat{x}$ and a represent the unit vector along x and the lattice constant, respectively. As explained in the orbital only model introduced in Eq. (61), the interactions between the orbitals explicitly depend on the bond direction. Due to the existence of the impurity, one of the interactions along the x direction,

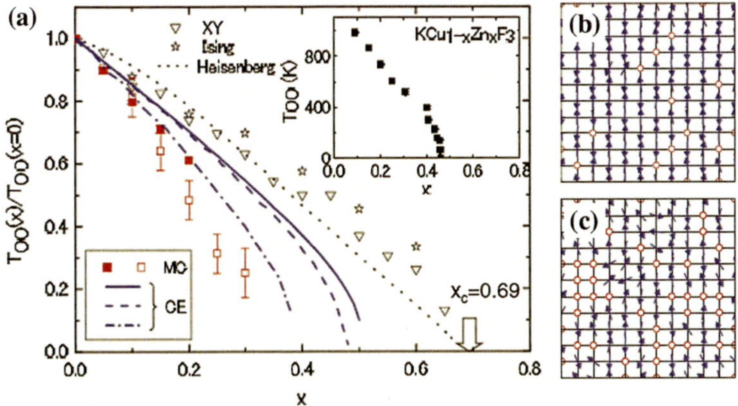

Fig. 10 (*Left*) Impurity-concentration dependence of the orbital-ordering temperature in the orbital only model. The ordering temperatures in the XY model, Ising model, and the Heisenberg model are also shown. The *inset* shows the Zn concentration dependence of the orbital ordering temperatures in $KCu_{1-x}Zn_xF_3$ obtained by the RXS experiments [131]. (*Right*) A snapshot in the Monte Carlo simulation for the PS configuration at $x = 0.1$ (*upper*), and that at $x = 0.3$ (*lower*). *Arrows* and *open circles* indicate PSs and the impurities, respectively. Data are excerpt from Ref. [136]

i.e. $J \sum_{\langle ij \rangle_x} (-\frac{1}{2} T_i^z + \frac{\sqrt{3}}{2} T_i^x)(-\frac{1}{2} T_i^z + \frac{\sqrt{3}}{2} T_i^x)$ vanishes. As a result, the PS tilts to gain the interaction energies in other bonds. This is unique in the orbital system and is highly in contrast to the diluted magnets.

3 RIXS and Orbital Excitation

3.1 Collective Orbital Excitation

It is well recognized that the collective excitations in crystalline solids contribute to a number of the low-energy physical properties, such as thermodynamics, transport, magnetism and so on. Owing to the Goldstone's theorem, when the continuous symmetry is broken spontaneously, a gapless excitation mode appears at zero momentum. When the orbital order is realized, the collective orbital excitation, termed the orbiton, is expected to appear. The realistic interacting orbital model does not have the continuous symmetry, as suggested in Sect. 2.1. This is highly in contrast to the interacting spin systems, in which several continuous symmetries, e.g. SU(2), or O(2), exist and the total spin angular momentum or its z component are conserved. In this article, we use "the dd excitation" for the electronic charge excitations in the local multiplets in the transition-metal compounds, where the ground state orbital is separately well from the excited ones due the large crystalline field effect, e.g. the high transition-temperature superconducting cuprates, and NiO. On the other hand, in the systems

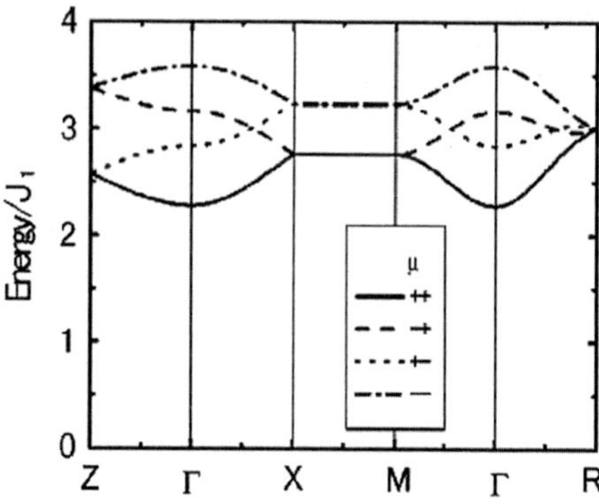

Fig. 11 A dispersion relation of the collective orbital excitation in the A-type antiferromagnetic phase [140]. Index μ represents the excitation modes

where the ground state orbitals are essentially degenerated within the energy scales of temperature and external fields, e.g. perovskite manganites and titanates, we use "the collective orbital excitation" for the charge excitations between these orbitals which can propages through the intersite interactions.

The collective orbital excitation in the long-range orbital ordered state was first proposed theoretically by Cyrot and Lyon-Caen [9], and Komarov et al. [138]. In Ref. [9], the spin-orbital model with the continuous symmetry, similar to the model in Eq. (12), was analyzed, and the gaplless orbital excitation modes are shown at the Brillouine zone center. The dispersion relations in a realistic orbital model given in Eq. (10) were first calculated in Refs. [17, 139, 140] as shown in Fig. 11. There, the JT-type lattice distortion was assumed to be frozen, and the orbital-lattice interaction was introduced as a static potential for the orbital degree of freedom. These are justified in the case that the orbital excitation energy is much larger than the phonon energy, and the lattice vibrations do not follow the coherent orbital motion. After the calculations in Refs. [17, 139, 140], several studies in the collective orbital excitations have been performed. The interaction between the orbital degree of freedom and the Einstein-type phonon was introduced by the perturbational approximation in Ref. [141]. Owing to the bilinear coupling between the orbiton and the phonon, an anticrossing behavior occurs in the dispersion relations. In Ref. [142, 143], the orbital excitation at an isolated octahedron was examined by taking the vibronic coupling into account. The results are shown in Fig. 12. The electronic orbital excitation associated with the multiple phonons were obtained by the adiabatic approximation, and the optical spectra from the vibronic excitations were calculated in the Franck-Condon scheme.

A unified treatment of the collective orbital excitations, in which both the inter-site exchange interactions and the local vibronic interaction due to the JT coupling are taken into account, were presented in Ref. [145]. The model is given by a sum of the inter-site exchange interaction introduced in Sect. 2.1 and the on-site JT coupling where the doubly degenerate e_g orbitals considered at each site. The JT coupling term is defined as

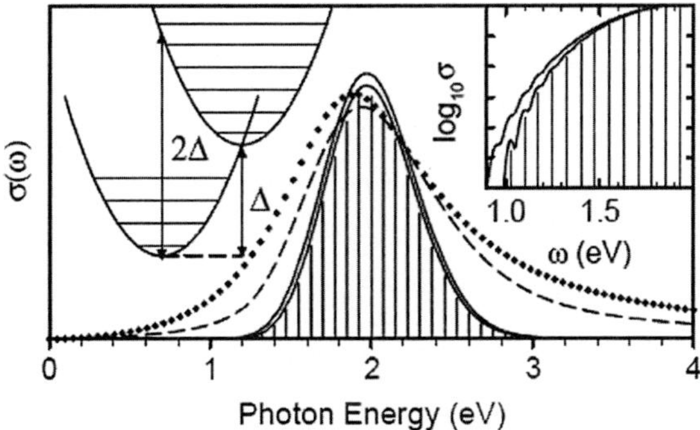

Fig. 12 Optical conductivity spectra in the orbital ordered LaMnO$_3$ due to the self-trapped electronic orbital excitation [142]. The points are the experimental data [144]. Excerpt from Ref. [142]

$$\mathcal{H}_{JT} = \sum_{i\, l=(u,v)} \left(-\frac{1}{2M}\frac{\partial^2}{\partial Q_{il}^2} + \frac{M\omega_0^2}{2} Q_{il}^2 \right) + 2A \sum_i \left(T_i^z Q_{iu} + T_i^x Q_{iv} \right),$$

$$= \sum_i \left\{ \omega_0 \sum_{l=(u,v)} b_i^{l\dagger} b_i^l - g T_i^z (b_i^{u\dagger} + b_i^u) + \sqrt{2/(M\omega_0)}\, A T_i^x (b_i^{v\dagger} + b_i^v) \right\},$$

$$\tag{62}$$

where Q_{iu} and Q_{iv} are the vibrational modes in an octahedron at site i, and M and ω_0 are the ion mass and the vibrational frequency, respectively. In the second line, the Hamiltonian is rewritten by introducing the phonon operators b_i^l. By applying the mean-field approximation to the inter-site exchange interactions, the Hamiltonian is reduced into the local interactions. By describing the transitions between the local eigen states by the boson operators, the Hamiltonian is obtained as $\mathcal{H} = \sum_{\mathbf{q}\eta} \Omega_{\mathbf{q}\eta} \alpha_{\mathbf{q}\eta}^\dagger \alpha_{\mathbf{q}\eta}$ with the vibronic excitation energies $\Omega_{\mathbf{q}\eta}$ at momentum $\mathbf{q}$ and mode η. The collective vibronic excitations in the $d_{3x^2-r^2}/d_{3y^2-r^2}$-type orbital order are shown in Fig. 13 for several values of the JT coupling. In the case without the JT coupling [see Fig. 13a], the well defined dispersion relations of the collective excitation modes are seen. These are the pure electronic-orbital excitations shown in Ref. [17]. With increasing the JT coupling constant, the excitation spectra are pushed up and become broad. In the large JT coupling, the discrete dispersion-less modes appear and their energy separation is given by about ω_0. These are interpreted as the multiple phonon excitations suggested in Ref. [143]. In the low energy region [see Fig. 13d], additional dispersive excitation modes appear. The dispersion relation has the similarity to those in the pure electronic orbital excitation shown in Fig. 13a. The low energy collective modes are identified as the "renormalized" orbiton mode. A schematic picture of the modes is also presented in Fig. 13. This is the collective vibronic mode where the local rotational mode on the lower adiabatic-potential plane owing to the JT coupling propagates through the inter-site exchange interactions.

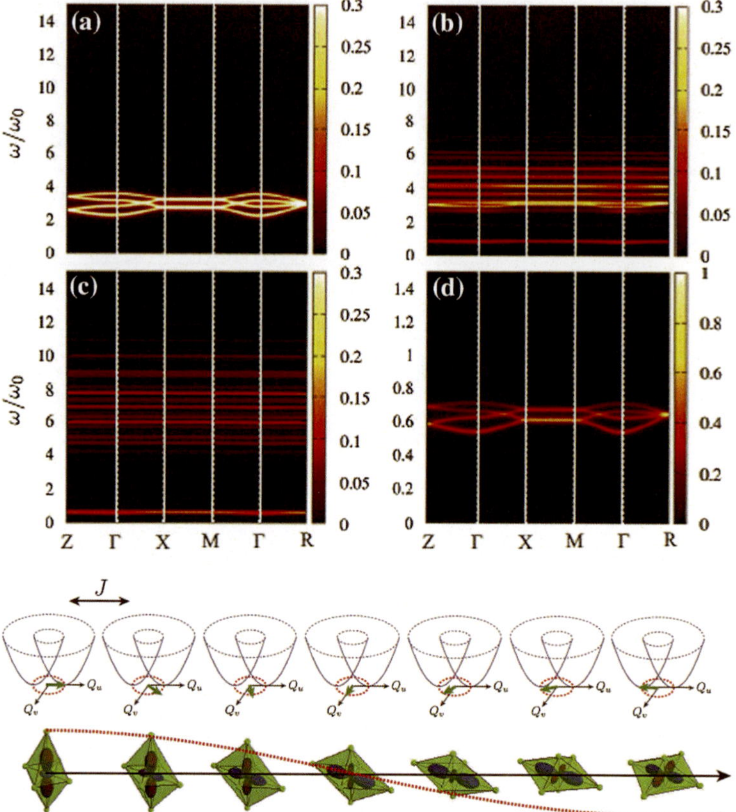

Fig. 13 (*Upper*) Vibronic excitation spectra at **a** $g/\omega_0 = 0$, **b** 1, and **c** 2. **d** An expansion of the low-energy region of (**c**) in the e_g orbital model with the JT coupling. A schematic picture for the collective vibronic excitation. (*Lower*) Adiabatic-potential planes in a lattice where *arrows* represent directions of a vector (Q_u, Q_v), and a schematic vibronic excitation modes. Excerpt from Ref. [145]

3.2 NIXS from Orbital Excitation

Direct experimental observations of the collective orbital excitation in the correlated electron systems are accepted as a challenging issue. The Raman scattering has been considered as a suitable tool to detect the collective orbital excitations, since the *dd* excitations are Raman active [146]. The Raman spectra were reported at the first time in the orbital ordered $LaMnO_3$ [147, 148]. After the first measurement, the experimental and theoretical studies for observation of the orbital excitations have been reported in several transition-metal oxides [149–154]. One limitation in the Raman scattering experiments is that the excitations at zero momentum transfer are only observed. The inelastic X-ray and neutron scatterings are the possible tools to observe the momentum dependent elementary excitations. In the case where the orbital excitations are associated with changes in the magnetic moments, the inelastic neutron scattering is available to detect them [151, 155]. In this section, before reviewing RIXS from the orbital excitations, we introduce NIXS as a probe to detect the dispersion relations of the orbital excitations.

The NIXS method can access to a variety of elementary excitations in crystalline solids and liquids. In the differential scattering cross section formulated in Sect. 2.3, we consider the case where the X-ray energy is far from any absorption edges. The relations $\omega_f \gg E_m - E_f$ and $\omega_i \gg E_m - E_i$ in Eq. (50) diminish the amplitude of the scattering due to S_1, and the scattering originating from S_2 in Eq. (51) is dominant. As shown in the definition of the charge density in Eq. (41), there are the off-diagonal matrix elements, i.e. $\gamma \neq \gamma'$ which describe the orbital excitations. As shown in Eq. (46), the polarization vectors appear in the scattering cross section as an inner product $\varepsilon_{\mathbf{k}_i \lambda_i} \cdot \varepsilon_{\mathbf{k}_f \lambda_f}$, i.e. no polarization dependence. The energy resolution is now less than 10 meV which is sufficiently enough for observation of the low-energy collective excitations. The physical interpretation is simple in comparison with that in RIXS derived from the higher order perturbational processes.

A possibility of the observation of the orbital excitation by the NIXS technique was suggested in Ref. [5], and the experimental measurements were carried out in KCuF$_3$, YTiO$_3$ and LaMnO$_3$ in Ref. [156]. Away from the orbital degenerated systems, the dd excitations was observed in NIXS in the transition-metal mono-oxides, CoO and NiO in Ref. [157]. As shown in Fig. 14, sharp peak structures were observed around 1–3 eV, and do not show remarkable dispersion relations. Through the comparison with the correlation functions calculated by the first-principle calculation and the random-phase approximation, these are assigned as the dd excitations in Ni and

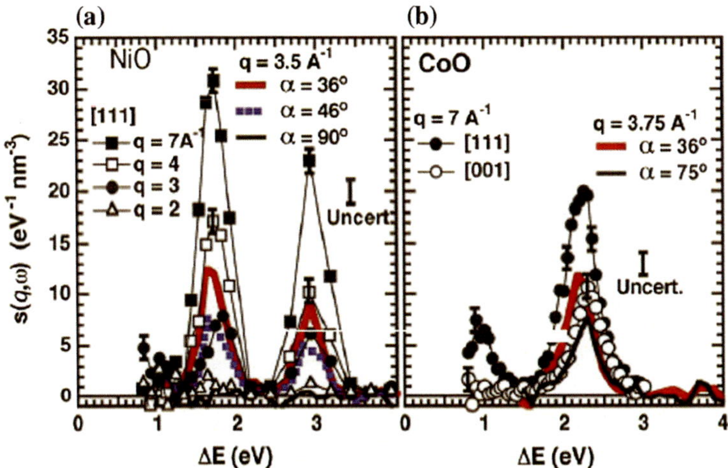

Fig. 14 (*Upper*) The NIXS spectra for the dd excitations in NiO and CoO. A symbol α represents the momentum q orientation angle between the 110 and 001 directions. Excerpt from Ref. [157]

Co ions. Motivated from the experiments, the scattering cross sections for the dd excitations in NIXS were formulated theoretically in Refs. [158–160], and were applied to the iron arsenides [161].

The scattering intensity of NIXS is given in Eq. (51) where the first term of Eq. (46) is responsible for the scattering. This originates from the first order perturbation of the A^2 term in the electron-photon interaction. The scattering cross section is related to the dynamical correlation function defined as [158–160]

$$S(\mathbf{K}, \omega) = \int dt e^{-i\omega t} \langle \rho(\mathbf{K}, t)\rho(-\mathbf{K}, 0)\rangle, \tag{63}$$

where $\rho(\mathbf{K}, t)$ represents the charge density operator defined by

$$\rho(\mathbf{K}) = \sum_{\gamma\gamma'} f_{\gamma\gamma'}(\mathbf{K}) \sum_{s\mathbf{q}} c^\dagger_{\mathbf{q}+\mathbf{K}\gamma s} c_{\mathbf{q}\gamma' s}, \tag{64}$$

where $f_{\gamma\gamma'}(\mathbf{K})$ is defined in Eq. (42). The charge operator is expanded by using the spherical harmonic function $Y_{lm}(\mathbf{q})$ as

$$\rho(\mathbf{K}) = \sum_{i} e^{i\mathbf{K}\cdot\mathbf{r}_i} \sum_{l=(0,2,4)} A_l(\mathbf{K}) \sum_{m=-l}^{l} Y^*_{lm}(\mathbf{K})w^l_m. \tag{65}$$

We define

$$A_l(\mathbf{K}) = i^l 4\pi \sqrt{\frac{2l+1}{4\pi}} P_l(|\mathbf{K}|)C^{20}_{l010}, \tag{66}$$

$$w_l = \sum_{\gamma\gamma's}\sum_{n_1,n_2} f^{(\gamma)*}_{n_1} f^{(\gamma')}_{n_2} C^{2n_1}_{lm2n_2} c^\dagger_{i\gamma s} c_{i\gamma's}, \tag{67}$$

and

$$P_l(|\mathbf{K}|) = \int dr r^2 |R_{3d}(r)|^2 j_l(|Kr|), \tag{68}$$

where $C^{\Gamma\gamma}_{\Gamma_1\gamma_1\Gamma_2\gamma_2}$ is the Clebsh–Gordon coefficient, $R_{3d}(r)$ is the Laguerre function, and $j_l(|qr|)$ is the spherical Bessel function. We consider the e_g orbital orders and introduce the coefficients $f^{(3z^2-r^2)}_0 = 1$, $f^{(x^2-y^2)}_2 = 1/\sqrt{2}$ and $f^{(x^2-y^2)}_{-2} = -1/\sqrt{2}$. When $\gamma \neq \gamma'$ in Eq. (67), the NIXS occurs from the orbital excitation from γ' to γ.

Here, we demonstrate the NIXS intensity from the collective orbital excitation in $KCuF_3$ where the $d_{z^2-x^2}$ and $d_{y^2-z^2}$ orbitals for holes are ordered as shown in Fig. 5. The four orbital excitation modes appear due to the four kinds of the orbitals in a unit cell. The scattering intensities in NIXS for each excitation mode are plotted as

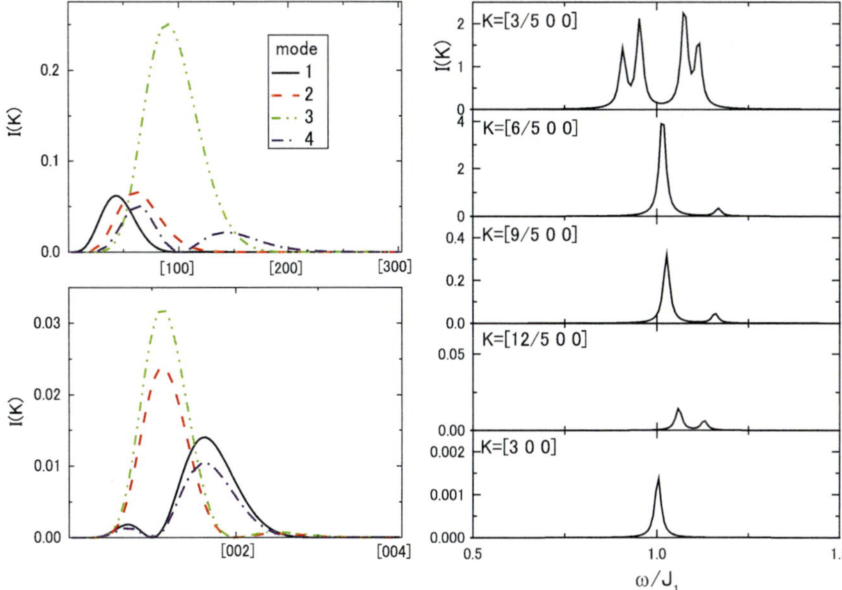

Fig. 15 (*Left*) Momentum dependences of the scattering intensities in NIXS from the orbital excitation in the ($d_{y^2-z^2}/d_{z^2-x^2}$) orbital ordered state. The momentum transfers are chosen along [$h00$] (*upper*) and [$00l$] (*lower*) directions. Each *line* represent the scattering intensity for each excitation mode. (*Right*) The NIXS spectra. The momentum transfer is chosen to be along [$h00$] [162]

functions of the momentum transfer **K** for different two directions. Several characteristics are seen in the results; (1) The intensity at **K** = 0 is zero, since $\langle\gamma|\gamma'\rangle = 0$ if $\gamma \neq \gamma'$. (2) The scattering intensities take their maximum around $|\mathbf{K}| \sim r_{3d}^{-1}$ where r_{3d} is the radius of the $3d$ orbitals. (3) The scattering intensity depends strongly on the direction of **K** and the modes of the orbital excitations. For example, when **K** is taken to be along [$h00$], the scattering intensity originating from the excitation mode 3 is dominant. The NIXS spectra from the collective orbital excitation are presented in the right panels in Fig. 15 where **K** is taken to be along [$h00$]. A broadening parameter value corresponding to the energy resolution is chosen to be 1 % of the exchange interaction. It is well demonstrated that the energy dispersions of the collective orbital excitations are able to be detected in the NIXS technique.

3.3 *RIXS from Orbital Excitation: Polarization Analyses*

3.3.1 RIXS in Orbital Ordered and Related Materials

The RIXS technique is now recognized widely as a powerful experimental probe to observe the momentum dependent charge, spin, orbital and structural excitations

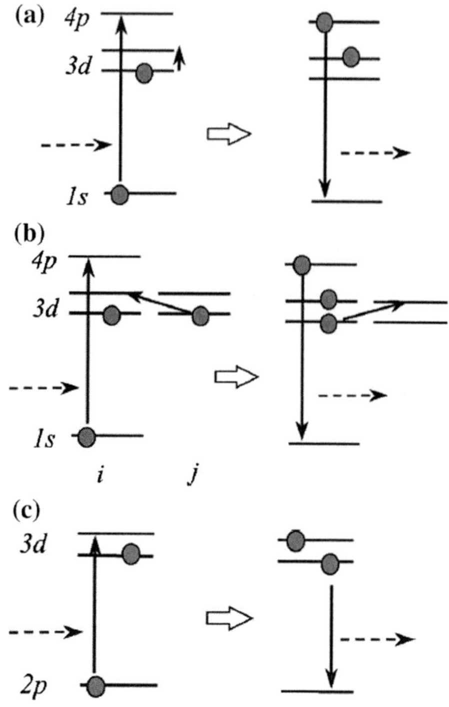

Fig. 16 **a, b** Possible orbital excitation processes in the K-edge RIXS, and **c** that in the L-edge RIXS for the $3d$ transition-metal compounds [140]. The *broken arrows* indicate the incoming and outgoing X-rays. A symbol j denotes one of the NN transition-metal sites or the NN ligand sites of i

in solids. This has been widely applied to a number of the correlated electron system. In this subsection, we introduce the studies in RIXS from the orbital excitation in the orbital ordered and related correlated electron systems. We focus on the X-ray polarization dependences of the RIXS intensity, which is crucially important to identify the scattering from the orbital excitations. Also see other reviews for RIXS in correlated electron systems [163–165].

The possible RIXS processes due to the orbital excitations are shown schematically in Fig. 16 for the K- and L-edge RIXS in the $3d$ transition-metal compounds [140]. Since one electron is excited from the $1s$ to $4p$ orbitals in the K-edge RIXS, additional excitation processes are necessary to induce the $3d$ orbital excitations. In the process shown in Fig. 16a, the orbital excitation is generated due to the Coulomb interaction between the electrons in the intermediate states. In the process shown in Fig. 16b, the charge transfer between the ligand p orbitals and the transition-metal $3d$ orbital is induced by the core hole potential. On the other hand, the orbital excitation is induced directly in the L-edge RIXS as shown in Fig. 16c. The incident X-ray excites an electron from the $2p$ to $3d_\gamma$ orbitals, and an electron is relaxed from the $3d_{\gamma'}$ orbital to the unoccupied $2p$ orbital with emitting the X-ray. When $\gamma \neq \gamma'$, this process implies the orbital excitation.

There are the selection rules between the symmetries of the orbital excitations in RIXS and the polarizations of the incoming and outgoing X-rays. This is seen in the following compact from of the cross section of RIXS and the group theoretical analyses [84]. The scattering cross section given in Eq. (49) with Eq. (50) is rewritten as

$$\frac{d^2\sigma}{d\Omega d\omega_f} = A\frac{\omega_f}{\omega_i} \sum_{\alpha\beta\alpha'\beta'} P_{\beta'\alpha'} P_{\beta\alpha} \Pi_{\beta'\alpha'\beta\alpha}(\omega, \mathbf{K}), \tag{69}$$

where $\Pi_{\beta'\alpha'\beta\alpha}(\omega, \mathbf{K})$ is the Fourier transformation of the correlation function of the polarizability operator. This is defined by

$$\Pi_{\beta'\alpha'\beta\alpha}(\omega, \mathbf{K}) = \frac{1}{2\pi} \int dt e^{i\omega t} \sum_{ll'} e^{-i\mathbf{K}\cdot(\mathbf{r}_{l'}-\mathbf{r}_l)} \langle i|\alpha^\dagger_{l'\beta'\alpha'}(t)\alpha_{l\beta\alpha}(0)|i\rangle, \tag{70}$$

with the momentum transfer $\mathbf{K} = \mathbf{k}_i - \mathbf{k}_f$ and the energy transfer $\omega = \omega_i - \omega_f$. The polarization part is defined as $P_{\beta\alpha} = \varepsilon^\beta_{\mathbf{k}_i \lambda_f} \varepsilon^\alpha_{\mathbf{k}_i \lambda_i}$. The polarizability operator is given by $\alpha_{l\beta\alpha}(t) = e^{iHt}\alpha^{(\omega)}_{l\beta\alpha}e^{-iHt}$ with the definition

$$\alpha^{(\omega)}_{l\beta\alpha} = i \int_{-\infty}^{0} dt e^{-i\omega t}[\mathbf{J}^\beta_l(0), \mathbf{J}^\alpha_l(t)], \tag{71}$$

where $\mathbf{J}^\alpha_l$ is the dipole operator introduced in Eq. (52) at site l. The similar expressions of the scattering cross section represented by the correlation functions were proposed in Refs. [166, 167]. The polarizability operator is expanded with respect to products of the PS operators from the viewpoint of the group theoretical analyses. We consider a case where the doubly degenerate e_g orbitals are introduced at each site in the cubic lattice. The coupling constants, reflecting the polarizations of X-rays, have the $T_{1u} \times T_{1u}$ symmetry, which is reducible to $A_{1g} + E_g + T_{1g} + T_{2g}$. The charge operator defined by $T^0_l \equiv \sum_{\gamma s} c^\dagger_{\gamma a}c_{\gamma s}$ has the A_{1g} symmetry, and the orbital PS operators $\mathbf{T}^\mu_l$ have the symmetries of E_{gv}, A_{2g}, and E_{gu} for $\mu = (x, y, z)$, respectively. Since the polarizability should have the A_{1g} symmetry, we have

$$\alpha_{l\beta\alpha} = \delta_{\alpha\beta} I_{A_{1g}} T^0_l + \delta_{\alpha\beta} I_{E_g} \left(\cos\frac{2\pi n_\alpha}{3} T^z_l - \sin\frac{2\pi n_\alpha}{3} T^x_l \right), \tag{72}$$

where $I_{A_{1g}}$ and I_{E_g} are the coupling constants being independent of the X-ray polarizations. The spin flip scattering is not taken into account. It is worthnoting that T^y_l does not appear, implying that the charge octupole moment is not observed by this method. By inserting Eq. (72) in Eq. (70), we have the expression as

$$\Pi_{\beta'\alpha'\beta\alpha}(\omega, \mathbf{K}) = \delta_{\beta'\alpha'}\delta_{\beta\alpha} \frac{1}{2\pi} \int dt e^{i\omega t} \sum_{ll'} e^{-i\mathbf{K}\cdot(\mathbf{r}_{l'}-\mathbf{r}_l)}$$
$$\times \sum_{\gamma\gamma'=(0,x,z)} I_{\gamma'\alpha'} I_{\gamma\alpha} \langle T^{\gamma'}_{l'}(t) T^\gamma_l(0)\rangle, \tag{73}$$

where $I_{0\alpha} = I_{A_{1g}}$, $I_{x\alpha} = I_{E_g}\cos(2\pi_\alpha/3)$, and $I_{z\alpha} = -I_{E_g}\sin(2\pi_\alpha/3)$. This expression provides phenomenologically the relation between the polarization dependences of the RIXS intensity and the excited orbitals.

The first experimental study for the polarization dependences of the RIXS intensity in the orbital-ordered correlated electron systems was performed in Ref. [112]. The K-edge RIXS experiments in LaMnO$_3$ observed broad spectra around 2.5 eV, which are attributable to the individual electronic excitations across the Mott gap. The electron excitations from the occupied band consisting of the $d_{3x^2-r^2}/d_{3y^2-r^2}$ orbitals to the unoccupied $d_{y^2-z^2}/d_{z^2-x^2}$ band are expected to be induced by the processes shown in Fig. 16a. The characteristic polarization dependence in the RIXS intensity was confirmed in the 2.5 eV peak, in which the polarization of the incident X-ray is chosen to be the π configuration, and the azimuthal angle (ψ) rotation was carried out. The intensity shows the two-fold oscillatory behavior as a function of ψ shown in Fig. 17. This is well reproduced by the theoretical calculations that the intensity is proportional to $\sin^2\psi$ in which the $d_{3x^2-r^2}/d_{3y^2-r^2}$-type orbital order is assumed [168]. The polarization dependence of the RIXS spectra was also measured in the doped manganites La$_{1-x}$Sr$_x$MnO$_3$ with x=0.2 and 0.4, where the long-range orbital order is not confirmed [113]. With increasing x, an energy gap around 2.5 eV observed in LaMnO$_3$ is filled and the low energy spectra emerge. This is consistent with the doping dependence of the optical conductivity spectra. The spectral intensity at 2 eV shows a similar azimuthal-angle dependence to that in LaMnO$_3$ as shown in Fig. 18. This result implies that the $d_{3x^2-r^2}/d_{3y^2-r^2}$-type short-range orbital correlation remains in the doped manganites. The intersite charge transfer excitations detected by RIXS in manganites were examined in Refs. [115, 117]. Motivated from these experimental results, the RIXS spectra in the orbital ordered manganites were

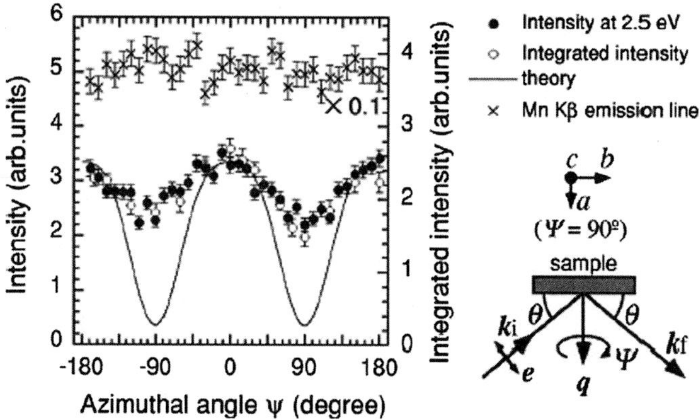

Fig. 17 Azimuthal angle (Ψ) dependence of the RIXS intensity in LaMnO$_3$. *Open* and *solid circles* show the integrated intensity of the 2.5 eV peak and the intensity at the energy transfer 2.5 eV, respectively. A *solid line* shows the integrated intensity of the theoretically calculated RIXS spectra. The experimental setup for the azimuthal angle dependence measurement is also shown. Excerpt from Ref. [112]

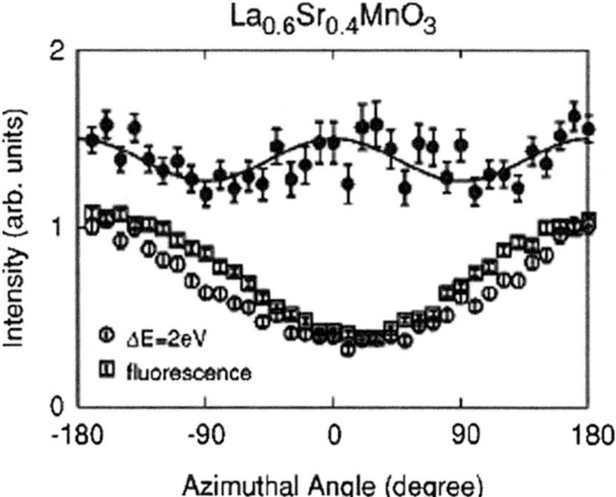

Fig. 18 Azimuthal angle dependence of the RIXS intensity in La$_{0.6}$Sr$_{0.4}$MnO$_3$. The *open circles* and *squares* are the scattering intensity of the energy transfer 2 eV at $Q = (2.7, 0, 0)$ and the fluorescence yield, respectively. The filled circles represent the scattering intensity divided by the fluorescence. The *solid lines* are the fitting result using a function of $A(1 + B \cos 2\psi)$. Excerpt from Ref. [113]

analyzed theoretically by the Hartree-Fock method [168] and the Keldysh Green function method [169] applied to the multi-band Hubbard model.

Away from the orbital degenerated materials, a number of the polarization analyses of the K-edge RIXS in the $3d$ transition-metal oxides have been applied to the electronic dd excitations. The RIXS studies at the Cu K-edge in the superconducting cuprates and Sr$_2$CuO$_2$Cl$_2$ [116, 170–173] are reviewed Chap. 6. The dd excitations in NiO and NiCl$_2$ were observed around 1–3 eV in the RIXS spectra at the Ni K-edge. The experimental results are shown in Fig. 19. The polarization of the incoming X-ray is perpendicular to that of the outgoing X-ray. Through the comparison with the theoretical calculations in a small-size cluster, these peaks are attributable to the excitations between the local dd multiplets in Ni ions. The excitations are interpreted to be due to the Coulomb interactions between the excited $4p$ electron and the valence $3d$ electrons [174].

Recently, the soft X-ray RXS technique has been greatly developed as a probe to detect the orbital excitation, as well as the dd excitations, in the $3d$ transition-metal oxides. The detailed theoretical study for the polarization analyses of RIXS and the orbital excitations was presented in Ref. [175, 176]. The RIXS at Ti L-edge was applied to YTiO$_3$ and LaTiO$_3$, where the t_{2g} orbital degree of freedom exist [177], and observed broad peak structures around 0.2–0.4 eV. Both the peak energies and intensities show the clear momentum dependence in the two materials. Through the comparison with the calculation based on the spin-orbital model with the crystalline field effect, these were interpreted to be the collective orbital excitations [178]. The clear dispersion relations of the dd excitations located around 1.6–3 eV were observed in an one dimensional Mott insulator Sr$_2$CuO$_3$. The relation between these dd excitations and magnetic excitations were discussed [179]. The dispersion relations were also observed in the RIXS experiments in CaCuO$_3$ with the ladder-type structure, and differences from the results in the one-dimensional Mott insulator were examined [180]. The RIXS from the dd and magnetic excitations in

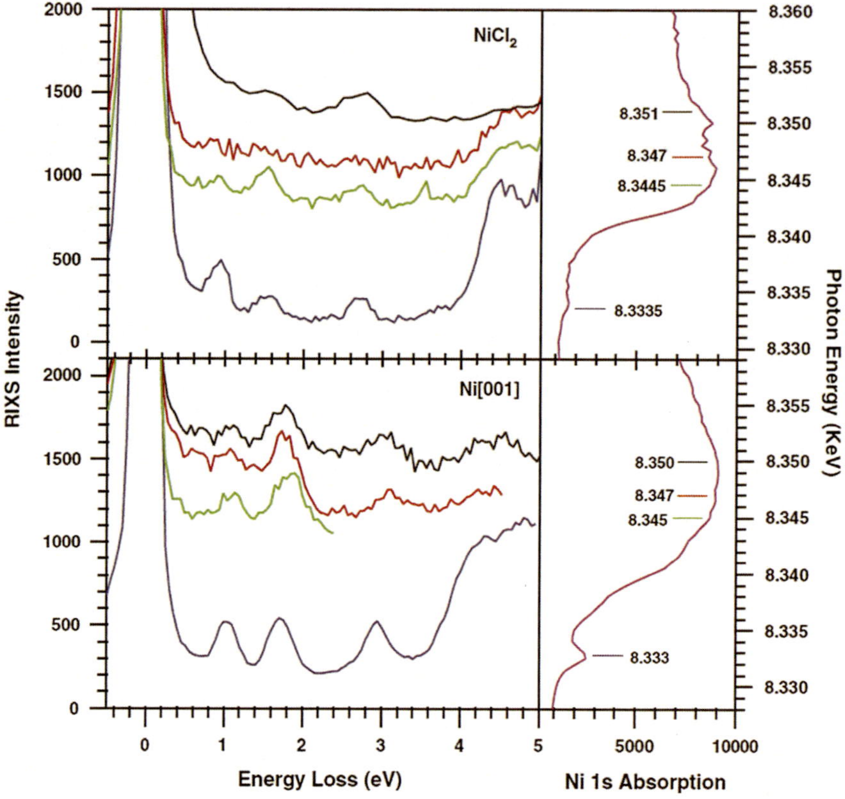

Fig. 19 Ni K-edge RIXS of NiO[001] and NiCl$_2$. The X-ray absorption is shown in the *right panel*. RIXS spectra for different excitation energy are shown in the *left panel*. Excerpt from Ref. [174]

the one-dimensional Mott insulators were analyzed theoretically in Refs. [181, 182]. This result is introduced in more detail in Chap. 6.

3.3.2 A Case Study of RIXS: Polarization Analyses in KCuF$_3$

In this subsection, we introduce the polarization dependences of the RIXS intensity in the orbital ordered KCuF$_3$ [118]. Observed typical RIXS spectra are shown in Fig. 20. The weak peak around 1.2 eV and the broad spectra above 5 eV are attributable to the Cu $3d$ orbital excitations and the one electron excitation across the Mott gap and the charge transfer gap, respectively. The low energy spectra below 3 eV were measured in several experimental configurations where the polarizations of the incident and scattered X-rays are varied. One example is shown in Fig. 20 where the experimental geometry is also shown in the inset. The clear polarization dependences are confirmed; two components at around 1 and 1.5 eV are seen, and the lower energy

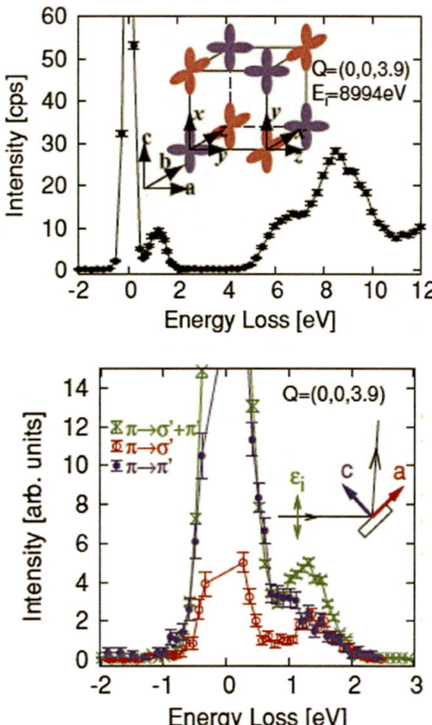

Fig. 20 (*Upper*) The RIXS spectrum of KCuF$_3$. (*Lower*) Polarization-analyzed RIXS spectra. Spectra with *filled* and *open circles* are measured in the $\pi \to \sigma$ and $\pi \to \pi$ conditions, respectively. The spectra without polarization analysis are also shown by *crosses*. Corresponding experimental geometries are shown in *insets*. Excerpt from Ref. [118]

component is pronounced in the $\pi \to \pi'$ configuration. The experimental results are introduced in Chap. 6 in more detail.

The polarization dependence of the RIXS spectra from the orbital excitation is phenomenologically interpreted from the symmetrical consideration introduced in Sect. 3.3.1. The local symmetry of the Cu atom in each orbital sublattice is approximated as D$_{4h}$. Owing to the group theoretical analyses, the RIXS can occurs, when the product representation

$$P_i \times P_f \times \Gamma_i \times \Gamma_f, \tag{74}$$

includes A$_{1g}$ in D$_{4h}$, where $P_{i(f)}$ and $\Gamma_{i(f)}$ are the irreducible representations for the incoming (outgoing) X-ray polarization, and the initial (final) electronic $3d$ orbital, respectively. The symmetries of $\Gamma_{d_{xy}} \times \Gamma_{d_{x^2-y^2}}$, $(\Gamma_{d_{yz}}, \Gamma_{d_{zx}}) \times \Gamma_{d_{x^2-y^2}}$, and $\Gamma_{d_{3z^2-r^2}} \times \Gamma_{d_{x^2-y^2}}$ are identified as A$_{2g}$, E$_g$, and B$_{1g}$, respectively. In the experimental geometry shown in the lower panel in Fig. 20, the incoming and outgoing X-ray polarizations $(\varepsilon_{\mathbf{k}_i \lambda_i}, \varepsilon_{\mathbf{k}_f \lambda_f})$, respectively in the two orbital sublattices are $(y + z, x)$ and $(x + y, z)$ in the $\pi \to \sigma'$ case, and are $(y + z, y - z)$ and $(x + y, x - y)$ in the $\pi \to \pi'$ cases. The symmetries of $P_i \times P_f$ are $\{A_{2g}, B_{2g}, E_g\}$ and $\{A_{1g}, A_{2g}, B_{1g}, E_g\}$ in the $\pi \to \sigma'$ and $\pi \to \pi'$ polarization cases, respectively. Therefore, the $t_{2g} \to e_g$ orbital excitations are possible in both the $\pi \to \sigma'$ and

$\pi \to \pi'$ cases, and the $e_g \to e_g$ excitations only occur in the $\pi - \pi'$ case in this experimental geometry. From these analyses, the spectra around 1 and 1.5 eV in Fig. 20 are ascribed to the $e_g \to e_g$ and $t_{2g} \to e_g$ excitations, respectively.

The microscopic calculations of the polarization dependences of the RIXS in KCuF$_3$ was performed in Refs. [162, 183] where the RIXS spectra were calculated by the exact diagonalization method in the small cluster. Also see the theoretical calculations in Ref. [184] where the RIXS spectra are calculated by the Keldysh Green function formula. The RIXS spectra were obtained based on the model where the $3d$ orbital part of the Hamiltonian given in Eq. (7), and the interactions between the $3d$ electrons and the $1s/4p$ electrons are taken into account. The latter is given as

$$
\begin{aligned}
\mathscr{H}_{3d-1s/4p} &= \sum_{i\gamma} \sum_{\sigma\sigma'} v^{sd} d_{i\gamma\sigma}^{\dagger} d_{i\gamma\sigma} s_{i\sigma'}^{\dagger} s_{i\sigma'} \\
&+ \sum_{i\gamma\gamma'\alpha\alpha'} \sum_{\sigma\sigma'} v^{pd}_{\gamma\gamma'\alpha\alpha'} d_{i\gamma\sigma}^{\dagger} d_{i\gamma'\sigma} p_{i\alpha\sigma'}^{\dagger} p_{i\alpha'\sigma'},
\end{aligned}
\tag{75}
$$

where $s_{i\sigma}$ and $p_{i\alpha\sigma}$ are the annihilation operators for the $1s$ core hole and the $4p$ electron with orbital $\alpha(=x, y, z)$, respectively, and $v^{sd}_{\gamma\gamma'}$ and $v^{pd}_{\gamma\gamma'\alpha\alpha'}$ represent the Coulomb interactions. A diagramatic representation of the K-edge RIXS process from the orbital excitation is given in Fig. 21.

The calculated X-ray absorption spectra (XAS) near the K-edge are shown in Fig. 22a. The lower and higher energy peak structures in XAS are attributed to the $(3d)^{10}$ and $(3d)^9$ final states, respectively, corresponding to the well-screened and poor-screened states. In the RIXS spectra shown in Fig. 22b, a broad spectra around $5t < \omega < 10t$ and sharp peaks below $2t$ are identified as the electronic excitations across the Mott gap, and the orbital excitations, respectively. The RIXS spectra for several X-ray energies are shown in Fig. 22c where both the polarizations of the incoming and outgoing X-rays are chosen to be $(\varepsilon_{\mathbf{k}_i\lambda_i}, \varepsilon_{\mathbf{k}_f\lambda_f}) = (x, x)$. The charge excitation across the Mott gap and the orbital excitations, respectively, are

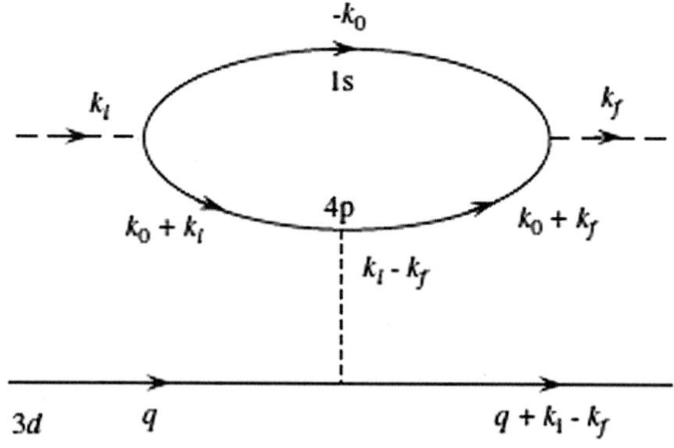

Fig. 21 A diagramatic representation of the K-edge RIXS process. *Broken* and *dotted lines* indicate the X-rays and the Coulomb interactions, respectively [183]

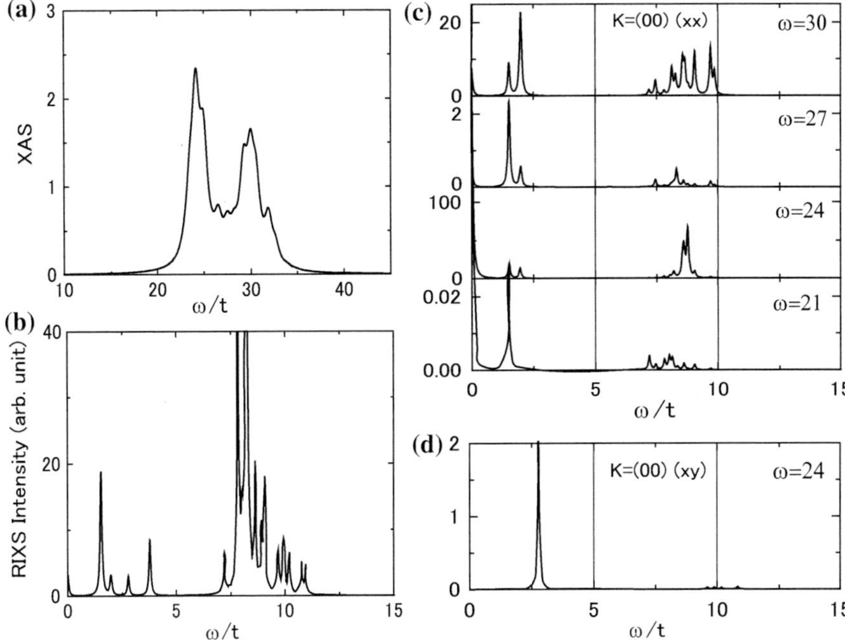

Fig. 22 **a** The XAS and **b** RIXS spectra calculated for KCuF$_3$. The polarizations of the incident and scattered X-rays are chosen to be along the x axis [162]. **c** The RIXS spectra for several incident X-ray energies with $(\varepsilon_{\mathbf{k}_i \lambda_i}, \varepsilon_{\mathbf{k}_f \lambda_f}) = (x, x)$, and **d** the spectra with $(\varepsilon_{\mathbf{k}_i \lambda_i}, \varepsilon_{\mathbf{k}_f \lambda_f}) = (x, y)$

pronounced around $\omega/t = 24$ and 30 corresponding to the $(3d)^{10}$ and $(3d)^9$ final states in XAS. These results are understood that the charge excitations across the Mott gap and the orbital excitations are induced through the processes by the charge transfer excitation and the Coulomb interactions, respectively, shown in Fig. 16. The RIXS spectra with the cross polarizations, i.e. $(\varepsilon_{\mathbf{k}_i \lambda_i}, \varepsilon_{\mathbf{k}_f \lambda_f}) = (x, y)$, are also shown in Fig. 22d. The lower and higher energy peaks in the orbital excitations only appear in the cases of $(\varepsilon_{\mathbf{k}_i \lambda_i}, \varepsilon_{\mathbf{k}_f \lambda_f}) = (x, x)$ and (x, y), respectively. These results support the assignment of the experimentally observed low-energy RIXS spectra given in Ref. [118].

Finally, we demonstrate a combination effect between the X-ray polarizations and the momentum transfer studied in Refs. [183, 185]. This breaks the selection rule based on the group theoretical analyses given in Eq. (74), and may explain some discrepancies between the measurements of the polarization analyses in KCuF$_3$ [118] and the selection rules introduced in Eq. (74). In the case of $\mathbf{K} \neq 0$, instead of Eq. (74), $\langle 3d_\gamma, \varepsilon_{\mathbf{k}_i \lambda_i}^\alpha | e^{i \mathbf{K} \cdot \mathbf{r}} | 3d_{\gamma'}, \varepsilon_{\mathbf{k}_f \lambda_f}^\beta \rangle$ should have the A$_{1g}$ symmetry. This combination effect originates microscopically from the following processes: consider an intermediate electronic state of the RIXS processes where the X-ray is absorbed at site i in a cubic lattice. When the incident X-ray polarization is parallel to the x axis, the $4p_x$ orbital is occupied by an electron, and the charge transfers excitation from the

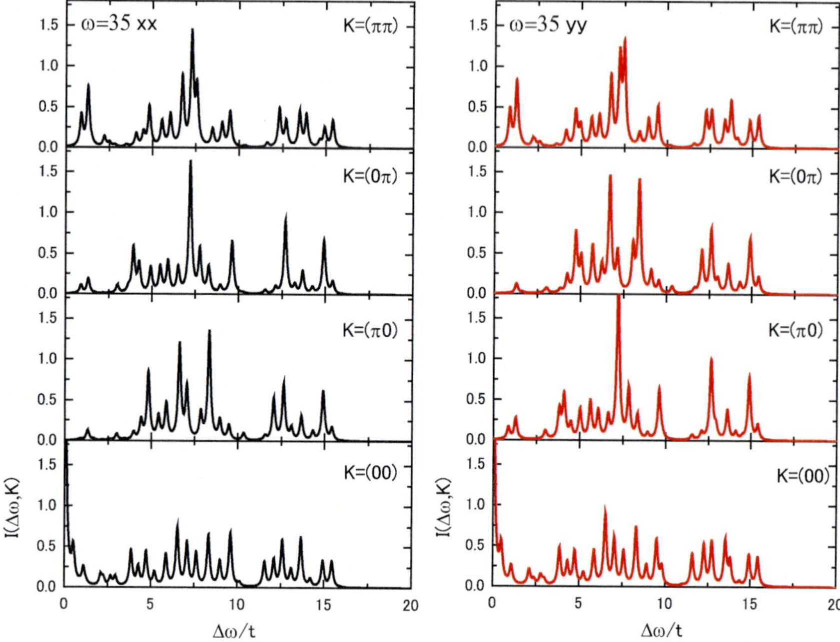

Fig. 23 (*Left*) The RIXS spectra for several momentum transfers. The $d_{x^2-y^2}$ orbital is occupied at each site in a square lattice. The polarizations of the X-rays are taken to be (*left*) $(\varepsilon_{\mathbf{k}_i\lambda_i}, \varepsilon_{\mathbf{k}_f\lambda_f}) = (x, x)$, and (*right*) (y, y) [162, 183]

$\mathbf{r}_i \pm a\hat{x}$ sites are not equivalent to those from $\mathbf{r}_i \pm a\hat{y}$. This is seen in the diagramatic representation of the RIXS process in Fig. 21, and is represented by the momentum specified interaction given as

$$\mathcal{H}_{3d-1s/4p}(\mathbf{K}) = \sum_{i\gamma} \sum_{\sigma\sigma'} v^{sd}(\mathbf{K}) d^\dagger_{\mathbf{q}+\mathbf{K}\gamma\sigma} d_{\mathbf{q}\gamma\sigma} s^\dagger_{\mathbf{k}_0-\mathbf{K}\sigma'} s_{\mathbf{k}_0\sigma'}$$

$$+ \sum_{i\gamma\gamma'} \sum_{\sigma\sigma'} v^{pd}_{\gamma\gamma'\alpha\alpha'}(\mathbf{K}) d^\dagger_{\mathbf{q}+\mathbf{K}\gamma\sigma} d_{\mathbf{q}\gamma'\sigma} p^\dagger_{\mathbf{k}_0-\mathbf{K}\alpha\sigma'} p_{\mathbf{k}_0\alpha'\sigma'}. \tag{76}$$

A simple demonstration of this combination effect is shown in the K-edge RIXS spectra presented in Fig. 23, where the $d_{x^2-y^2}$ orbital is occupied at each site in a square lattice. The polarizations are taken to be $(\varepsilon_{\mathbf{k}_i\lambda_i}, \varepsilon_{\mathbf{k}_f\lambda_f}) = (x, x)$ and (y, y) in the right and left panels, respectively. It is shown that fine structures in the RIXS spectra at $\mathbf{K} = (\pi, 0)$ with the (x, x) polarization are different from the spectra at $(\pi, 0)$ with the (y, y) polarization.

4 Summary and Perspective

We have reviewed, in this chapter, the RXS and RIXS studies in correlated electron systems with the orbital degree of freedom. After more than 15 years since the first experimental observation of RXS due to the orbital order in perovskite manganites, the RXS and related techniques have been applied to a wide variety of correlated electron materials, and are now developed as the indispensable experimental tool to study the electronic structure in solids, to be on a par with the angular-resolved photoemission spectroscopy and the inelastic neutron scattering. Along with the recent great progresses in the light source, including the X-ray free-electron laser, the situations of the RXS and RIXS experiments are now taking a new stage. The following are the new perspectives of the RXS and RIXS researches in the correlated electron systems with the orbital degree of freedom.

- The orbital concept in condensed matter physics now spreads to a wide variety of materials. The RXS and RIXS will uncover the exotic electronic structure and dynamics in such novel "orbital systems" with strong electron correlations. The dimer-type organic molecular solids, such as $(TMTTF)_2AsF_6$ and κ-(BEDT-TTF)$_2$Cu$_2$(CN)$_3$, are the candidate materials. Connections of the electronic degree of freedom inside the molecular dimers, recognized as the "orbital", to the superconductivity, the spin liquid state, and the dielectric anomalies have attracted much attention [186, 187]. The collective electronic excitations was also examined by the optical measurements as well as the theoretical calculations [188, 189]. Another target material is the excitonic insulator. The excitonic insulator can be retraced back the studies in 1960s, where narrow gap semiconductors and semimetals were examined as the candidate materials. The recent researches using the modern experimental and theoretical techniques clarify that this phenomenon is in the category of the orbital physics [190]. The experimental confirmations of the collective excitations using RIXS may review a new attractive force for the superconductivity mediated by the excitionic fluctuation.
- Studies in the non-equilibrium and transient properties of the orbital degree of freedom are now greatly developed. Informations obtained from the optical pump-probe measurements are limited at the zero momentum, and the time- and momentum-resolved probes which can access directly to the orbital degree of freedom are strongly required. The transient measurement can disentangle the strongly interacting multiple degrees of freedom, such as the orbital and JT-type lattice distortion, and can identify the main interaction dominating the macroscopic phenomena. Along this new directions, a discovery of the "hidden orbital state", which cannot be realized in the thermal equilibrium state, is expected. New states of matter is possible owing to a certain number of metastable states under the multiple degree of freedom.

Acknowledgements We thank T. Arima, A.R. Baron, Y. Endoh, K. Hirota, D.J. Huang, T. Inami, K. Ishii, T. Matsumura, S. Maekawa, J. Mizuki, Y. Murakami, H. Nakao, J. Nasu, H. Sawa and Y. Wakabayashi for their collaborations and helpful discussions. This work was supported by MEXT KAKENHI Grant Nos. 26287070 and 15H02100. Some of the numerical calculations were performed using the facilities of the Supercomputer Center, the Institute for Solid State Physics, the University of Tokyo.

References

1. Y. Murakami, H. Kawada, H. Kawata, M. Tanaka, T. Arima, H. Moritomo, Y. Tokura, Phys. Rev. Lett. **80**, 1932 (1998)
2. Y. Murakami, J.P. Hill, D. Gibbs, M. Blume, I. Koyama, M. Tanaka, H. Kawata, T. Arima, Y. Tokura, K. Hirota, Y. Endoh, Phys. Rev. Lett. **81**, 582 (1998)
3. S. Ishihara, S. Maekawa, Rep. Prog. Phys. **65**, 561 (2002)
4. S. Maekawa, T. Tohyama, S.E. Barnes, S. Ishihara, W. Koshibae, G. Khaliullin, *Physics of Transition Metal Oxides* (Springer, Berlin, 2004)
5. S. Ishihara, Y. Murakami, T. Inami, K. Ishii, J. Mizuki, K. Hirota, S. Maekawa, Y. Endoh, New J. Phys. **7**, 119 (2005)
6. T. Matsumura, H. Nakao, Y. Murakami, J. Phys. Soc. Jpn. **82**, 021007 (2013)
7. L.M. Roth, Phys. Rev. **149**, 306 (1966)
8. K.I. Kugel, D.I. Khomskii, Sov. Phys. JETP Lett. **15**, 446 (1972)
9. M. Cyrot, C. Lyon-Caen, J. Phys. **36**, 253 (1975)
10. S. Inagaki, J. Phys. Soc. Jpn. **39**, 596 (1975)
11. D.I. Khomskii, *Transition Metal Compounds* (Cambridge University Press, Cambridge, 2014)
12. L.I. Korovin, E.K. Kudinov, Sov. Phys. Solid State **16**, 1666 (1975)
13. J.D. Dunitz, L.E. Orgel, J. Phys. Chem. Solids **3**, 12S (1957)
14. J. Kanamori, J. Appl. Phys. **31**, S14 (1960)
15. M.D. Kaplan, B.G. Vekhter, *Cooperative Phenomena in Jahn-Teller Crystals* (Plenum, NewYork, 1995)
16. P.M. Levy, Sol. Stat. Comm. **7**, 1813 (1969)
17. S. Ishihara, J. Inoue, S. Maekawa, Phys. Rev. B **55**, 8280 (1997)
18. L.F. Feiner, A.M. Oleś, J. Zaanen Phys. Rev. **78**, 2799 (1997)
19. R. Shiina, T. Nishitani, H. Shiba, J. Phys. Soc. Jpn. **66**, 3159 (1997)
20. Y. Yamashita, N. Nishitani, K. Ueda, Phys. Rev. B **58**, 9114 (1998)
21. Y.Q. Li, M. Ma, D.N. Shi, F.C. Zhang, Phys. Rev. B **60**, 12781 (1999)
22. I. Affleck, Nucl. Phys. B **265**, 409 (1986)
23. M. Kataoka, J. Kanamori, J. Phys. Soc. Jpn. **32**, 113 (1972)
24. A.J. Millis, Phys. Rev. B **53**, 8434 (1996)
25. S. Okamoto, S. Ishihara, S. Maekawa, Phys. Rev. B **65**, 144403 (2002)
26. S. Ishihara, M. Yamanaka, N. Nagaosa, Phys. Rev. B **56**, 686 (1997)
27. G. Khaliullin, V. Oudovenko, Phys. Rev. B **56**, R14243 (1997)
28. L.F. Feiner, A.M. Oleś, Phys. Rev. B **59**, 3295 (1999)
29. Z. Nussinov, M. Biskup, L. Chayes, J. van den Brink, Europhys. Lett. **67**, 990 (2004)
30. T. Tanaka, S. Ishihara, Phys. Rev. B **79**, 035109 (2009)
31. A.I. Liechtenstein, V.I. Anisimov, J. Zaanen, Phys. Rev. B **52**, R5467(R) (1995)
32. I. Leonov, N. Binggeli, D. Korotin, V.I. Anisimov, N. Stojić, D. Vollhardt, Phys. Rev. Lett. **101**, 096405 (2008)
33. E. Pavarini, E. Koch, A.I. Lichtenstein, Phys. Rev. Lett. **101**, 266405 (2008)
34. Y. Ito, J. Akimitsu, J. Phys. Soc. Jpn. **40**, 1333 (1976)
35. J. Akimitsu, H. Ichikawa, N. Eguchi, T. Miyano, M. Nishi, K. Kakurai, J. Phys. Soc. Jpn. **70**, 3475 (2001)

36. H. Ichikawa, L. Kano, M. Saitoh, S. Miyahara, N. Furukawa, J. Akimitsu, T. Yokoo, T. Matsumura, M. Takeda, K. Hirota, J. Phys. Soc. Jpn. **74**, 1020 (2005)
37. M. Takata, E. Nishibori, K. Kato, M. Sakata, Y. Moritomo, J. Phys. Soc. Jpn. **68**, 2190 (1999)
38. T. Itakura, T. Nakao, H. Kimura, Y. Noda, Y. Ishikawa, Y. Tokura, S. Miyasaka, Meeting abstracts of the Phys. Soc. Jpn. **67**, 592 (2012)
39. T. Fujii, T. Sugawara, T. Higuchi, N. Katayama, H. Sawa, T. Shishido, K. Sugimoto, M. Takada, T. Arima, N. Taguchi, Y. Tokura, Meeting abstracts of the Phys. Soc. Jpn. **70**, 2422 (2015)
40. A.N. Mariano, R.E. Hanneman, J. Appl. Phys. **34**, 384 (1963)
41. R.J. Wakelin, E.L. Yates, Proc. Phys. Soc. B **66**, 221 (1953)
42. Y. Waseda, S. Tamaki, Phil. Mag. **32**, 951 (1975)
43. D.H. Templeton, L.K. Templeton, Acta. Cryst. A **36**, 237 (1980)
44. D.H. Templeton, L.K. Templeton, Acta. Cryst. A **38**, 62 (1982)
45. V.E. Dmitrienko, Acta. Cryst. A **39**, 29 (1983)
46. S. Ishihara, S. Maekawa, Phys. Rev. B **58**, 13442 (1998)
47. L. Paolasini, C. Vettier, F. de Bergevin, F. Yakhou, D. Mannix, A. Stunault, W. Neubeck, Phys. Rev. Lett. **82**, 4719 (1999)
48. M. Fabrizio, M. Altarelli, M. Benfatto, Phys. Rev. Lett. **80**, 3400 (1998)
49. Y. Endoh, K. Hirota, S. Ishihara, S. Okamoto, Y. Murakami, A. Nishizawa, T. Fukuda, H. Kimura, H. Nojiri, K. Kaneko, S. Maekawa, Phys. Rev. Lett. **82**, 4328 (1999)
50. M.V. Zimmermann, J.P. Hill, D. Gibbs, M. Blume, D. Casa, B. Keimer, Y. Murakami, Y. Tomioka, Y. Tokura, Phys. Rev. Lett. **83**, 4872 (1999)
51. Y. Wakabayashi, Y. Murakami, I. Koyama, T. Kimura, Y. Tokura, Y.A. Moritomo, K. Hirota, Y. Endoh, J. Phys. Soc. Jpn. **69**, 2731 (2000)
52. J. García, M.C. Sanchez, J. Blasco, G. Subias, M.G. Proietti, J. Phys. Condens. Matter **13**, 3243 (2001)
53. M.V. Zimmermann, C.S. Nelson, J.P. Hill, D. Gibbs, M. Blume, D. Casa, B. Keimer, Y. Murakami, C.-C. Kao, C. Venkataraman, T. Gog, Y. Tomioka, Y. Tokura, Phys. Rev. B **64**, 195133 (2001)
54. Y. Wakabayashi, Y. Murakami, I. Koyama, T. Kimura, Y. Tokura, Y. Moritomo, Y. Endoh, K. Hirota, J. Phys. Soc. Jpn. **72**, 618 (2003)
55. S.B. Wilkins, P.D. Spencer, T.A.W. Beale, P.D. Hatton, M.V. Zimmermann, S.D. Brown, D. Prabhakaran, A.T. Boothroyd, Phys. Rev. B **67**, 205110 (2003)
56. H. Ohsumi, Y. Murakami, T. Kiyama, H. Nakao, M. Kubota, Y. Wakabayashi, Y. Konishi, M. Izumi, M. Kawasaki, Y. Tokura, J. Phys. Soc. Jpn. **72**, 1006 (2003)
57. S. Di Matteo, T. Chatterji, Y. Joly, A. Stunault, J.A. Paixao, R. Suryanarayanan, G. Dhalenne, A. Revcolevschi, Phys. Rev. B **68**, 024414 (2003)
58. Y. Sua, K. Istomina, D. Wermeilleb, A. Fattaha, P. Foucarta, P. Meuffelsa, D. Hupfelda, T. Brueckela, J. Mag. Mag. Mat. **272–276**, E291 (2004)
59. S. Grenier, J.P. Hill, D. Gibbs, K.J. Thomas, M.V. Zimmermann, C.S. Nelson, V. Kiryukhin, Y. Tokura, Y. Tomioka, D. Casa, T. Gog, C. Venkataraman, Phys. Rev. B **69**, 134419 (2004)
60. J.H. Martín, J. Garcia, G. Subías, J. Blasco, M.C. Sanchez, Phys. Rev. B **70**, 024408 (2004)
61. S.S. Dhesi, A. Mirone, C. De, Nadaï P. Ohresser, P. Bencok, N.B. Brookes, P. Reutler, A. Revcolevschi, A. Tagliaferri, O. Toulemonde, G. van der Laan. Phys. Rev. Lett. **92**, 056403 (2004)
62. H. Nakao, Y. Wakabayashi, T. Kiyama, Y. Murakami, M.V. Zimmermann, J.P. Hill, D. Gibbs, S. Ishihara, Y. Taguchi, Y. Tokura, Phys. Rev. B **66**, 184419 (2002)
63. H. Nakao, M. Tsubota, F. Iga, K. Uchihira, T. Nakano, T. Takabatake, K. Kato, Y. Murakami, J. Phys. Soc. Jpn. **73**, 2620 (2004)
64. M. Kubota, H. Nakao, Y. Murakami, Y. Taguchi, M. Iwama, Y. Tokura, Phys. Rev. B **70**, 245125 (2004)
65. M. Tsubota, F. Iga, K. Uchihira, T. Nakano, S. Kura, T. Takabatake, S. Kodama, H. Nakao, Y. Murakami, J. Phys. Soc. Jpn. **74**, 3259 (2005)

66. H. Nakao, S. Kodama, K. Kiyoto, Y. Murakami, M. Tsubota, F. Iga, K. Uchihira, T. Takabatake, H. Ohsumi, M. Mizumaki, N. Ikeda, J. Phys. Soc. Jpn. **75**, 094706 (2006)
67. M. Noguchi, A. Nakazawa, S. Oka, T. Arima, Y. Wakabayashi, H. Nakao, Y. Murakami, Phys. Rev. B **62**, R9271(R) (2000)
68. D. Bizen, N. Shirane, T. Murata, H. Nakao, Y. Murakami, J. Fujioka, T. Yasue, S. Miyasaka, Y. Tokura, J. Phys. Conf. Ser. **150**, 042010 (2009)
69. T.A.W. Beale, R.D. Johnson, Y. Joly, S.R. Bland, P.D. Hatton, L. Bouchenoire, C. Mazzoli, D. Prabhakaran, Phys. Rev. B **82**, 024105 (2010)
70. H. Nakao, T. Murata, D. Bizen, Y. Murakami, K. Ohoyama, K. Yamada, S. Ishiwata, W. Kobayashi, I. Terasaki, J. Phys. Soc. Jpn. **80**, 023711 (2011)
71. J. Fujioka, Y. Yamasaki, H. Nakao, R. Kumai, Y. Murakami, M. Nakamura, M. Kawasaki, Y. Tokura, Phys. Rev. Lett. **111**, 027206 (2013)
72. U. Staub, G.I. Meijer, F. Fauth, R. Allenspach, J.G. Bednorz, J. Karpinski, S.M. Kazakov, L. Paolasini, F. d'Acapito, Phys. Rev. Lett. **88**, 126402 (2002)
73. V. Scagnoli, U. Staub, A.M. Mulders, M. Janousch, G.I. Meijer, G. Hammerl, J.M. Tonnerre, N. Stojic, Phys. Rev. B **73**, 100409(R) (2006)
74. R. Caciuffo, L. Paolasini, A. Sollier, P. Ghigna, E. Pavarini, J. van den Brink, M. Altarelli, Phys. Rev. B **65**, 174425 (2002)
75. C. Mazzoli, L. Paolasini, F. de Bergevin, P. Ghigna, R. Caciuffo, Phys. B Condens. Matter **378–380**, 563 (2006)
76. J.C.T. Lee, S. Yuan, S. Lal, Y.I. Joe, Y. Gan, S. Smadici, K. Finkelstein, Y. Feng, A. Rusydi, P.M. Goldbart, S.L. Cooper, P. Abbamonte, Nat. Phys. **8**, 63 (2012)
77. J. García, G. Subias, M.G. Proietti, H. Renevier, Y. Joly, J.L. Hodeau, J. Blasco, M.C. Sanchez, J.F. Berar, Phys. Rev. Lett. **85**, 578 (2000)
78. D.J. Huang, H.-J. Lin, J. Okamoto, K.S. Chao, H.-T. Jeng, G.Y. Guo, C.-H. Hsu, C.-M. Huang, D.C. Ling, W.B. Wu, C.S. Yang, C.T. Chen, Phys. Rev. Lett. **96**, 096401 (2006)
79. S. Ishihara, S. Maekawa, Phys. Rev. Lett. **80**, 3799 (1998)
80. I.S. Elfimov, V.I. Anisimov, G.A. Sawatzky, Phys. Rev. Lett. **82**, 4264 (1999)
81. M. Takahashi, J. Igarashi, P. Fulde, J. Phys. Soc. Jpn. **68**, 2530 (1999)
82. M. Benfatto, Y. Joly, C.R. Natoli, Phys. Rev. Lett. **83**, 636 (1999)
83. S. Ishihara, S. Maekawa, Phys. Rev. B **62**, 5690 (2000)
84. S. Ishihara, S. Maekawa, Phys. Rev. B **62**, R9252 (2000)
85. C. Castleton, M. Altarelli, Phys. Rev. B **62**, 1033 (2000)
86. P. Benedetti, J. van den Brink, E. Pavarini, A. Vigliante, P. Wochner, Phys. Rev. B **63**, 060408 (2001)
87. M. Takahashi, J. Igarashi, Phys. Rev. B **64**, 075110 (2001)
88. I. Elfimov, N. Skorikov, V. Anisimov, G. Sawatzky, Phys. Rev. Lett. **88**, 015504 (2001)
89. S.W. Lovesey, K.S. Knight, D.S. Sivia, Phys. Rev. B **65**, 224402 (2002)
90. Y. Tanaka, T. Inami, T. Nakamura, H. Yamauchi, H. Onodera, K. Ohoyama, Y. Yamaguchi, J. Phys. Condens. Matter **11**, L505 (1999)
91. K. Hirota, N. Oumi, T. Matsumura, H. Nakao, Y. Wakabayashi, Y. Murakami, Y. Endoh, Phys. Rev. Lett. **84**, 2706 (2000)
92. H. Nakao, K. Magishi, Y. Wakabayashi, Y. Murakami, K. Koyama, K. Hirota, Y. Endoh, S. Kunii, J. Phys. Soc. Jpn. **70**, 1857 (2001)
93. D. McMorrow, K. McEwen, U. Steigenberger, H. Ronnow, F. Yakhou, Phys. Rev. Lett. **87**, 057201 (2001)
94. T. Matsumura, N. Oumi, K. Hirota, H. Nakao, Y. Wakabayashi, Y. Murakami, J. Phys. Soc. Jpn. **71**, 91 (2002)
95. T. Matsumura, N. Oumi, K. Hirota, H. Nakao, Y. Murakami, Y. Wakabayashi, T. Arima, S. Ishihara, Y. Endoh, Phys. Rev. B **65**, 094420 (2002)
96. J. Paixao, C. Detlefs, M. Longfield, R. Caciuffo, P. Santini, N. Bernhoeft, J. Rebizant, G. Lander, Phys. Rev. Lett. **89**, 187202 (2002)
97. D. Mannix, Y. Tanaka, D. Carbone, N. Bernhoeft, S. Kunii, Phys. Rev. Lett. **95**, 117206 (2005)
98. T. Nagao, J. Igarashi, J. Phys. Soc. Jpn. **70**, 2892 (2001)

99. H. Kusunose, Y. Kuramoto, J. Phys. Soc. Jpn. **74**, 3139 (2005)
100. T. Nagao, J. Igarashi, J. Phys. Soc. Jpn. **75**, 247 (2006)
101. T. Kiyama, Y. Wakabayashi, H. Nakao, H. Ohsumi, Y. Murakami, M. Izumi, M. Kawasaki, Y. Tokura, J. Phys. Soc. Jpn. **72**, 785 (2003)
102. H. Nakao, J. Nishimura, Y. Murakami, A. Ohtomo, T. Fukumura, M. Kawasaki, T. Koida, Y. Wakabayashi, H. Sawa, J. Phys. Soc. Jpn. **78**, 024602 (2009)
103. H. Nakao, T. Sudayama, M. Kubota, J. Okamoto, Y. Yamasaki, Y. Murakami, H. Yamada, A. Sawa, K. Iwasa, Phys. Rev. B **92**, 245104 (2015)
104. S.B. Wilkins, P.D. Spencer, P.D. Hatton, S.P. Collins, M.D. Roper, D. Prabhakaran, A.T. Boothroyd, Phys. Rev. Lett. **91**, 167205 (2003)
105. K.J. Thomas, J.P. Hill, S. Grenier, Y.-J. Kim, P. Abbamonte, L. Venema, A. Rusydi, Y. Tomioka, Y. Tokura, D.F. McMorrow, G. Sawatzky, M. van Veenendaal, Phys. Rev. Lett. **92**, 237204 (2004)
106. J.H. Martín, J. García, G. Subías, J. Blasco, M.C. Sánchez, S. Stanescu, Phys. Rev. B **73**, 224407 (2006)
107. S.B. Wilkins, N. Stoji, T.A.W. Beale, N. Binggeli, C.W.M. Castleton, P. Bencok, D. Prabhakaran, A.T. Boothroyd, P.D. Hatton, M. Altarelli, Phys. Rev. B **71**, 245102 (2005)
108. H. Wadati, J. Geck, E. Schierle, R. Sutarto, F. He, D.G. Hawthorn, M. Nakamura, M. Kawasaki, Y. Tokura, G.A. Sawatzky, New J. Phys. **16**, 033006 (2014)
109. B.J. Kim, H. Ohsumi, T. Komesu, S. Sakai, T. Morita, H. Takagi, T. Arima, Science **323**, 1329 (2009)
110. K. Ohgushi, J. Yamaura, H. Ohsumi, K. Sugimoto, S. Takeshita, A. Tokuda, H. Takagi, M. Takata, T. Arima, Phys. Rev. Lett. **110**, 217212 (2013)
111. M.M. Sala, M. Rossi, S. Boseggia, J. Akimitsu, N.B. Brookes, M. Isobe, M. Minola, H. Okabe, H.M. Rønnow, L. Simonelli, D.F. McMorrow, G. Monaco. Phys. Rev. B **89**, 121101(R) (2014)
112. T. Inami, T. Fukuda, J. Mizuki, S. Ishihara, H. Kondo, H. Nakao, T. Matsumura, K. Hirota, Y. Murakami, S. Maekawa, Y. Endoh, Phys. Rev. B **67**, 045108 (2003)
113. K. Ishii, T. Inami, K. Ohwada, K. Kuzushita, J. Mizuki, Y. Murakami, S. Ishihara, Y. Endoh, S. Maekawa, K. Hirota, Y. Moritomo, Phys. Rev. B **70**, 224437 (2004)
114. Y. Kim, J.P. Hill, F.C. Chou, D. Casa, T. Gog, C.T. Venkataraman, Phys. Rev. B **69**, 155105 (2004)
115. S. Grenier, J. Hill, V. Kiryukhin, W. Ku, Y.-J. Kim, K. Thomas, S.-W. Cheong, Y. Tokura, Y. Tomioka, D. Casa, T. Gog, Phys. Rev. Lett. **94**, 047203 (2005)
116. J. Kim, D.S. Ellis, T. Gog, D. Casa, Y.-J. Kim, Phys. Rev. B **81**, 073109 (2010)
117. F. Weber, S. Rosenkranz, J.-P. Castellan, R. Osborn, J.F. Mitchell, H. Zheng, D. Casa, J.H. Kim, T. Gog, Phys. Rev. B **82**, 085105 (2010)
118. K. Ishii, S. Ishihara, Y. Murakami, K. Ikeuchi, K. Kuzushita, T. Inami, K. Ohwada, M. Yoshida, I. Jarrige, N. Tatami, S. Niioka, D. Bizen, Y. Ando, J. Mizuki, S. Maekawa, Y. Endoh, Phys. Rev. B **83**, 241101(R) (2011)
119. X. Liu, T.F. Seman, K.H. Ahn, M. van Veenendaal, D. Casa, D. Prabhakaran, A.T. Boothroyd, H. Ding, J.P. Hill, Phys. Rev. B **87**, 201103 (2013)
120. C.S. Nelson, J.P. Hill, D. Gibbs, F. Yakhou, F. Livet, Y. Tomioka, T. Kimura, Y. Tokura, Phys. Rev. B **66**, 134412 (2002)
121. J.J. Turner, K.J. Thomas, J.P. Hill, M.A. Pfeifer, K. Chesnel, Y. Tomioka, Y. Tokura, S.D. Kevan, New J. Phys. **10**, 053023 (2008)
122. H. Ehrke, R.I. Tobey, S. Wall, S.A. Cavill, M. Forst, V. Khanna, N. Th Garl, D. Stojanovic, A.T. Prabhakaran, M. Boothroyd, A. Gensch, P. Mirone, A. Reutler, S.S. Revcolevschi, A.Cavalleri Dhesi, Phys. Rev. Lett. **106**, 217401 (2011)
123. P. Beaud, A. Caviezel, S.O. Mariager, L. Rettig, G. Ingold, C. Dornes, S.-W. Huang, J.A. Johnson, M. Radovic, T. Huber, T. Kubacka, A. Ferrer, H.T. Lemke, M. Chollet, D. Zhu, J.M. Glownia, M. Sikorski, A. Robert, H. Wadati, M. Nakamura, M. Kawasaki, Y. Tokura, S.L. Johnson, U. Staub, Nat. Mater. **13**, 923 (2014)
124. M. Blume, in *Resonant Anomalous X-ray Scattering*, ed. by G. Materlik, C.J. Sparks, K. Fischer (Elsevier Science, Amsterdam, 1994)

125. S.W. Lovesey, S.P. Collins, *X-ray Scattering and Absorption by Magnetic Materials* (Oxford University Press, Oxford, 1995)
126. J.J. Sakurai, *Advanced Quantum Mechanics* (Addison-Wesley, Massachusetts, 1967)
127. M. Azuma, Y. Fujishiro, M. Takano, M. Nohara, H. Takagi, Phys. Rev. B **55**, R8658 (1997)
128. M. Sigrist, A. Furusaki, J. Phys. Soc. Jpn. **65**, 2385 (1996)
129. S.B. Oseroff, S.-W. Cheong, B. Aktas, M.F. Hundley, Z. Fisk, L.W. Rupp Jr., Phys. Rev. Lett. **74**, 1450 (1995)
130. M. Hase, K. Uchinokura, R.J. Birgeneau, K. Hirota, G. Shirane, J. Phys. Soc. Jpn. **65**, 1392 (1996)
131. N. Tatami, Y. Ando, S. Niioka, H. Kira, M. Onodera, H. Nakao, K. Iwasa, Y. Murakami, T. Kakiuchi, Y. Wakabayashi, H. Sawa, S. Itoh, J. Mag. Mag. Mater. **310**, 787 (2007)
132. N. Tatami: Master thesis, Tohoku University, 2004
133. S. Ohtani, Y. Watanabe, M. Saito, N. Abe, K. Taniguchi, H. Sagayama, T. Arima, M. Watanabe, Y. Noda, J. Phys. Condens. Matter **22**, 176003 (2010)
134. R.B. Stinchcombe, in *Phase Transition in Critical Phenomena*, vol. 7, ed. by C. Domb, J.L. Lebowitz (Academic Press, London, 1983)
135. D.J. Breed, K. Gilijamse, J.W.E. Sterkenburg, A.R. Miedema, J. Appl. Phys. **41**, 1267 (1970)
136. T. Tanaka, M. Matsumoto, S. Ishihara, Phys. Rev. Lett. **98**, 256402 (2007)
137. T. Tanaka, S. Ishihara, Phys. Rev. B **78**, 153106 (2008)
138. A.G. Komarov, L.I. Korovin, E.K. Kudinov, Sov Phys. Solid State **17**, 1531 (1975)
139. S. Ishihara, J. Inoue, S. Maekawa, Phys. C **263**, 130 (1996)
140. S. Ishihara, S. Maekawa, Phys. Rev. B **62**, 2338 (2000)
141. J. van den Brink, Phys. Rev. Lett. **87**, 217202 (2001)
142. P.B. Allen, V. Perebeinos, Phys. Rev. Lett. **83**, 4828 (1999)
143. V. Perebeinos, P.B. Allen, Phys. Rev. B **64**, 085118 (2001)
144. J.H. Jung, K.H. Kim, D.J. Eom, T.W. Noh, E.J. Choi, J. Yu, Y.S. Kwon, Y. Chung, Phys. Rev. B **55**, 15 489 (1997)
145. J. Nasu, S. Ishihara, Phys. Rev. B **88**, 205110 (2013)
146. J. Inoue, S. Okamoto, S. Ishihara, W. Koshibae, Y. Kawamura, S. Maekawa, Phys. B **237–238**, 51 (1997)
147. E. Saitoh, K.T. Okamoto, K. Tobe, K. Yamamoto, T. Kimura, S. Ishihara, S. Maekawa, Y. Tokura, Nature **418**, 40 (2001)
148. S. Okamoto, S. Ishihara, S. Maekawa, Phys. Rev. B **66**, 014435 (2002)
149. M. Gruninger, R. Ruckamp, M. Windt, P. Reutler, C. Zobel, T. Lorenz, A. Freimuth, A. Revcolevshi, Nature **418**, 39 (2002)
150. R. Kruger, B. Schulz, S. Naler, R. Rauer, D. Budelmann, J. Backstrom, K.H. Kim, S.-W. Cheong, V. Pereveinos, M. Rubhausen, Phys. Rev. Lett. **92**, 097203 (2004)
151. S. Ishihara, Phys. Rev. B **69**, 075118 (2004)
152. S. Miyasaka, S. Onoda, Y. Okimoto, J. Fujioka, M. Iwama, N. Nagaosa, Y. Tokura, Phys. Rev. Lett. **94**, 076405 (2005)
153. C. Ulrich, A. Gossling, M. Gruninger, M. Guennou, H. Roth, M. Cwik, T. Lorenz, G. Khaliullin, B. Keimer, Phys. Rev. Lett. **97**, 157401 (2006)
154. G. Khaliullin, S. Okamoto, Phys. Rev. Lett. **89**, 167201 (2002)
155. D. Kawana, Y. Murakami, T. Yokoo, S. Ito, A.T. Savich, G.E. Granrth, K. Ikeuchi, H. Nakao, K. Iwasa, R. Fukuta, S. Miyasaka, S. Taiima, S. Ishihara, Y. Tokura, Meeting Abstract of Japanese Physical Society Meeting **67**, 675 (2012)
156. Y. Tanaka, A.Q. Baron, Y.-J. Kim, K.J. Thomas, J.P. Hill, Z. Honda, F. Iga, S. Tsutui, D. Ishihkawa, C.S. Nelson, New J. Phys. **6**, 161 (2004)
157. B.C. Larson, W. Ku, J.Z. Tischler, C.-C. Lee, O.D. Restrepo, A.G. Eguiluz, P. Zschack, K.D. Finkelstein, Phys. Rev. Lett. **99**, 026401 (2007)
158. M. van Veenendaal, M.W. Haverkort, Phys. Rev. B **77**, 224107 (2008)
159. W. Haverkort, A. Tanaka, L.H. Tjeng, G.A. Sawatzky, Phys. Rev. Lett. **99**, 257401 (2007)
160. W. Haverkort, A. Tanaka, L.H. Tjeng, G.A. Sawatzky, Phys. Rev. B **77**, 224107 (2008)
161. K. Tsutsui, E. Kaneshita, T. Tohyama, Phys. Rev. B **92**, 195103 (2015)

162. S. Ishihara: (unpublished)
163. L.J.P. Ament, M. van Veenendaal, T.P. Devereaux, J.P. Hill, J. van den Brink, Rev. Mod. Phys. **83**, 705 (2011)
164. C. Jia, K. Wohlfeld, Y. Wang, B. Moritz, T.P. Devereaux, Phys. Rev. X **6**, 021020 (2016)
165. K. Ishii, T. Tohyama, J. Mizuki, J. Phys. Soc. Jpn. **82**, 021015 (2013)
166. J. van den Brink, M. van Veenendaal, Europhys. Lett. **73**, 121 (2006)
167. L.J.P. Ament, F. Forte, J. van den Brink, Phys. Rev. B **75**, 115118 (2007)
168. H. Kondo, S. Ishihara, S. Maekawa, Phys. Rev. B 64, 014414 (2001); Erratum. Phys. Rev. B **64**, 179904 (2001)
169. T. Semba, M. Takahashi, J. Igarashi, Phys. Rev. B **78**, 155111 (2008)
170. J.P. Hill, C.C. Kao, W.A.L. Caliebe, M. Matsubara, A. Kotani, J.L. Peng, R.L. Greene, Phys. Rev. Lett. **80**, 4967 (1998)
171. K. Hamalainen, J.P. Hill, S. Huotari, C.-C. Kao, L.E. Berman, A. Kotani, T. Ide, J.L. Peng, R.L. Greene, Phys. Rev. B **61**, 1836 (2000)
172. L. Lu, J.N. Hancock, G. Chabot-Couture, K. Ishii, O.P. Vajk, G. Yu, J. Mizuki, D. Casa, T. Gog, M. Greven, Phys. Rev. B **74**, 224509 (2006)
173. G. Chabot-Couture, J.N. Hancock, P.K. Mang, D.M. Casa, T. Gog, M. Greven, Phys. Rev. B **82**, 035113 (2010)
174. M. van Veenendaal, X. Liu, M.H. Carpenter, S.P. Cramer, Phys. Rev. B **83**, 045101 (2011)
175. M. van Veenendaal, Phys. Rev. Lett. **96**, 117404 (2006)
176. L.J.P. Ament, G. Ghiringhelli, M.M. Sala, L. Braicovich, J. van den Brink, Phys. Rev. Lett. **103**, 117003 (2009)
177. C. Ulrich, L.J.P. Ament, G. Ghiringhelli, L. Braicovich, M. Moretti, Sala, N. Pezzotta, T. Schmitt, G. Khaliullin, J. van den Brink, H. Roth, T. Lorenz, B. Keimer. Phys. Rev. Lett. **103**, 107205 (2009)
178. L.J.P. Ament, G. Khaliullin, Phys. Rev. B **81**, 125118 (2010)
179. J. Schlappa, K. Wohlfeld, K.J. Zhou, M. Mourigal, M.W. Haverkort, V.N. Strocov, L. Hozoi, C. Monney, S. Nishimoto, S. Singh, A. Revcolevschi, J.-S. Caux, L. Patthey, H.M. Ronnow, J. van den Brink, T. Schmitt, Nature **485**, 82 (2012)
180. V. Bisogni, K. Wohlfeld, S. Nishimoto, C. Monney, J. Trinckauf, K. Zhou, R. Kraus, K. Koepernik, C. Sekar, V. Strocov, B. Buchner, T. Schmitt, J. van den Brink, J. Geck, Phys. Rev. Lett. **114**, 096402 (2015)
181. K. Wohlfeld, M. Daghofer, S. Nishimoto, G. Khaliullin, J. van den Brink, Phys. Rev. Lett. **107**, 147201 (2011)
182. K. Wohlfeld, S. Nishimoto, M.W. Haverkort, J. van den Brink, Phys. Rev. B **88**, 195138 (2013)
183. S. Ishihara, S. Ihara, J. Phys. Chem. Sol. **69**, 3184 (2008)
184. T. Nomura, J. Phys. Soc. Jpn. **83**, 064707 (2014)
185. P. Abbamonte, C.A. Burns, E.D. Isaacs, P.M. Platzman, L.L. Miller, M.V. Klein, Phys. Rev. Lett. **83**, 860 (1999) ibid.preprint
186. M. Naka, S. Ishihara, J. Phys. Soc. Jpn. **79**, 063707 (2010)
187. C. Hotta, Phys. Rev. B **82**, R31044 (2010)
188. M. Naka, S. Ishihara, J. Phys. Soc. Jpn. **82**, 023701 (2013)
189. K. Itoh, H. Itoh, M. Naka, S. Saito, I. Hosako, N. Yoneyama, S. Ishihara, T. Sasaki, S. Iwai, Phys. Rev. Lett. **110**, 106401 (2013)
190. J. Nasu, T. Watanabe, M. Naka, S. Ishihara, Phys. Rev. B **93**, 205136 (2016)

Resonant X-ray Scattering in $3d$ Electron Systems

Hironori Nakao

1 Introduction

The resonant X-ray scattering (RXS) technique had been used to study local symmetry around the target ion using the anisotropy of the anomalous scattering factor of the atomic scattering factor (ASF) near the absorption energy [1–4]. In 1998, Murakami et al. applied the RXS technique to study orbital and charge ordering. The e_g orbital ordering of Mn^{3+} and the charge ordering of Mn ions in $La_{0.5}Sr_{1.5}MnO_3$ were determined [5]. In the case of $LaMnO_3$, the e_g orbital ordering of Mn^{3+} was clearly determined, and a scattering mechanism was proposed [6]. These results demonstrated the potential for measuring the order parameter of the ordered state. These charge- and orbital-ordered states are important parameters for understanding the origin of various intriguing physical properties in strongly correlated electron systems (SCES) [7, 8]. As a result, RXS studies have been extended to many types of materials [9, 10]. The charge and orbital states of e_g electrons have been systematically studied in manganites [11–19] and thin films [20–24]. Orbital and charge ordering of the t_{2g} electron system has been studied in titanates [25–29] and vanadates [30–38]. Moreover, orbital and charge ordering has been investigated in Fe [39–46], Co [47–50], Ni [51–53], and Cu [54–56] systems.

Historical RXS investigations of manganites are detailed in Chap. 1. Herein, we review studies of orbital ordering in perovskite-type titanate as a typical example. First, we summarize the theoretical framework of RXS in Sect. 2. Next, in Sect. 3, we explain in detail the RXS studies of titanates and discuss the origin of RXS signals. Then, we establish RXS as a technique to determine orbital and charge ordering. Thereafter, we investigate the RXS technique experimentally under extreme conditions; experimental examples are presented in Sect. 4.

H. Nakao (✉)
Condensed Matter Research Center and Photon Factory, IMSS, KEK,
Tsukuba, Ibaraki 305-0801, Japan
e-mail: hironori.nakao@kek.jp

© Springer-Verlag Berlin Heidelberg 2017
Y. Murakami and S. Ishihara (eds.), *Resonant X-Ray Scattering
in Correlated Systems*, Springer Tracts in Modern Physics 269,
DOI 10.1007/978-3-662-53227-0_2

2 Theoretical Framework

The scattering intensity $I(E, \mathbf{Q})$ of X-ray diffraction is proportional to the square of the structure factor $F(E, \mathbf{Q})$:

$$I(E, \mathbf{Q}) = |F(E, \mathbf{Q})|^2 L(E, \mathbf{Q}), \tag{1}$$

where $L(E, \mathbf{Q})$ is a correction term, including the Lorentz factor and sample absorption effect. The structure factor is expressed as

$$F(E, \mathbf{Q}) = \sum_j f_j(E, \mathbf{Q}) \exp(-i\mathbf{Q} \cdot \mathbf{r}_j). \tag{2}$$

$f_j(E, \mathbf{Q})$ is the ASF of the j-th atom at position $\mathbf{r}_j$ in a unit cell. The ASF $f(E, \mathbf{Q})$ is given by

$$f(E, \mathbf{Q}) = f_0(\mathbf{Q}) + f'(E) + if''(E), \tag{3}$$

where $f_0(\mathbf{Q})$ is the Thomson scattering factor and $f'(E)$ and $f''(E)$ are real and imaginary terms of the anomalous scattering factor, respectively. The anomalous scattering factor changes remarkably near the absorption-edge energy E_a. Hence, we can obtain structural and electronic information of only a target ion using the energy dependence of $I(E, \mathbf{Q})$ near E_a.

2.1 Anomalous Scattering Factor: Valence State

The absorption-edge energy, E_a, is quite sensitive to the valence of an adsorbed ion. An energy shift due to valence difference is the so-called *chemical shift*. An RXS signal reflecting charge ordering can be observed only near E_a according to the following explanation. Hence, RXS is a powerful technique for determining the charge-ordering structure.

To calculate the structure factor using Eq. (2), we need the ASF and atomic position, $\mathbf{r}$, of all atoms in a unit cell. The value of the Thomson scattering factor is known, and we can use tabulated ASF values, except near the E_a [57–59]. Therefore, the ASFs, $f'(E)$ and $f''(E)$, of charge-ordered ions are needed to calculate the structure factor of the charge-ordering state.

The imaginary term of the ASF $f''(E)$ can be directly obtained from the absorption coefficient $\mu(E)$ using the following equation.

$$f''(E) = \frac{m_e E}{2Nhe^2} \mu(E), \tag{4}$$

where m_e is the electron mass, e is the electron charge, N is the atomic number density, and h is Planck's constant [57, 60]. The real term of the ASF $f'(E)$ can be calculated using Kramers–Krönig transformation of $f''(E)$:

$$f'(E) = \frac{2}{\pi} P \int_0^\infty \frac{E' f''(E')}{E^2 - E'^2} dE', \tag{5}$$

$$f''(E) = \frac{2E}{\pi} P \int_0^\infty \frac{f'(E')}{E^2 - E'^2} dE'. \tag{6}$$

Next, we present an example to determine the ASF of V ions in vanadium compounds [31]. Here, NaV_2O_5 has an average valence of $V^{4.5+}$ at room temperature, and charge-ordering transitions of V^{4+} and V^{5+} occur at approximately 35 K. To determine the charge-ordering structure, the ASFs of V^{4+} and V^{5+} were estimated using CaV_2O_5 and V_2O_5, which have the same crystal structure as NaV_2O_5 and contain only V^{4+} and V^{5+} ions, respectively. The absorption spectra of the vanadium compounds were obtained, as shown in Fig. 1. The absorption spectrum and absorption coefficient are expressed as follows:

$$I(E) = I_0(E) \exp(-\mu(E)t), \tag{7}$$

$$\mu(E)t = log(I_0(E)/I(E)), \tag{8}$$

where $I_0(E)$ and t are the intensity of incident x-rays and the sample thickness, respectively. E_a of V_2O_5 (V^{5+}) is approximately 1.8 eV higher than that of CaV_2O_5 (V^{4+}), i.e., the chemical shift. The peak at around 5.47 keV is the pre-edge $1s \rightarrow 3d$ transition energy, which is strongly observed as a dipole transition owing to the lack of local inversion symmetry at a vanadium site [61]. A similar feature was observed in the V_2O_3 system [30]. E_a of NaV_2O_5 is placed between the E_a of V_2O_5 and CaV_2O_5,

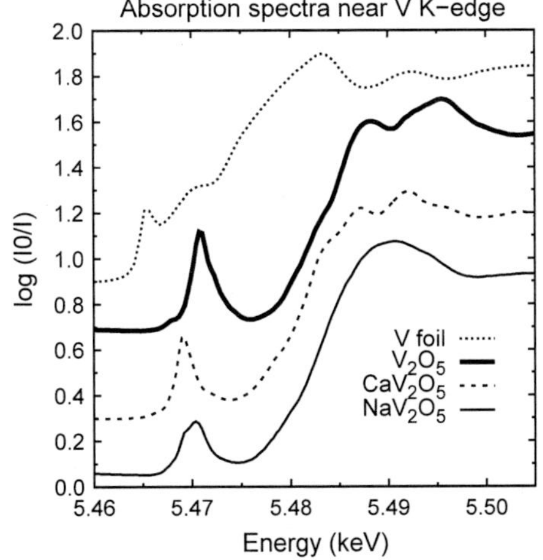

Fig. 1 Absorption spectra of V foil (*dotted line*) and powdered samples of CaV_2O_5 (V^{4+}) (*broken line*), V_2O_5 (V^{5+}) (*thick solid line*), and NaV_2O_5 (*thin solid line*). NaV_2O_5 has an average valence of $V^{4.5+}$ at room temperature, and it shows a charge-ordering transition of V^{4+} and V^{5+} at about 35 K. Part of data were taken from Ref. [31]

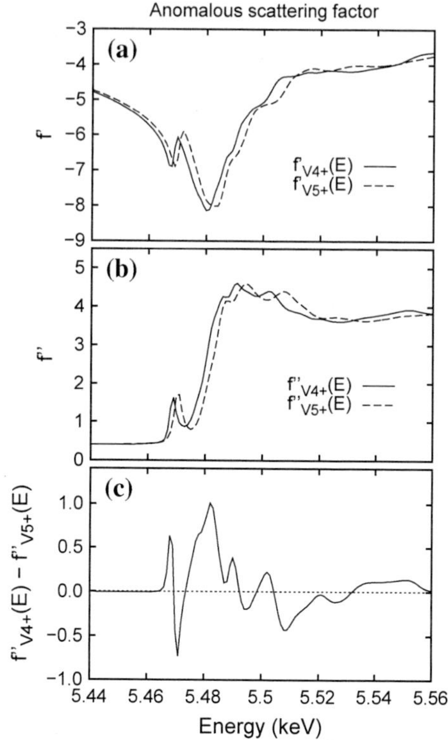

Fig. 2 Energy dependence of ASFs **a** $f'(E)$ and **b** $f''(E)$ for V^{4+} and V^{5+} shown by *solid* and *broken lines*, respectively. **c** Difference between ASFs of V^{4+} and V^{5+}, $f''_{V4+}(E) - f''_{V5+}(E)$. Data were taken from Ref. [31]

and the pre-edge peak broadens, reflecting the existence of V^{4+} and V^{5+}. In contrast, E_a of V foil is considerably lower than those of other compounds. The imaginary parts of the ASFs, $f''(E)$, of V^{4+} and V^{5+} are calculated according to Eq. 4 using $\mu(E)$ of CaV_2O_5 and V_2O_5. Then, $f'(E)$ of V^{4+} and V^{5+} are calculated according to the Kramers–Krönig transformation of $f''(E)$ using the program DIFFKK [60]. The obtained ASFs $f'(E)$ and $f''(E)$ of V^{4+} and V^{5+} are shown in Fig. 2a, b, respectively. The difference between the ASFs of V^{4+} and V^{5+} is small at $E \ll E_a$, whereas it is critically enhanced with a large modulation at E_a. Hence, the difference of one electron between the V^{4+} and V^{5+} sites can be easily detected as the RXS signal $I(E, \mathbf{Q})$, which remarkably changes at E_a. In fact, the charge-ordered state in NaV_2O_5 was clarified using the RXS technique [31, 33].

2.2 Anomalous Scattering Factor: Orbital State

Reflecting the anisotropy of orbitals, the ASF becomes a tensor near the absorption energy. This acts as the origin of an RXS signal in orbital-ordering systems. As a result, the RXS intensity indicates a unique azimuthal angle and polarization dependence, which is caused by the tensorized ASF, whereas the usual Thomson scattering shows no such dependence. In studies of orbital ordering in $3d$ electron systems, RXS at the main edge is mainly used; the $1s \rightarrow 4p$ transition at the main edge is a dipole process. The $4p$ energy level splits because of $3d$ orbital ordering. Then, the ASF becomes a tensor that reflects the anisotropic $4p$ orbital. Two possible scenarios for this lifting have been proposed. One is on-site Coulomb interaction between the $4p$ and $3d$ orbitals, which raises the $4p$ energy level lying parallel to the direction of the extension of the $3d$ orbital (Coulomb mechanism) [63–65]. The other is the anisotropic hybridization of the $4p$ orbital with the p orbital of neighboring ions due to the so-called Jahn–Teller distortion (JTD) (JT mechanism) [66–68]. The origin of $4p$ energy-level splitting depends on the type of system, and it remains controversial as to which of the mechanisms is more significant from the viewpoint of giving rise to RXS [69]. Here, we consider the $3x^2 - r^2$ orbital, as shown in Fig. 3b. In the case of the Coulomb mechanism, $4p$ energy levels are split by on-site Coulomb interaction. Hence, the $4p_x$ energy level is raised, and the $4p_{y,z}$ energy level is lowered. In contrast, in the case of the JT mechanism, the local structure (octahedron) is distorted, reflecting the anisotropic $3d$ orbital, i.e., the JTD. The octahedron is extended along the x direction. Therefore, the $4p_x$ energy level is lowered, and the $4p_{y,z}$ energy level is raised, as shown in Fig. 3c. However, in both cases, an RXS signal provides information about orbital ordering, and the tensor is described as follows:

$$\hat{f} = \begin{pmatrix} f_\parallel(E) & 0 & 0 \\ 0 & f_\perp(E) & 0 \\ 0 & 0 & f_\perp(E) \end{pmatrix}, \tag{9}$$

where $f_\parallel(E)$ $(f_\perp(E))$ corresponds to the ASF when the polarization of the incident X-ray $\hat{\varepsilon}_i$ is parallel to the x (y, z) direction. As explained here, the tensor has energy

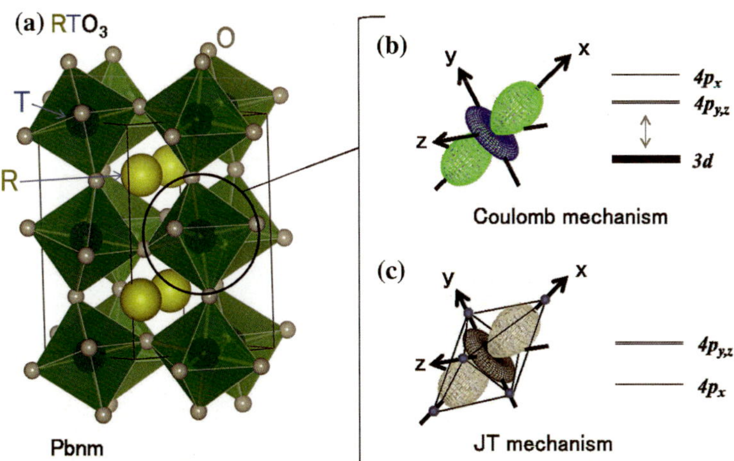

Fig. 3 **a** Perovskite structure with space group $Pbmn$, so called GdFeO$_3$-type structure. **b, c** $3x^2 - r^2$ of $3d$ electron orbital. Schematic energy level diagram of $4p_{x,y,z}$ states, which depend on the scenario for the lifting of the degeneracy of $4p$ orbitals

dependence but no Q dependence because the transition is a dipole process. In contrast, the RXS signal at the pre-edge contains the quadrupole transition process, which is the $1s \rightarrow 3d$ transition. In this case, the tensor has Q dependence. The scattering framework is explained in depth in previous studies [10, 62].

Here, we consider the tensor of the structure factor of RTO_3 (T: transition metal with orbital degree of freedom, R: rare earth or Y) perovskite structure with the space group $Pbnm$, as shown in Fig. 3a. There are four T ion sites in the unit cell. The structure factor is expressed by

$$F(E, \mathbf{Q}) = \sum_{j}^{4} \hat{\varepsilon}_i \hat{f}_j(E) \hat{\varepsilon}_s \exp(-i\mathbf{Q} \cdot \mathbf{r}_j) + F_o, \tag{10}$$

where $\hat{\varepsilon}_s$ is the polarization vector of the scattered X-ray and F_o is the scalar component of the structure factor. F_o includes the components of R, O atoms and the Thomson scattering factor of T atoms. The tensor in Eq. (9) is defined using xyz coordinates, where the principal axes are aligned with the oxygen direction in the TO_6 octahedron. There is a tilt of the octahedron in the crystal structure with $Pbnm$ (GdFeO$_3$-type structure), as shown in Fig. 3. Hence, the xyz coordination of the tensor depends on T-ion sites. The coordination of the tensor $\hat{f}_j(E)$ is adjusted for calculation using a rotation matrix. On the basis of the structure factor (Eq. (10)), the azimuthal angle dependence of RXS was calculated for titanate and vanadate systems [25, 32, 36].

Let us consider the RXS experimental geometry shown in Fig. 4. The scattering plane defined by $\mathbf{k}_i$ and $\mathbf{k}_f$ is the vertical plane, i.e., the yz plane, where $\mathbf{k}_i$ and $\mathbf{k}_f$ are the wave vectors of incident and scattered X-rays, respectively. The scattering

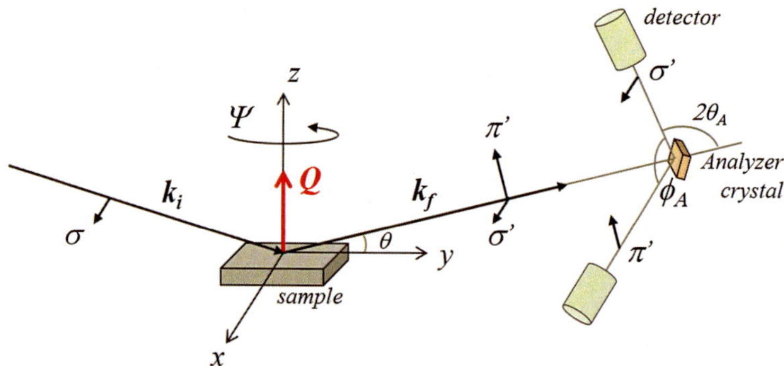

Fig. 4 Schematic geometry of RXS measurement. The scattering plane defined by $\mathbf{k}_i$ and $\mathbf{k}_f$ is vertical, and the scattering vector $\mathbf{Q} = \mathbf{k}_i - \mathbf{k}_f$ is perpendicular to the sample surface in this example, where $\mathbf{k}_i$ and $\mathbf{k}_f$ are the wave vectors of the incident and scattered X-rays, respectively. X-ray polarization is denoted as σ, and π; σ (π) is perpendicular (parallel) to the scattering plane; $\sigma = \sigma' = (1, 0, 0)$ and $\pi' = (0, -\sin\theta, \cos\theta)$. X-ray polarization is analyzed by rotating an analyzer crystal through angle ϕ_A, and $2\theta_A$ is the scattering angle

vector $\mathbf{Q} = \mathbf{k}_i - \mathbf{k}_f$ is perpendicular to the sample surface in this example. The azimuthal angle Ψ, i.e., the angle corresponding to a rotation around the scattering vector $\mathbf{Q}$, is defined as the angle between the scattering plane and reference vector. The incident polarization vector is denoted by σ, and the polarization vector σ' (π') of the scattered beam is perpendicular (parallel) to the scattering plane. The polarization of the scattered X-ray is analyzed by rotating the analyzer crystal by an angle ϕ_A. The analyzer crystal is selected such that the scattering angle $2\theta_A$ is close to $90°$. The σ'-polarized beam is detected by the analyzer with $\phi_A = 0°$, and the structure factor is expressed by $F_{\sigma\sigma'}$ with $\hat{\varepsilon}_i = \sigma$ and $\hat{\varepsilon}_s = \sigma'$. The π'-polarized beam is detected at $\phi_A = 90°$, and the structure factor is expressed by $F_{\sigma\pi'}$.

3 Orbital Ordering in Perovskite Titanate

Perovskite titanate, $YTiO_3$, is a t_{2g} electron system. One electron occupies the triply degenerated t_{2g} orbitals, and it shows typical t_{2g} orbital ordering. Hence, the orbitally ordered state was investigated in depth using the RXS technique, and this became a target next to the e_g orbital-ordering study. This study not only acted as evidence of t_{2g} orbital ordering but also facilitated a discussion of the RXS mechanism. In fact, substances having the e_g electron system such as manganites exhibit considerable JTD owing to a strong electron–lattice coupling; then the scattering mechanism of Coulomb and JT mechanisms became the controversial problem as described above. In contrast, the t_{2g} electron system has a weak electron–lattice coupling. As a result, it is expected that the JT mechanism in the t_{2g} system is less effective in giving rise to RXS when compared with the e_g system. In this section, RXS studies of t_{2g} orbital ordering in $YTiO_3$ [25] and $Y_{1-x}Ca_xTiO_3$ [26, 28, 29] are presented.

3.1 *YTiO₃*

$YTiO_3$ is a ferromagnetic (FM) insulator with $T_C \sim 30\,K$. The Ti^{3+} ion has one t_{2g} electron, and it shows orbital ordering. Orbital ordering is expected to be the origin of its ferromagnetism and has been studied extensively both theoretically and experimentally. Unrestricted Hartee–Fock calculation [70, 71] and generalized gradient approximation [72] have been employed to predict the wave functions of the orbitally ordered state to be a linear combination of two t_{2g} orbitals at sites $1-4$, as shown in Fig. 5. By polarized neutron scattering, spin-density distribution was evaluated, and the wave function of the ordered orbital was determined [73, 74]. The wave function was also estimated by performing an NMR experiment [75]. Both sets of experimental results were consistent with theoretical predictions. However, these techniques can only be used to observe orbital ordering when accompanied by magnetic ordering. Hence, an observation of the orbitally ordered state without magnetic ordering was strongly desired.

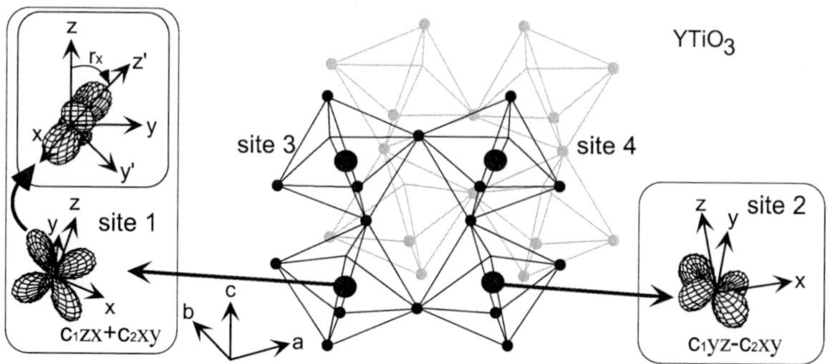

Fig. 5 Orbital ordering in YTiO$_3$. Four Ti sites exist in the unit cell. Ti atoms numbered 1, 2, 3, and 4 are located at $(0, \frac{1}{2}, 0)$, $(\frac{1}{2}, 0, 0)$, $(0, \frac{1}{2}, \frac{1}{2})$, and $(\frac{1}{2}, 0, \frac{1}{2})$, respectively. Wave functions at sites 1, 2, 3, and 4 are $c_1 zx + c_2 xy$, $c_1 yz - c_2 xy$, $c_1 zx - c_2 xy$, and $c_1 yz + c_2 xy$, respectively. The x, y, and z axes are taken along the directions of Ti–O in the TiO$_6$ octahedron. The inset of site 1 shows an alternate coordinate system with its axes aligned with the orbital, i.e., the $xy'z'$ coordinates, defined as the xz' plane in which the t_{2g} orbital is elongated. The angle between the z and z' axes following a rotation about the x axis is defined as r_x

To investigate the orbital state of the Ti $3d$ electron above T_C, we performed an RXS study of YTiO$_3$ at room temperature. First, we recorded the X-ray absorption spectrum near the Ti K-edge, as shown in Fig. 6a. We found two large peaks at $E = 4.974$ (main edge) and 4.987 keV, which correspond to the $1s \rightarrow 4p$ dipole transition. The spectrum shows a shoulder feature at $E \sim 4.963$ keV (pre-edge), which corresponds to the $1s \rightarrow 3d$ quadrupole transition energy.

We searched for RXS in the (1 0 0), (0 0 1), and (0 1 1) reflections, which are forbidden reflections on the *Pbnm* space group. We can expect to observe RXS reflecting orbital ordering at peak positions, as shown in Fig. 5. We measured energy dependence at several azimuthal angles to avoid contamination due to multiple scattering (MS). Strong MS intensity was observed sometimes, and the spectrum was strongly affected, as shown around 4.96 keV in Fig. 7c. The MS intensity can be suppressed by changing the azimuthal angle and incident X-ray energy. In contrast, the real RXS signal changes gradually depending on the azimuthal angle. This is an important point for distinguishing an RXS signal from a spurious signal owing to MS. The data, free of MS, are shown in Fig. 6b–d. The energy spectra depend strongly on the observed reflections. The RXS spectrum at (1 0 0) shows three large resonant peaks at 4.974, 4.986, and 4.999 keV, while that at (0 0 1) shows two large peaks at 4.972 and 4.984 keV. At the pre-edge energy, a small resonant signal was observed for both reflections. Moreover, both RXS spectra have only a $\sigma \rightarrow \pi'$ component. At (0 1 1), both scattering components, $\sigma \rightarrow \sigma'$ and $\sigma \rightarrow \pi'$, were observed simultaneously. Both the components show the same energy dependence, exhibiting two large resonant peaks at 4.975 and 4.986 keV and a weak peak at the pre-edge. In each reflection, several resonant peaks were observed above the main edge energy. In contrast, strong RXS was observed at only one resonant energy (main edge) in the manganite system [5, 6]. The ASF tensor at the main edge is expected to have no

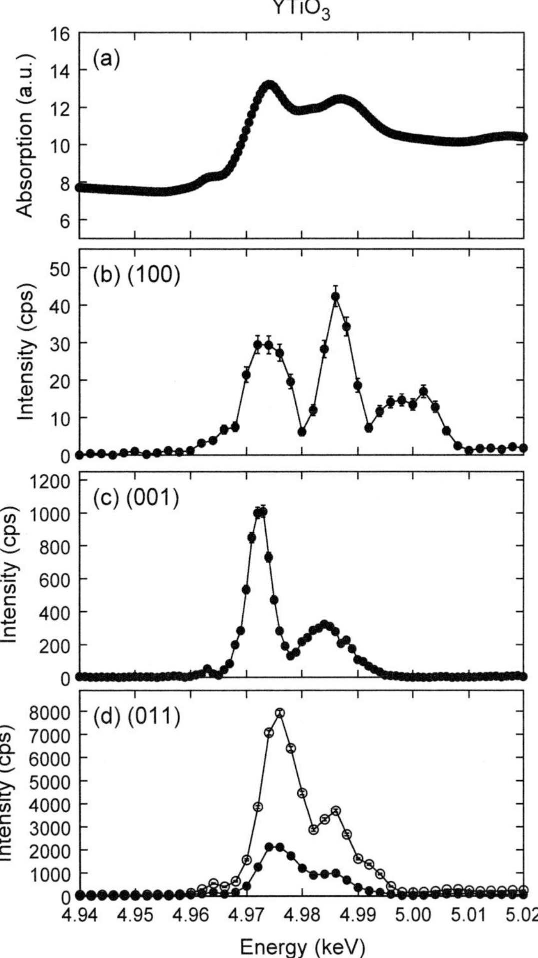

Fig. 6 a Absorption spectrum near Ti K-edge energy. **b–d** Energy dependence of RXS intensities at forbidden reflections. Part of data were taken from Ref. [25]. The scattering components $\sigma \to \sigma'$ and $\sigma \to \pi$ are shown by *open* and *filled circles*, respectively

radial direction dependence of the scattering vector Q because it is a dipole transition, i.e., the $1s \to 4p$ transition, as expressed in Sect. 2.2; on the other hand, RXS indicates a marked azimuthal dependence. To clarify the property of RXS, we recorded the energy spectra corresponding to the (0 0 1) and (0 0 3) reflections, which have the same direction of Q and different Q lengths, as shown in Fig. 7a. The energy spectra are quite similar to each other, and RXS intensities are comparable. The quantitative estimation of the scattering intensity is quite difficult near the absorption energy because sample absorption is strong. The spectra at (1 0 0) and (3 0 0) and at (0 1 1) and (0 3 3) are similar to each other, as shown in Fig. 7b, c, respectively. Therefore the ASF tensor at the main edge has no radial direction dependence of Q, which is as expected.

Fig. 7 Energy dependence of RXS intensities for forbidden reflections: **a** (0 0 h), **b** (h 0 0), and **c** (0 h h). To evaluate the radial direction dependence of RXS, the RXS spectra corresponding to the reflections of $h = 1$ are compared with those corresponding to $h = 3$. **c** shows strong MS at around 4.96 keV. Part of data were taken from Ref. [25]

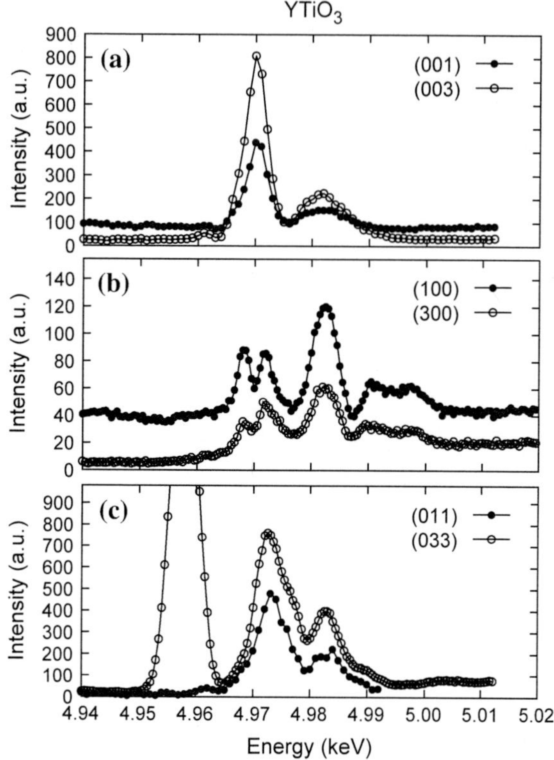

An important feature of RXS is azimuthal angle dependence; the azimuthal angle, Ψ, is a rotation around the scattering vector, as shown in Fig. 4. Azimuthal angle dependence gives direct information about the ASF tensor. Hence, the azimuthal angle dependence of RXS intensity was observed for forbidden reflections at the main edge. At each azimuthal angle, we performed $\theta - 2\theta$ scans. The resulting integrated intensities at (1 0 0), (0 0 1), and (0 1 1) were normalized by those of the fundamental peak of (2 0 0), (0 0 2), and (0 2 2), respectively, to correct any variations due to the sample shape. The structure factors of the fundamental peaks were calculated using crystal parameters [76]. The results were finally normalized by the intensity of (0 2 2) as a standard. The azimuthal angle dependence of the (1 0 0) and (0 0 1) reflections exhibits twofold symmetry, as shown in Fig. 8a, b. For the (0 1 1) reflection, the azimuthal angle dependence of the $\sigma \rightarrow \sigma'$ component shows fourfold symmetry, whereas that of the $\sigma \rightarrow \pi'$ component has a period of 360°, as shown in Fig. 8c, d. Azimuthal angle dependence at the pre-edge was also observed for the (0 0 1) and (0 1 1) reflections, as indicated by open circles in Fig. 8; it was the same as that at the main edge. Moreover, the intensities were normalized by the fundamental peak at the main edge. Hence, correction for the energy dependence of the fundamental peak intensity is needed for a quantitative discussion.

Fig. 8 Azimuthal angle dependence of RXS intensities at (1 0 0) (**a**), (0 0 1) (**b**), and (0 1 1) (**c, d**). Data were taken from Ref. [25]. The intensities at the main edge are denoted by *filled circles*, and those at the pre-edge multiplied by 20 are shown by *open circles*. Ψ of the azimuthal angle is defined as follows: **a, b** $\Psi = 0°$ at $\sigma \parallel b$; **c, d** $\Psi = 0°$ at $\sigma \parallel a$

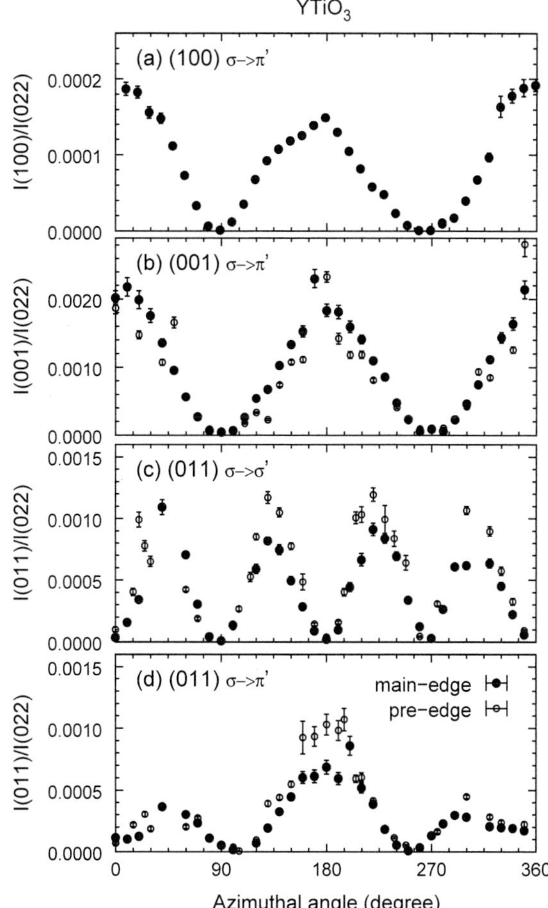

The origin of RXS is the splitting of the Ti $4p$ energy levels, as indicated in Sect. 2.2. However, the origin of the splitting remains controversial. RXS was calculated on the basis of the Coulomb and JT mechanisms. In the case of the Coulomb mechanism, the Ti $4p$ energy levels are split by on-site Coulomb interaction between the ordered $3d$ orbital and $4p$ orbital of a Ti ion. The wave function, $c_1 zx + c_2 xy$, of the ordered orbital is expected at site 1, as shown in Fig. 5. The wave function is represented by xz' by definition if the principal axes based on xyz coordinates are exchanged for those based on $xy'z'$ coordinates. That is, the t_{2g} orbital is elongated in the xz' plane but not in the y' direction. Because of on-site Coulomb interaction, the $4p_{y'}$ energy level is lowered, and the $4p_x$ and $4p_{z'}$ energy levels are raised. Hence, the tensor at site 1 can be described as follows:

$$\begin{pmatrix} f_a & 0 & 0 \\ 0 & f_a + \Delta f_a & 0 \\ 0 & 0 & f_a \end{pmatrix} \text{ in the } xy'z' \text{ coordinates,}$$

where Δf_a is the anisotropic strength and f_a is the isotropic strength of the ASF. The angle between the z and z' axes following a rotation about the x axis is defined as r_x. Thus, the tensor is described with only two parameters: r_x and Δf_a. The tensors at the other three sites are determined in the same manner. These tensors satisfy the space group of the crystal structure $Pbnm$. If the tensors do not satisfy the space group, the azimuthal angle and polarization dependence values calculated from such tensors are completely different from those obtained in the experimental results presented herein. In fact, the azimuthal angle dependence was reported to drastically change in the case of RVO$_3$ when the space group of the crystal structure was changed, reflecting V $3d$ orbital ordering [36]. Thus, restrictions pertaining to the space group are very strict. Moreover, sites 1–4 have different xyz coordinates owing to the tilting of the TiO$_6$ octahedron, the so-called GdFeO$_3$-type structural distortion. In our model calculation, this distortion was considered in terms of the difference of the xyz coordinates among the sites, as explained in Sect. 2.2. On the basis of these tensors, the structure factor can be calculated. The calculated $\sigma \rightarrow \pi'$ components of the intensity at each reflection with $\Psi = 180°$ as a function of r_x are shown in Fig. 9. The intensity ratio among the RXS peaks depends strongly on the value of r_x. The experimentally observed RXS intensities are also plotted on the right ordinate in the figure. In the model calculation, the range of r_x for which the order among the RXS intensities is observed is $30° \sim 70°$. By comparing the measured intensity ratio among the RXS peaks with the calculated value, it is found that $r_x = 45° \pm 10°$, as indicated by an arrow in Fig. 9. The value of $r_x = 45°$ is used to calculate the azimuthal and polarization dependence values of the three reflections. The results of these calculations are shown by lines in Fig. 10. The periodicity and intensity of

Fig. 9 Rotation angle, r_x, dependence of RXS intensities of the $\sigma \rightarrow \pi'$ component at $\Psi = 180°$. Data were taken from Ref. [25]. The relative intensities observed in the experiments are plotted on the right ordinate

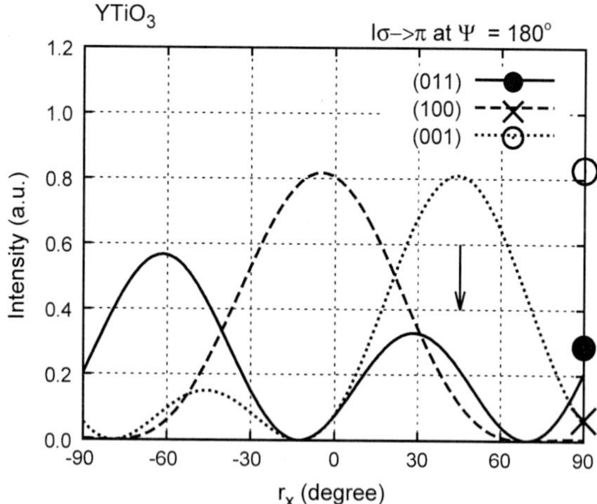

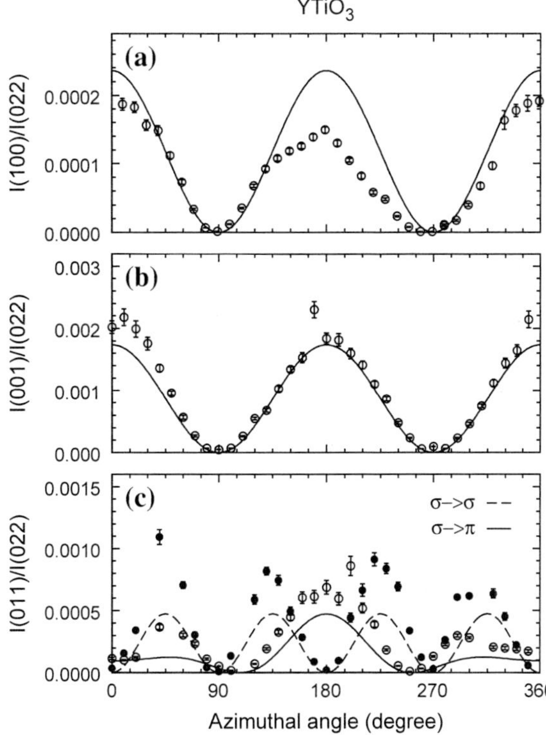

Fig. 10 Azimuthal angle dependence of RXS intensities at the main edge is compared with model calculations using $r_x = 45°$. Data were taken from Ref. [25]. The *solid line* (*open circle*) and *broken line* (*filled circle*) denote the $\sigma \to \pi'$ and $\sigma \to \sigma'$ components, respectively

Table 1 Lists of wave function parameters, c_1, determined experimentally and theoretically. The wave function at site 1 is defined as $c_1 zx + c_2 xy \ (c_1^2 + c_2^2 = 1)$

	c_1
Theory	0.8 [70], 0.71 [72]
Neutron	0.77 [73, 74]
NMR	0.8 [75]
RXS	0.71 [25]

the azimuthal angle dependence were well reproduced by the calculations. Thus, the azimuthal angle and polarization dependence and the intensity ratio among the RXS peaks can be explained by the model calculation that is based on the Coulomb mechanism. It was concluded that the parameters of the wave function of the Ti $3d$ orbital, c_1 and c_2, were $c_1 = 0.71 \pm 0.11$ and $c_1^2 + c_2^2 = 1$, respectively. These values agreed well with the results of previous theoretical and experimental studies within our experimental error bar, as can be inferred from Table 1.

In the case of the JT mechanism, the splitting of the $4p$ energy levels is a result of oxygen movement that distorts the octahedron surrounding the Ti ion in the orbitally ordered state. Then, the tensor is that of the Coulomb mechanism without rotations, i.e., $r_x = 0 \ (c_2 = 0)$. It is obvious that $r_x = 0$ required by this simplest version of the JT mechanism cannot explain our results. Hence, it is expected that RXS is mainly caused by the Coulomb mechanism, although a weak RXS signal resulting from the

JT mechanism may exist. However, for quantitative estimation, not only the JTD effect but also the effect of neighboring octahedra should be considered as the origin of RXS, as discussed in the literature [78].

The anisotropic strength of the tensor, Δf_a, determines the intensity ratio between the RXS peak and fundamental peak, while r_x controls the intensity ratio among the RXS peaks. By comparing RXS with the fundamental intensities, Δf_a was determined to be 1.3 ± 0.1. This value was estimated without correction for any extinction effect. For further quantitative estimation, a data-analysis-like structural analysis is required. In addition, the aforementioned Δf_a value was used for the solid curves shown in Fig. 10. It was noted that this Δf_a value was larger than that of $LaMnO_3$ ($\Delta f_a \sim 0.3$), although the JTD was smaller for $YTiO_3$ than that for $LaMnO_3$. Δf_a of $LaMnO_3$ was estimated roughly on the basis of the result in Ref. [6]. This also indicates that the Coulomb mechanism is a plausible model of RXS in $YTiO_3$.

The tensors at the Ti site resulting from the JT and Coulomb mechanisms used in our model calculation are JTD and a simple linear combination of two t_{2g} orbitals, respectively. Thus, we only examined which of these mechanisms is more significant in terms of the origin of RXS. These observations were supportive of the Coulomb mechanism. Ishihara et al. derived a general form for the scattering cross-section of RXS and approached our RXS results as a linear combination of three t_{2g} orbitals [77]. Their wave function parameters better fit our results than those obtained using our simple model calculation. In contrast, Takahashi et al. investigated RXS intensities corresponding to the forbidden reflections using band-structure calculation combined with local density approximation, as shown in Fig. 11 [78]. Accordingly, the calculated RXS intensities arise from not only the JTD but also the tilts of neighboring TiO_6 octahedra. In our JT mechanism model, only the tilt of the octahedra at the site where X-rays are absorbed is considered. The energy spectra obtained theoretically are similar to our experimental results. However, the RXS intensities are comparable in the theoretical calculations of Takahashi et al. Namely the experimental intensity ratio cannot be explained by their theory. There the RXS intensity due to the Coulomb mechanism is expected to be combined with their theory. Moreover, their model leads to a metallic ground state without orbital ordering. Therefore, the experimentally observed RXS intensities at the pre-edge cannot be reproduced by their theory. Finally, this study supports the Coulomb mechanism in the main edge, although other mechanisms also may influence RXS.

We note RXS intensity at the pre-edge. Theoretically, this intensity is expected to directly reflect the $3d$ orbital state, although the precise scattering mechanism is not well understood. In $LaMnO_3$, RXS due to a weak dipole transition arising from the hybridization of Mn $4p$ with neighboring orbitally ordered Mn $3d$s was predicted [66, 68]. In V_2O_3, in contrast, strong RXS was observed at the pre-edge owing to broken local inversion symmetry; on the other hand, no RXS was observed at the main edge [30, 79]. In this study, in $YTiO_3$, the azimuthal angle, polarization, and Q-position dependence at the pre-edge are the same as those at the main edge, and there is inversion symmetry. Therefore, RXS at the pre-edge probably arose from a dipole transition caused by the hybridization of Ti $4p$ with neighboring orbitally ordered Ti $3d$s. Within the experimental error bound, the ratio between

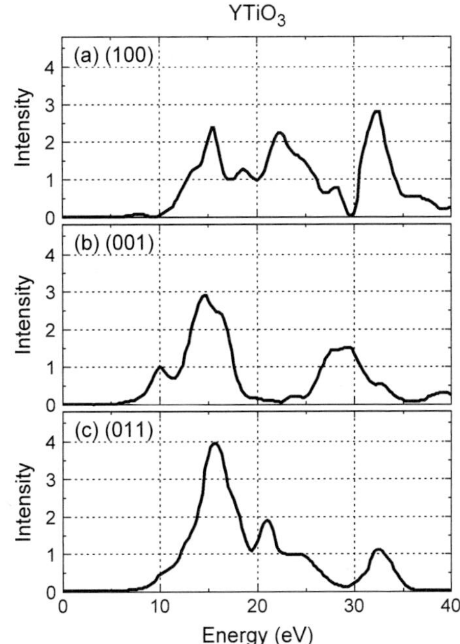

Fig. 11 Theoretical calculations of RXS spectra for (1 0 0), (0 0 1), and (0 1 1) reflections. The origin of the energy corresponds to the photon energy required for exciting an electron from the 1s state to the Fermi level. Data were taken from Ref. [78]

intensities at the pre-edge and main edge is almost independent of the reflections. As a result, $r_x \sim 45°$ at the pre-edge was immediately determined. The anisotropic strength of the ASF, $\Delta f_a \sim 0.2$, at the pre-edge was obtained by comparing RXS and the fundamental intensities with correction for the absorption effect. Thus, the ASF tensor at the pre-edge was quantitatively determined.

3.2 $Y_{1-x}Ca_xTiO_3$

In the study of RXS in $YTiO_3$, we discussed the orbital ordering of the 3d electron of Ti^{3+}. However, it was expected that the RXS intensity appears to reflect not only the orbital state but also the structural character. Hence, the origin of RXS remains controversial both experimentally and theoretically. The origin of RXS is important for a qualitative discussion of an orbital state. Herein we noted $Y_{1-x}Ca_xTiO_3$, in which orbital ordering is suppressed by Ca substitution [26, 28]. Here, we can discuss the origin of RXS because we expect that RXS signals reflecting the orbital state and structural character show different Ca substitution dependence values. The RXS signal at the Y K edge ($1s \rightarrow 5p$ transition energy) was also noted [29]. There we expect that only the RXS intensity due to the octahedral tilt can be observed. By combining the RXS signals at the Ti K edge and Y K edge, we discuss the origin of RXS. The result elucidates the potential to determine the wave function of the ordered orbital using this RXS technique.

In $Y_{1-x}Ca_xTiO_3$, the number of $3d$ electrons in the Ti ion changes from 1 to 0 on Ca substitution [80–84]. FM ordering disappears at $x_{FP} \sim 0.15$ with increasing x, whereas the insulating state is preserved up to x_{MI}. A marked metal–insulator transition was observed at x_{MI}. Then, the paramagnetic (PM)–metal phase stabilized above x_{MI}. Hence, it was expected that the orbital degree of freedom plays an important role in these phases. The orbitally ordered state was systematically investigated in $Y_{1-x}Ca_xTiO_3$ ($0 \leq x \leq 0.75$) using the RXS technique [26, 28].

RXS intensities at the forbidden reflections, (0 0 1), (1 0 0), and (0 1 1), were searched for near the Ti K-edge energy on the basis of the result obtained using $YTiO_3$. Energy dependence was measured at several azimuthal angles to avoid contamination due to MS. The data, free of MS, are shown in Fig. 12. The (0 0 1),

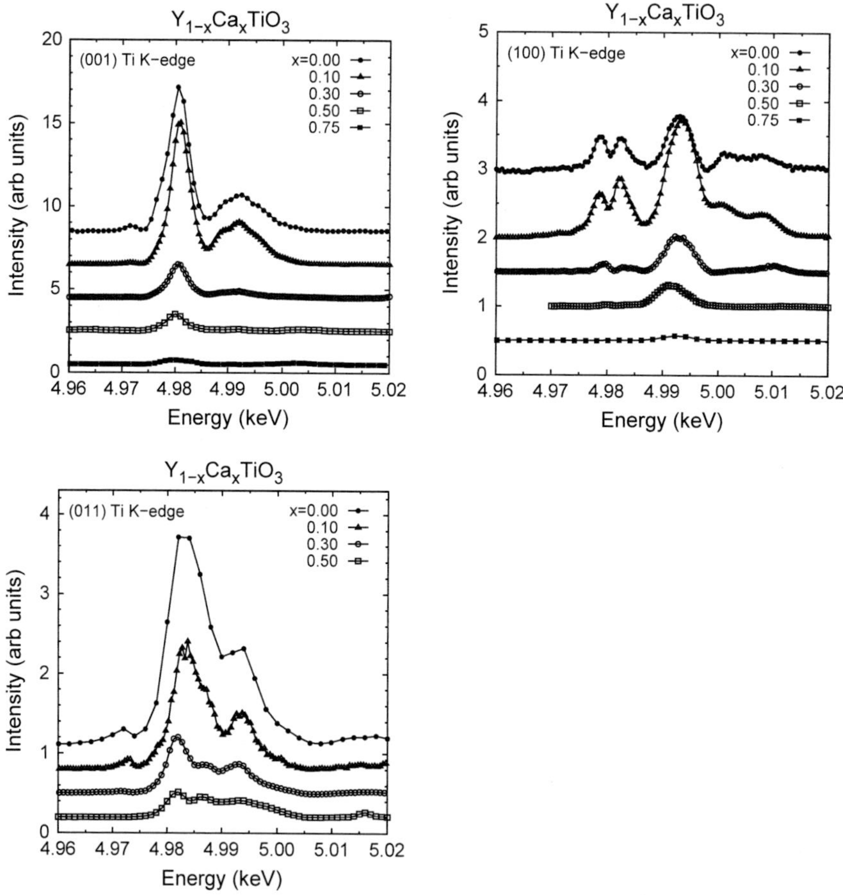

Fig. 12 Energy dependence of RXS intensities of $Y_{1-x}Ca_xTiO_3$ for the forbidden reflections (0 0 1), (1 0 0), and (0 1 1). Data were taken from Refs. [25, 26, 28]. Baselines were shifted for clarity

(1 0 0), and (0 1 1) intensities were normalized by the fundamental reflections, (0 0 2), (2 0 0), and (0 2 2), respectively, to correct any error due to the sample shape. To compare the intensities with those obtained for other Ca concentrations, the intensities were also normalized by the respective structure factors of the fundamental peaks, thus reflecting structural changes with Ca doping [28]. The RXS intensities at (0 0 1) and (0 1 1) decreased monotonically, and the energy spectra indicated no remarkable change with increasing Ca concentration, x. In contrast, the energy spectra at (1 0 0) changed depending on x. The spectra above the main edge changed drastically, while the intensities decreased at the main edge and pre-edge energies. The main edge and pre-edge energies were approximately 4.980 and 4.972 eV, respectively. These energies were different from those defined in Sect. 3.1 because of differences in energy calibration.

Figure 13 shows the Ca concentration dependence of an RXS intensity at the main edge. The RXS intensities decreased gradually with increasing x. The intensity ratio among these peaks was almost the same as a function of x. Next, RXS at the pre-edge, which is expected to be the order parameter of orbital ordering as discussed in Sect. 3.1, is noted. The Ca concentration dependence of the RXS intensities at (0 0 1) and (0 1 1) is shown in Fig. 14b. The intensities decreased rapidly in the concentration range of $0 < x < 0.15$ and could barely be discerned at $x = 0.30$ at the insulator phase. Moreover, the intensity almost disappeared at the metal phase ($x \geq 0.5$). In other words, the orbital ordering of Ti 3d was strongly suppressed in the $0 < x < 0.15$ concentration range but barely remained up to x_{MI}. This result is consistent with the results of a structural analysis based on powder X-ray diffraction [28]. Here, the Ca concentration dependence of the RXS intensity at the main edge, as shown in Fig. 14a,

Fig. 13 RXS intensities at main edge energy as a function of x. RXS intensities were normalized by observed and calculated fundamental peak intensities. Data were taken from Ref. [28]

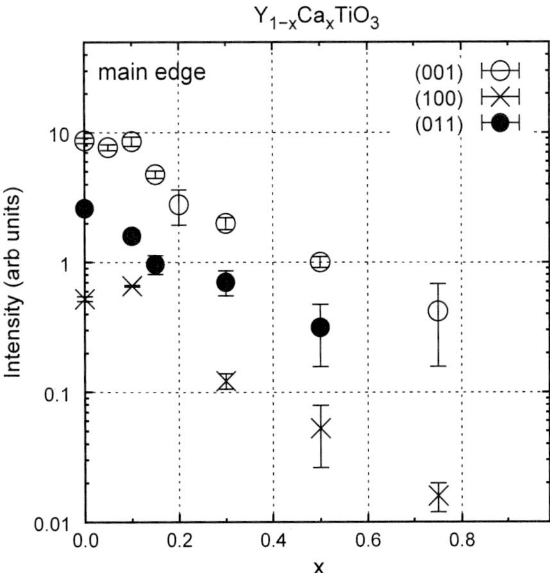

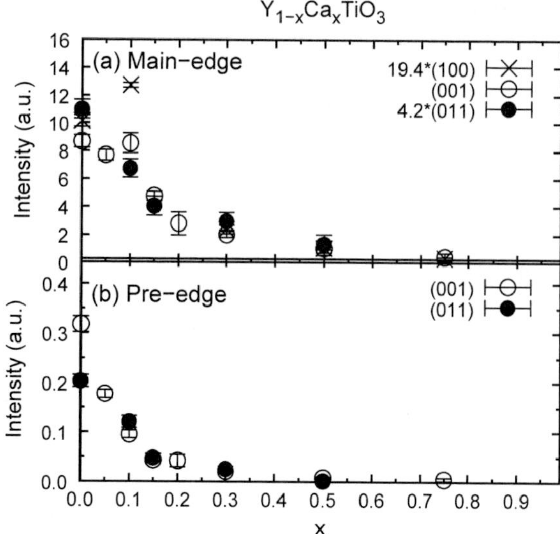

Fig. 14 Ca concentration dependence of RXS intensities **a** at main edge and **b** at pre-edge. Data were taken from Refs. [26, 28]

is considered. The RXS intensity declined with increasing x, and the rate of decrease slowed at $x \sim 0.2$. The dependence is similar to that at the pre-edge. Hence, it can be expected that the RXS at the main edge mainly reflects orbital ordering. However, this RXS intensity remains even at the metal phase ($x > x_{MI}$). The intensity is about 1/10th that of YTiO$_3$ ($x = 0$), and it may reflect the local structure around the Ti ion. As a result, we believe that the Coulomb mechanism is in effect on the main edge because RXS reflects mainly the order parameter of orbital ordering. Moreover, it is important that the other mechanisms have a small effect on RXS.

Precise RXS measurement can be used to quantitatively determine the ASF tensors, r_x and Δf_a. This is purely an experimental result and independent of the scattering mechanism. The intensity ratio among the RXS peaks changed drastically as a function of r_x (Fig. 9). In the case of Y$_{1-x}$Ca$_x$TiO$_3$, the intensity ratio was nearly independent of Ca concentration, as shown in Fig. 13. That is, r_x is about 45°, the same value as that reported in the case of $x = 0$. Then, Δf_a was estimated in the case of YTiO$_3$ because the intensity ratio between the RXS peak and fundamental peak reflects the anisotropic strength of the ASF, Δf_a. The Ca concentration dependence of Δf_a was estimated, as shown in Fig. 15. The anisotropic strength of the ASF decreased gradually with increasing x, and it remained in the metallic region ($x > x_{MI}$). At the metal phase, the orbital disordered state was expected. At that state, RXS cannot be explained by the Coulomb and JT mechanisms. The tilt of neighboring octahedra was proposed as the origin of an RXS signal at the metal phase [26, 28]. The Ca concentration dependence of RXS intensity, reflected in the tilt of neighboring octahedra, was expected to be estimated as a function of octahedral tilting.

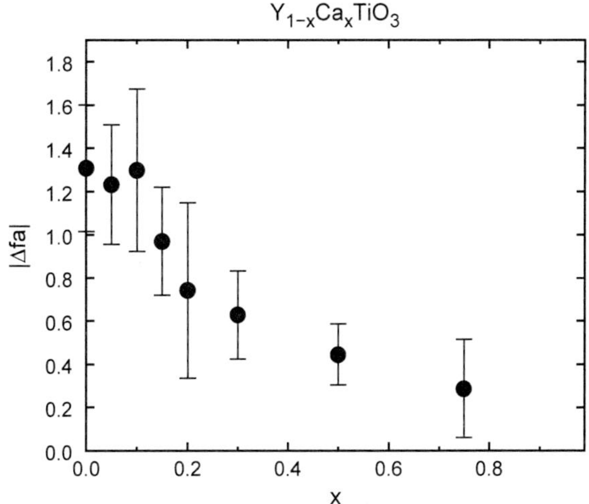

Fig. 15 Anisotropic strength of ASF, Δf_a, as a function of Ca concentration dependence. Data were taken from Ref. [28]

To evaluate RXS, which is reflected in the tilting of the TiO$_6$ octahedra in a crystal, experimentally, the RXS signal at the Y K edge was noted [29]. In this case, the $1s \rightarrow 5p$ transition was used. Hence, the local environment around the Y ion was detected using the RXS at Y K edge. It was reported that RXS at the La L edge is mainly caused by the tilting of the MnO$_6$ octahedra in LaMnO$_3$ [16, 85], which has the same crystal structure with the space group, $Pbnm$, as this perovskite titanate system [28, 86]. Therefore, it can be expected that only the RXS intensity due to octahedral tilting can be observed at the Y K edge, whereas RXS at the main edge of Ti K edge reflects both orbital ordering and octahedral tilting.

We searched for the RXS signal at several forbidden reflection positions near the Y K-edge energy to detect RXS reflecting the octahedral tilting. The absorption spectrum of YTiO$_3$ was obtained at the Y K edge, as shown in Fig. 16a. The energy dependence showed clear absorption at the K edge energy. The spectrum was detected to resonate near the Y K edge only at (5 0 0), as shown in Fig. 16b. The intensity had a $\sigma \rightarrow \pi'$ component and not a $\sigma \rightarrow \sigma'$ one. The azimuthal angle dependence of the intensity was measured at $E = 17.033\,\text{keV}$. The integrated intensities at (5 0 0) were normalized by that of the fundamental peak of (6 0 0). The azimuthal angle dependence exhibited twofold symmetry ($\propto \cos^2 \Psi$), as shown in Fig. 17, and was consistent with the space group.

The energy spectra of RXS were obtained at (5 0 0) as a function of Ca concentration, x, as shown in Fig. 18. The spectra did not change markedly with increasing x. Then, the Ca dependence of the RXS intensity at the azimuthal angle $\Psi = 0°$ was estimated, as shown in Fig. 19a. To avoid error due to MS and sample shape, the intensities at $\Psi = 0°$ were estimated according to the fitting function, $A \cos^2 \Psi$. The fitted result at $x = 0$ is shown in Fig. 17. The obtained intensities normalized by the respective structure factors of the fundamental peak (6 0 0), reflecting the structural change with Ca doping and the number of Y ions. The obtained RXS intensities are

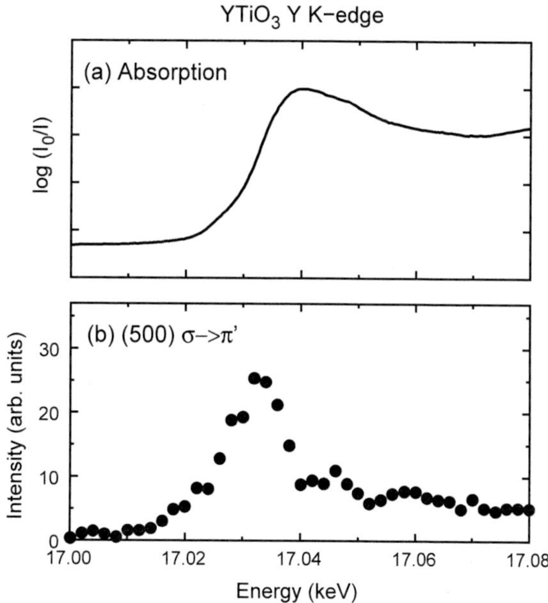

Fig. 16 a Absorption spectrum of the YTiO$_3$ powder sample. **b** Energy dependence of RXS intensity at (5 0 0) reflection with $\sigma \rightarrow \pi'$ polarization. Data were taken from Ref. [29]

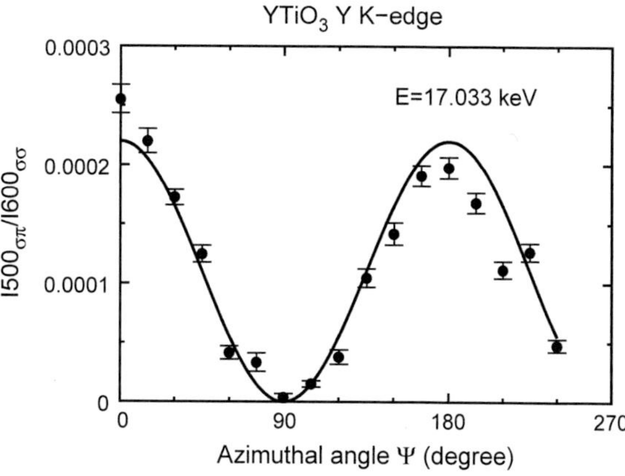

Fig. 17 Azimuthal angle dependence of the scattering intensity with $\sigma \rightarrow \pi'$ polarization at (5 0 0) in YTiO$_3$. Data were taken from Ref. [29]. The azimuthal angle Ψ is defined as $\Psi = 0°$ when the polarization vector σ is parallel to the b axis. The plotted function, $A \cos^2 \Psi$, is represented by the *solid line*

shown in Fig. 19a. The RXS intensities were almost constant in the entire Ca concentration region, whereas the RXS intensities at the Ti K edge decreased markedly with increasing x, as shown in Fig. 19b, c. Figure 19d shows the Ca dependence of the deviation from 180° of the Ti–O–Ti bond angle (Δ_{tilt}). Δ_{tilt} shows no remarkable change and decreases slightly with increasing x. Hence, the Ca dependence of the RXS intensity at Y K edge is similar to that of Δ_{tilt}. Finally, it was expected that the origin of RXS is related to the deviation of the Ti–O–Ti bond angle from 180°, as in the case of RXS at the La L edge in LaMnO$_3$. In the low Ca concentration region ($0 < x < 0.2$), the RXS intensity increased with increasing x, whereas Δ_{tilt}

Fig. 18 Energy dependence
of RXS intensities as a
function of Ca concentration,
x. Data were taken from
Ref. [29]. The *dotted lines*
are visual guides

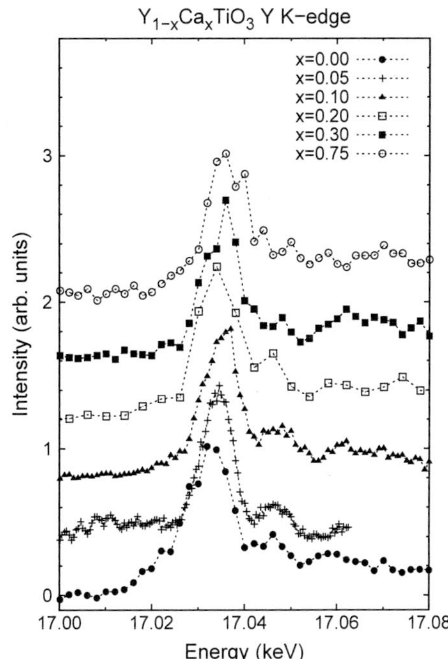

decreased monotonically. This cannot simply be explained by the RXS associated
with the deviation of the Ti–O–Ti bond angle from 180°. It may indicate a difference
in the local environment around only the Y ion, while parameters obtained by the
structural analysis are the averages of those at the Y and Ca sites. A theoretical study
is needed for a detailed discussion of the scattering mechanism.

Here, we focus on the Ca concentration dependence of the RXS intensity at the
main edge of the Ti K edge. The RXS intensity at the insulator phase ($0 \leq x < 0.4$)
is expected to reflect both orbital ordering and tilts of the octahedra (Δ_{tilt}). The
latter origin of RXS is the same as that of RXS at the Y K edge. Therefore, the Ca
concentration dependence of the RXS intensity at the main edge of the Ti K edge
originating from the Δ_{tilt} ($ITi_{tilt}(x)$) should be similar to that of the RXS intensity
at the Y K edge ($IY(x)$, Fig. 19a): the $ITi_{tilt}(x)$ is expected to be almost constant
as a function of Ca concentration. Here, the origin of RXS at the main edge of the
Ti K edge is discussed considering the following steps: (1) the Ca concentration
dependence of $ITi_{tilt}(x)$, (2) Ca concentration dependence of RXS reflecting the
orbitally ordered state ($ITi_{OO}(x)$), and (3) evaluation of RXS at the main edge of
the Ti K edge using the obtained $ITi_{OO}(x)$ and $ITi_{tilt}(x)$. As the first step, the
Ca concentration dependence of $ITi_{tilt}(x)$ was evaluated. The $ITi_{tilt}(x)$ reflects
the local crystal structure around the Ti ion, which is under the averaged effect
of the Y and Ca ions. In contrast, the RXS intensity at the Y K edge reflects the
local structure around only the Y ions. Therefore, the Ca concentration dependence
of $ITi_{tilt}(x)$ should be estimated not based on $IY(x)$ but based on the structural

Fig. 19 Ca concentration
dependence of the RXS
intensity at **a** Y K edge
($E \sim 17.033\,\mathrm{keV}$), **b**
pre-edge, and **c** main edge of
the Ti K edge. **d** Ti–O–Ti
bond angles in ab plane
(*open circles*) and along c
axis (*filled circles*) as
functions of Ca
concentration. **e** Ratio
between Ti-O$_{long}$ and
Ti-O$_{short}$ in TiO$_6$
octahedron. These structural
parameters were determined
via the structural analysis.
Data were taken from
Refs. [26, 28, 29]

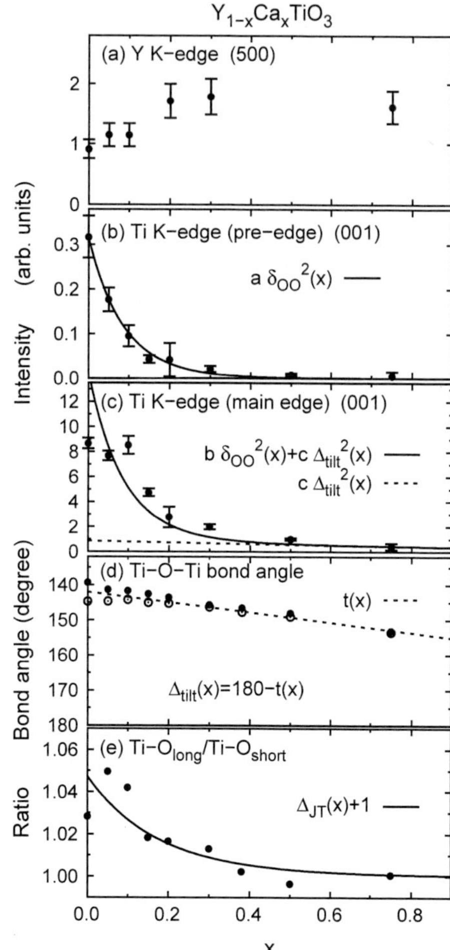

parameter Δ_{tilt}. The Ca concentration dependence values of the average Ti–O–Ti
bond angles in the ab plane and along the c axis were fitted by a linear function $t(x)$,
as shown in Fig. 19d. To estimate $ITi_{tilt}(x)$, we assumed that the RXS intensity
due to the octahedral tilt, $ITi_{tilt}(x)$, is proportional to $\Delta_{tilt}^2(x) = (180 - t(x))^2$;
the RXS intensity reflecting the JTD in LaMnO$_3$ was theoretically calculated to be
proportional to the square of the distortion [68]. The obtained function, $ITi_{tilt}(x)$,
is shown in Fig. 19c by a dotted line, which indicates no marked change over the
entire Ca concentration region. Second, we examined $ITi_{OO}(x)$ reflecting the $3d$
orbital state; $ITi_{OO}(x)$ was expected to be proportional to the square of the order
parameter of orbital ordering ($\delta_{OO}(x)$) [6]. In orbitally ordered systems, because of
the JT effect, the order parameter $\delta_{OO}(x)$ is of the same size as the JTD ($\Delta_{JT}(x)$),
i.e., $ITi_{OO}(x) \propto \delta_{OO}^2(x) \propto \Delta_{JT}^2(x)$. In this case, JT distortion was estimated from

the ratio of the longest Ti–O bond length (Ti-O_{long}) and the averaged short Ti–O bond length (Ti-O_{short}) of the TiO_6 octahedron. The Ca concentration dependence was fitted by the function $\Delta_{JT}(x) + 1$ ($\equiv \frac{(Ti-O_{long})-(Ti-O_{short})}{(Ti-O_{short})} + 1 = \frac{(Ti-O_{long})}{(Ti-O_{short})}$), as shown by a solid line in Fig. 19e. Thus, the Ca concentration dependence of the order parameter $\delta_{OO}(x)$ was obtained from the figure. The obtained $\delta_{OO}^2(x)$ agreed well with RXS dependence at the pre-edge of the Ti K edge, thus reflecting the anisotropy of the $3d$ orbital state directly, as shown by a solid line in Fig. 19b. Consequently, the Ca dependence values of RXS intensities reflecting the octahedral tilt and orbital ordering can be expressed by $\Delta_{tilt}^2(x)$ and $\delta_{OO}^2(x)$, respectively.

Finally, the Ca concentration dependence of the RXS intensity at the main edge of the Ti K edge was fitted by $b\delta_{OO}^2(x) + c\Delta_{tilt}^2(x)$, as shown in Fig. 19c by the solid line. The experimental data can be explained sufficiently by the fitted curve. The fitting parameters b and c reflect the extents of energy-level splitting due to orbital ordering and octahedral tilt, respectively. Both parameters have positive values. Therefore, the Ti $4p$ orbital reflecting the on-site Coulomb interaction between the $3d - 4p$ orbitals of the Ti ion and that originating from the tilts of neighboring TiO_6 octahedra may have the same energy-level scheme. At $YTiO_3$ ($x = 0.0$), $ITi_{tilt}(x) = c\Delta_{tilt}^2(x)$ shown by the dotted line in the figure is much smaller than $ITi_{OO}(x) = b\Delta_{JT}^2(x)$. Thus, RXS reflecting orbital ordering was mainly observed at $x = 0.0$, and the wave function of the ordered orbital was determined successfully from the RXS intensity at the main edge of the Ti K edge in $YTiO_3$. Using the obtained $ITi_{OO}(x)$, we can discuss the wave function of ordered orbitals over the entire Ca concentration region. However, there remain open questions for our assumptions: whether the Ca concentration dependence of $ITi_{tilt}(x)$ is the same as that of $IY(x)$, whether $ITi_{tilt}(x)$ is proportional to $\Delta_{tilt}^2(x)$, and whether the Ca concentration dependence of $ITi_{OO}(x)$ is the same as that of the RXS intensity at the pre-edge of the Ti K edge. To establish the scattering mechanism of RXS, a theoretical study of this mechanism is needed.

Consequently, the origin of RXS at the main edge of the Ti K edge is understood experimentally. It is well explained by considering two origins of RXS, i.e., the orbitally ordered state and structural characteristic. Strong RXS reflecting orbital ordering exists in the case of $YTiO_3$, and the wave function of the ordered $3d$ orbital can be estimated. Therefore, RXS studies of titanate elucidate that the RXS technique is useful for estimating the orbital state, although the origin of RXS needs to be understood.

4 Applications of RXS Measurements

As established above, RXS can be used for determining orbital and charge ordering. As the next step, the RXS technique has been applied to experiments at extreme conditions. Moreover, researchers have attempted to determine new types of electronic ordering states using this technique. Therefore, in this section, we describe experiments on thin-film systems and high-pressure experiments using a diamond anvil cell as challenging tasks. Observation of the spin-state ordering is provided as a new direction for RXS studies.

4.1 Charge and Orbitally Ordered States in Thin-Film Systems

Pulsed laser deposition has rapidly developed as a technique for fabricating artificial lattices. It can now be used to atomically control the artificial lattice of a perovskite system. In the perovskite structure family, various physical properties, high–temperature superconductivity, multiferroics and colossal magnetoresistance (CMR), have been reported. Hence, many groups have attempted to fabricate super-lattices combining different physical properties to create new physical properties and new functions. Strong coupling among charges, spins, and orbitals of $3d$ electrons and lattice degrees of freedom played important roles in those studies. Hence, the study of charge and orbitally ordered states in thin-film systems gained importance, and the RXS technique was applied to various film systems [20–24, 51].

4.1.1 Mn Valence States in $[(LaMnO_3)_m(SrMnO_3)_m]_n$ Thin Film

Perovskite manganites show various interesting phenomena such as CMR because of the close interplay among charges, orbitals, spins, and lattice degrees of freedom. The $[(LaMnO_3)_m(SrMnO_3)_m]_n$ superlattice was fabricated as a stage for artificially controlling the valence of Mn [87]. The films were composed of the same number of $LaMnO_3$ and $SrMnO_3$ layers; thus, the average Mn valence was maintained at 3.5+. The Mn valence values in the $LaMnO_3$ and $SrMnO_3$ layers were 3+ and 4+, respectively. The valence distribution was expected to be determined by the stacking sequence of $LaMnO_3/SrMnO_3$ layers. Recently, a new CMR effect that has never been realized in the $(La, Sr)MnO_3$ alloy was discovered in this superlattice system [88]. To microscopically understand the physical properties of the $[(LaMnO_3)_m(SrMnO_3)_m]_n$ superlattice, an evaluation of the Mn valence state in the superlattice is very important. Therefore, we attempted to evaluate the valence distribution of Mn ions using the RXS technique on the basis of $I(E, \mathbf{Q})$ near E_a of the Mn ions [24].

For estimating the Mn valence distribution, the ASFs of Mn^{3+} and Mn^{4+} are required, as discussed in Sect. 2. However, it is difficult to determine the ASFs of Mn^{3+} and Mn^{4+} of thin films because fluorescence signals from thin films are quite weak. First, we measured $\mu(E)$ of the bulk $LaMnO_3$ and $SrMnO_3$ samples to estimate the chemical shift between Mn^{3+} and Mn^{4+}. The observed $\mu(E)$ was transformed into the imaginary part of the ASF $f''(E)$ using Eq. (4). The obtained $f''(E)$ spectra are shown in Fig. 20a. E_a of $SrMnO_3$ is approximately 3 eV higher than that of $LaMnO_3$; this represents a chemical shift. These spectra have different energy

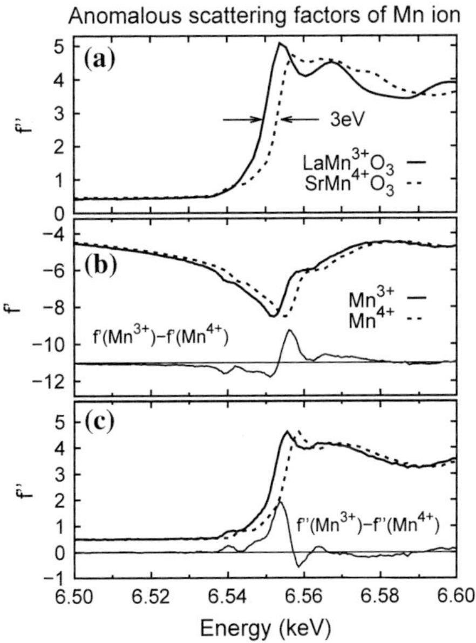

Fig. 20 **a** Anomalous scattering factor $f''(E)$ of LaMnO$_3$ and SrMnO$_3$ bulk. The difference in the absorption-edge energy (E_a) is indicated by *arrows*. **b** $f'(\mathrm{Mn}^{3+})$, $f'(\mathrm{Mn}^{4+})$, and $f'(\mathrm{Mn}^{3+}) - f'(\mathrm{Mn}^{4+})$. **c** $f''(\mathrm{Mn}^{3+})$, $f''(\mathrm{Mn}^{4+})$, and $f''(\mathrm{Mn}^{3+}) - f''(\mathrm{Mn}^{4+})$. Data were taken from Ref. [24]

profiles above E_a, which reflects the difference in the crystal structure. The ASFs $f'(\mathrm{Mn}^{3.5+})$ and $f''(\mathrm{Mn}^{3.5+})$ of La$_{0.5}$Sr$_{0.5}$MnO$_3$ grown on a SrTiO$_3$ (0 0 1) substrate were obtained from the fluorescence spectra in Ref. [21]. On the basis of the chemical shift of 3 eV, f'' of Mn^{3+} (Mn^{4+}) was obtained from the energy shift of -1.5 eV ($+1.5$ eV) using $f''(\mathrm{Mn}^{3.5+})$. $f'(\mathrm{Mn}^{3+})$ and $f'(\mathrm{Mn}^{4+})$ were calculated according to the Kramers–Krönig transformation of $f''(\mathrm{Mn}^{3+})$ and $f''(\mathrm{Mn}^{4+})$, respectively [57–59]. The obtained ASFs of Mn^{3+} and Mn^{4+} are shown in Fig. 20b, c, respectively. The differences between the ASFs of Mn^{3+} and Mn^{4+}, i.e., $\Delta f' = f'(\mathrm{Mn}^{3+}) - f'(\mathrm{Mn}^{4+})$ and $\Delta f'' = f''(\mathrm{Mn}^{3+}) - f''(\mathrm{Mn}^{4+})$, are strongly enhanced near the Mn K-edge energy, as shown in the figures. Hence, it is expected that a strong signal reflecting the Mn valence distribution will be observed near the energy at superlattice peaks, where the structure factor includes the difference terms $\Delta f'$ and $\Delta f''$.

The crystal structures of the superlattices were evaluated from the Q_z dependence of the scattering intensity, $I(Q_z)$. $I(Q_z)$ around the (0 0 1) reflection of $[(\mathrm{LaMnO}_3)_8(\mathrm{SrMnO}_3)_8]_6$ is shown in Fig. 21. The superlattice reflections are denoted as $q = \pm1, \pm2$, and ±3 from fundamental reflections, and the peak positions are consistent with $m \sim 8$. The peaks due to the Laue function reflecting $n = 6$ were observed clearly between the superlattice peaks. The peak position at (0 0 1) reflects the lattice constants of the average perovskite structure. The sharp peak at $Q_z = 1.61$ Å^{-1} is the (0 0 1) reflection of the SrTiO$_3$ substrate. $I(Q_z)$ was calculated on the basis of the crystal structure model reported in Ref. [24]. In this case, the superlattice peak positions cannot be explained by periodicities $m = 8$. Hence, the noninteger periodicity $m = 7.6$ was considered. By introducing a noninteger value of m, a La/Sr mixed layer is generated, and the mixing ratio at the interface depends

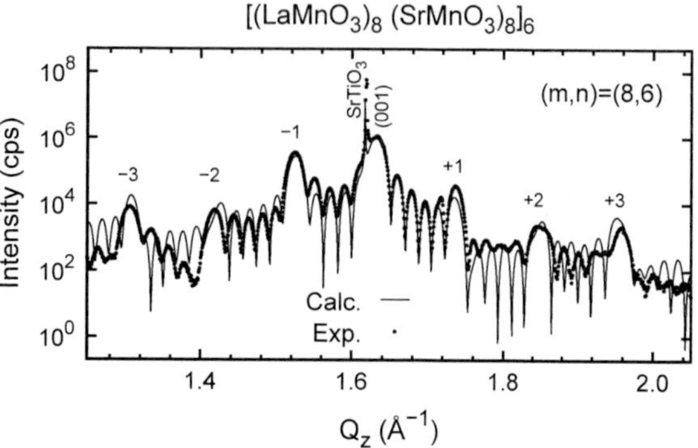

Fig. 21 Diffraction pattern around (0 0 1) of [(LaMnO$_3$)$_8$(SrMnO$_3$)$_8$]$_6$. Data were taken from Ref. [24]. The model calculation is denoted by the *solid line*

on the parameter m. The out-of-plane lattice constants c_{La} and c_{Sr} were considered in the LaMnO$_3$ and SrMnO$_3$ layers, respectively. Moreover, the relaxation of the lattice constant near the interface between the LaMnO$_3$ and SrMnO$_3$ layers was considered, i.e., the out-of-plane lattice constants expand or contract near the interface to relax lattice distortion. Finally, the parameters used to calculate the scattering intensity of the superlattices are the periodicities m and n; lattice constants in the La/Sr layer c_{La} and c_{Sr}; and lattice relaxation parameter α. The intensity ratios and peak positions are well explained by the model calculation, as indicated by the solid line in Fig. 21. Hence, the crystal structure of the superlattices, which provides sufficient accuracy to estimate the valence state of Mn, was determined.

Then, the energy dependence of the scattering intensity, $I(E, \mathbf{Q})$, was measured at the reflection positions to estimate the valence distribution of Mn ions. $I(E, \mathbf{Q})$ was observed at the fundamental and superlattice reflections shown in Fig. 22. All energy profiles showed strong energy dependence near the Mn K edge. This means that the structure factor includes the ASF of Mn ions. To evaluate energy dependence, four models of Mn valence distribution were considered, as discussed in Ref. [24]. (1) *rect* model: Mn valence in LaMnO$_3$ (SrMnO$_3$) is 3+ (4+), and the valence at the interface has an intermediate value. In the structural model, the periodicity m has a noninteger value, and a La/Sr mixed layer exists. Mn valence at the border of the mixed layer is determined by the La/Sr ratio. Mn valence distribution transforms into an almost rectangular wave. (2) α model: Mn valence is determined by the volume of the perovskite unit cell, the stacking sequence of which is determined by the lattice parameter α. The intermediate valence state occurs in six layers near the interface. (3) *sin* model: Mn valence distribution is a sinusoidal wave from 3+ to 4+, defined as a function of the number of stacking layers. (4) *no* model: All Mn valence values are 3.5+, and there is no charge modulation. On the basis of these valence distribution models, the energy dependence values $I(E, \mathbf{Q})$ were calculated and are shown by the lines in Fig. 22. Valence modulations of the *alpha* and *sin* models are similar to each other in the case of [(LaMnO$_3$)$_8$(SrMnO$_3$)$_8$]$_6$. Hence, calculation using the *sin* model is not shown here. The structure factor at the fundamental peak (0 0 1) shown in Fig. 22a reflects the average perovskite structure. Hence, the $\Delta f'$

Fig. 22 Energy dependence values of the scattering intensity at **a** (0 0 1), **b** (0 0 1)−1, **c** (0 0 1)−2, and **d** (0 0 1)−3 in [(LaMnO₃)₈(SrMnO₃)₈]₆. Data were taken from Ref. [24]. Model calculations are shown by the *thick solid line* (*rect* model), *dashed line* (*α* model), and *solid line* (*no* model). *no* model: All Mn valence values are 3.5+, and there is no charge modulation. *rect* model: Mn valence in the LaMnO₃ (SrMnO₃) layer is 3+ (4+), and the valence at the interface has an intermediate value. The Mn valence distribution almost transforms into a rectangular wave. *α* model: Valence distribution that reflects lattice relaxation near the interface is considered

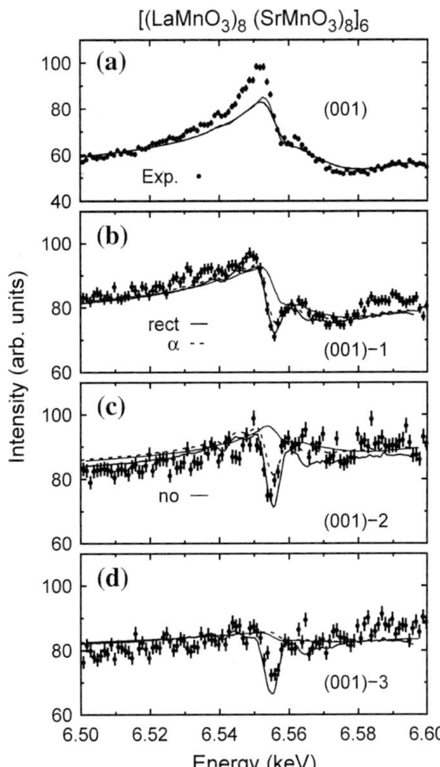

and $\Delta f''$ terms are canceled in the structure factor, and $I(E, \mathbf{Q})$ is independent of the valence distribution model. In the *no* model, the ASFs of only $Mn^{3.5+}$ are used, whereas in the *rect* and *α* models, the ASFs of Mn^{3+} and Mn^{4+} are used. Hence, the energy spectrum at the absorption energy of the *no* model is sharper than that of the *rect* model. Next, the energy dependence values of the superlattice peak intensity $I(E, \mathbf{Q})$ were compared with those obtained from model calculations. The resonant peaks near the Mn K edge reflected the existence of the $\Delta f'$ and $\Delta f''$ terms in the structure factor. For all distribution models except the *no* valence distribution model, $I(E, \mathbf{Q})$ at (0 0 1)−1 showed similar energy dependence. With increasing q, the dependence of the *α* model quickly approached that of the *no* model. In contrast, the *rect* model showed a strong resonant peak near the Mn K-edge energy, even at $q = -3$. Thus, peak height is important for evaluating the valence distribution. Finally, Mn valence distribution in thin-film systems were effectively evaluated using this RXS technique.

4.1.2 Observation of Orbital Superlattice

The observation of orbital states in thin-film systems using the RXS technique has recently become an issue. The structural and physical properties of $La_{1-x}Sr_xMnO_3$ ($x = 0.40$, 0.55) single-layer films and the superlattices of these films have been investigated systematically, and it was proposed that an orbital superlattice (Fig. 23b) can be realized in their films [89]. The in-plane lattices of both $La_{0.45}Sr_{0.55}MnO_3$ and $La_{0.60}Sr_{0.40}MnO_3$ layers grown epitaxially on $SrTiO_3$ (0 0 1) substrates expanded to match that of the substrates as $a = 0.391$ nm. The out-of-plane lattices shortened elastically as $c = 0.379$ and 0.384 nm for the $La_{0.45}Sr_{0.55}MnO_3$ and $La_{0.60}Sr_{0.40}MnO_3$ layers, respectively. The $La_{0.45}Sr_{0.55}MnO_3$ film exhibited antiferromagnetic and metallic ground states, and the $d_{x^2-y^2}$-type orbitals were considered to be occu-

Fig. 23 **a** X-ray diffraction pattern of $[(La_{0.45}Sr_{0.55}MnO_3)_{10} (La_{0.60}Sr_{0.40}MnO_3)_3]_{20}]$ superlattices along the (1 0 L) direction in reciprocal space. **b** Schematic of e_g orbital states in $[(La_{0.45}Sr_{0.55}MnO_3)_{10} (La_{0.60}Sr_{0.40}MnO_3)_3]_{20}]$ orbital superlattices. $(x^2 - y^2)$-type orbitals in the $x = 0.55$ layers and $(x^2 - y^2)/(3z^2 - r^2)$ disordered orbitals in the $x = 0.40$ layers are shown. Reprinted with permission from Ref. [20] (Copyright 2002 The Physical Society of Japan)

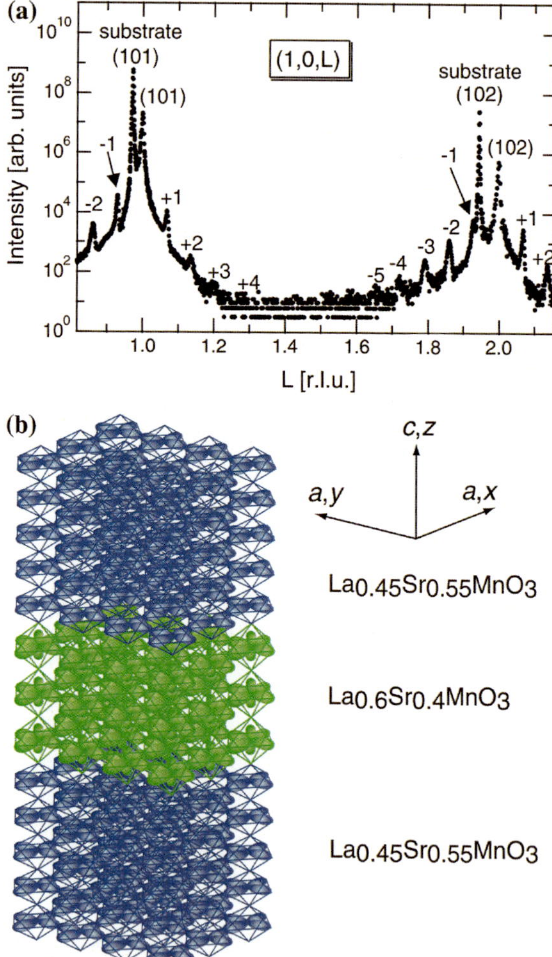

pied through the distortion. By contrast, a disordered orbital state was expected to be realized in the $La_{0.60}Sr_{0.40}MnO_3$ film, which showed metallic conductivity and three-dimensional FM ordering below $T_C = 330-340$ K. These results suggest that these layers preserve their structural and physical properties even in superlattices as in thin films. Therefore, the superlattice is expected to form orbital superlattices, as shown in Fig. 23b, with layered units of different orbital states stacked alternately. To elucidate the orbital superstructure, the interference technique was proposed to observe RXS reflections from ferro-type orbital ordering in the superlattices [20].

The diffraction pattern of $[(La_{0.45}Sr_{0.55}MnO_3)_{10}(La_{0.60}Sr_{0.40}MnO_3)_3]_{20}$ along the $(1\ 0\ L)$ direction is shown in Fig. 23a. The satellite peaks between the fundamental peaks $(1\ 0\ 1)$ and $(1\ 0\ 2)$ indicate successful construction of the superlattices. The satellite peaks are denoted as $\ldots, -1, 0, +1, \ldots$ from fundamental reflections in the stacking direction. To detect the RXS signal reflecting the orbital state, we used X-ray energy near the Mn K-absorption-edge energy E_a. Here, we noted the polarization and azimuthal angle dependence values of the scattering intensity, which indicate that the ASF becomes a tensor near E_a. However, the RXS signal is considerably weaker than that of the Thomson scattering, reflecting the superstructure; in other words, the intensity shown in Fig. 23a mainly reflects the Thomson scattering. The interference technique was used here. The scattering formula in the case of using a polarization analyzer is given as

$$\begin{aligned}
I &\propto |F_{\sigma\sigma'}\cos\phi_A - F_{\sigma\pi'}\sin\phi_A|^2 \\
&= |F_{\sigma\sigma'}|^2\cos^2\phi_A - 2Re\{F_{\sigma\sigma'}F_{\sigma\pi'}^*\}\cos\phi_A\sin\phi_A \\
&\quad + |F_{\sigma\pi'}|^2\sin^2\phi_A.
\end{aligned} \tag{11}$$

$F_{\sigma\sigma'}$ mainly reflects the Thomson scattering. $F_{\sigma\pi'}$ reflects the tensor of the ASF, namely, the orbital state, and it contains a $\sin\Psi$ term; Ψ is the azimuthal angle, which is defined as $\Psi = 0°$ at $\sigma \perp c$. ϕ_A is the analyzer rotation angle, as shown in Fig. 4. In a weak-$F_{\sigma\pi'}$ case, it is difficult to detect the third term. Hence, the RXS signal can be measured effectively using the second interference term $F_{\sigma\sigma'}F_{\sigma\pi'}^*$. Then, the interference term can be derived by subtracting the intensity at $\phi_A = 97.5°$ from that at $\phi_A = 82.5°$. The energy dependence of the interference term at $(1\ 0\ 2)+1$ indicates a resonating feature near E_a, as shown in Fig. 24a. Reflecting the existence of the $\sin\Psi$ term, the spectrum at $\Psi = \pm90°$ is inverted and that at $\Psi = 0°$ loses its signal. The azimuthal angle dependence of the interference term is shown in Fig. 24b. The dependence is explained clearly by the $\sin\Psi$ term, denoted by the solid line. These energy, polarization, and azimuthal angle dependence values are consistent with the orbital superlattice structure, as shown in Fig. 23b. The magnitude of the $4p$ energy split ($\Delta E = E_{4p_{xy}} - E_{4p_z}$) is also estimated using the energy spectrum of the interference term. Finally, the values of ΔE in $La_{0.45}Sr_{0.55}MnO_3$ and $La_{0.60}Sr_{0.40}MnO_3$ layers were calculated to be -1.3 and -0.9 eV, respectively. This is clear evidence of the existence of an orbital superlattice in which layers with different orbital states are stacked alternately on the atomic scale. This study became

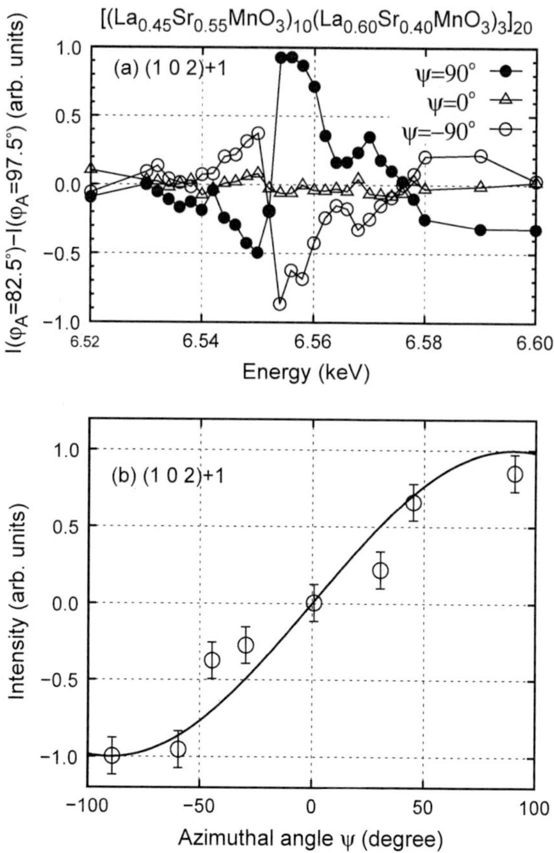

Fig. 24 **a** Energy dependence of interference term at (1 0 2)+1 reflection with azimuthal angles $\Psi = 90°$, 0° and 90°. **b** Azimuthal angle dependence of intensity of interference term at (1 0 2)+1 reflection. Data were taken from Ref. [20]

not only the first experiment to determine orbital ordering using the interference term but also direct observation of a $4p$ energy level. It led to a study of RXS scattering mechanism based on the proposed technique [21, 22].

4.2 RXS Study Under High Pressure

Under high pressure, various intriguing physical properties were discovered in SCES. In this case, the strong correlation among the orbital, charge, and spin degrees of freedom plays an important role. Hence, studying orbital and charge states is very important. In the RXS experiment, however, X-ray energy cannot be selected freely because it is determined by the absorption energy of the target ion. The K-absorption energy of a $3d$ transition metal is low for a high-pressure experiment because X-rays are absorbed strongly by high-pressure cells. Hence, RXS measurement under high pressure is generally difficult [34, 90]. Here, the Devil's Staircase-type phase transition in NaV_2O_5 under high pressure is presented.

The exciting phase diagram-Devil's flower-was reproduced perfectly in a temperature-pressure phase diagram of NaV_2O_5 [91]. Application of high pressure leads to changes in lattice modulation along the *c*-axis of NaV_2O_5. A precise X-ray diffraction experiment clarified that a large number of transitions occur successively among higher-order commensurate phases. Moreover, NaV_2O_5 is well described as a quarter-filled two-leg ladder system, and all vanadium ions have an average valence state of $V^{4.5+}$. In other words, the system has a charge degree of freedom. Then, NaV_2O_5 undergoes a novel cooperative phase transition associated with its charge disproportionation ($2V^{4.5+} \rightarrow V^{4+} + V^{5+}$) at $T_C = 34$ K [31]. Therefore, the relationship between vanadium charge ordering and the Devil's flower-type phase diagram is important.

To elucidate charge ordering state under high pressure, an RXS experiment was performed [34]. Since 5.47 keV (energy of the K-absorption edge of vanadium) X-rays cannot penetrate diamond anvils, a He-gas-driven diamond anvil cell (DAC) using a Be gasket was developed, as shown in Fig. 25. X-rays with an energy of 5.47 keV enter and exit the sample chamber through the Be gasket. As a result, RXS spectra were measured successfully using the DAC, as shown in Fig. 26. A sharp peak structure was observed at approximately 5.468 keV, which corresponds to the $1s \rightarrow 3d$ transition energy (pre-edge). The peak at approximately 5.475 keV corresponds to the main edge energy. At the pre-edge, the peak intensity decreases considerably, while that at the main edge remains almost constant. Here, the intensity at the main edge is dominated by the order parameter δ of charge ordering ($V^{4.5+\delta} + V^{4.5-\delta} \leftarrow 2V^{4.5+}$). This indicates that full charge disproportionation is achieved even in the

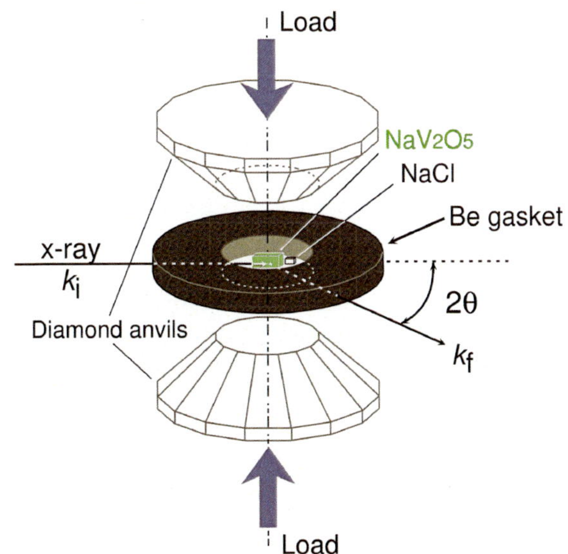

Fig. 25 Layout of RXS-DAC. Four windows were opened on the DAC jacket, each with a 70° equatorial, to allow for the collection of high-angle diffraction of NaV_2O_5 through the Be gasket. Our target reflection (7.5, 0.5, L) was located at $2\theta \sim 100°$ when the incident X-ray energy was 5.47 keV

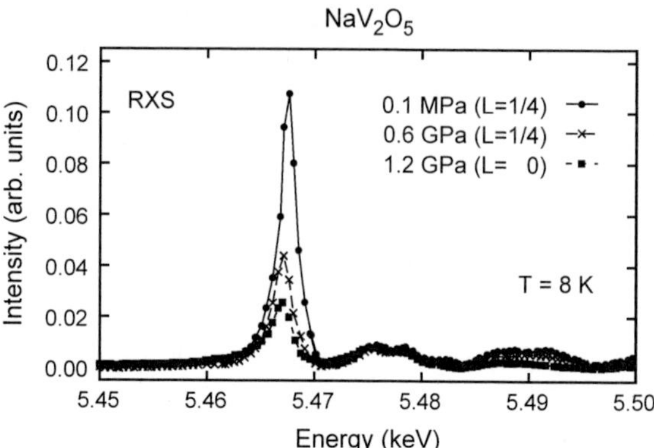

Fig. 26 Energy spectra at $Q = (15/2, 1/2, L)$ observed at 0.1 MPa ($L = 1/4$, $C_{1/4}$ phase), 0.6 GPa ($L = 1/4$, $C_{1/4}$ phase), and 1.2 GPa ($L = 0$, C_0 phase) at 8 K. Data were taken from Ref. [34]

phase under high pressure. In contrast, the intensity at the pre-edge sensitively reflects the breaking size of the local centrosymmetry around the V ion. This indicates that the breaking size of local centrosymmetry decreases with increasing pressure, which is consistent with the obtained crystal structure. This study clarified the possibility of RXS studies under high pressure by using the developed DAC.

4.3 Spin State Study by RXS

The ordered states of charge, spin, and orbital degrees of freedom play important roles in the appearance of many interesting physical properties in SCES. Moreover, spin-state degrees of freedom such as low-spin (LS), high-spin (HS), and intermediate-spin (IS) state emerge additionally depending on the type of ion. In cobalt oxides, a wide variety of physical properties associated with the spin state are expected. In LaCoO$_3$, the ground state of the Co^{3+} ion is a nonmagnetic LS state (t_{2g}^6, $S = 0$). The compound exhibits a gradual transition to a PM phase with increasing temperature [92]. The spin state in the PM phase has been debated, that is, it is controversial whether the moment arises from the IS ($t_{2g}^5 e_g^2$, $S = 1$) [93, 94] or HS ($t_{2g}^4 e_g^2$, $S = 2$) state of the Co^{3+} ion [95–97]. The possibility of the e_g orbital ordering of the IS state was suggested in the PM phase [94]. e_g orbital ordering provides crucial evidence of the existence of an IS state because only the IS state has e_g orbital degrees of freedom. However, there is still no clear evidence for e_g orbital ordering, and the existence of the IS state is a highly controversial issue.

Sr$_3$YCo$_4$O$_{10.5}$ is an unusual room-temperature ferrimagnet with $T_C \sim 370$ K, which is the highest T_C among perovskite Co oxides [98–100]. The Curie constant indicates the existence of the IS and/or HS state of the Co^{3+} ion. The crystal structure has been investigated by powder X-ray diffraction, and orbital ordering of the IS state of Co^{3+} has been proposed as the origin of its magnetism [101]. Therefore, this material is a good target for studying the spin states of the Co ion, and an RXS experiment has been performed [48].

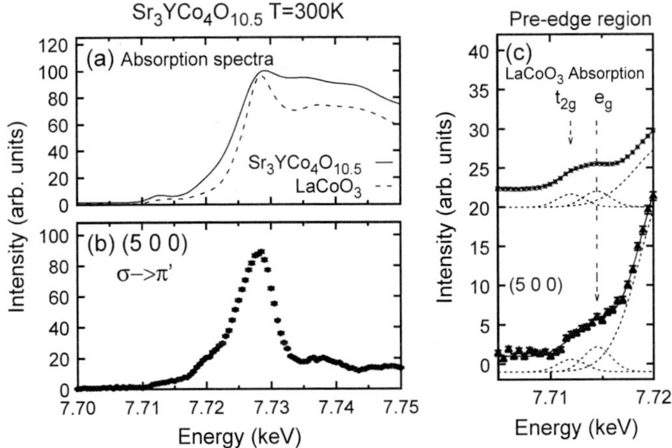

Fig. 27 **a** Absorption spectra of Sr$_3$YCo$_4$O$_{10.5}$ (*solid line*) and LaCoO$_3$ (*dashed line*). **b** Energy dependence of scattering intensity at (5 0 0) for $\varepsilon_\sigma \parallel b$. **c** *Bottom* energy dependence of (5 0 0) reflection magnified near pre-edge; *top* absorption spectrum of LaCoO$_3$. Data were taken from Ref. [48]. The *solid lines* show the results fitted with three peaks (*dashed line*). The base lines are shifted for clarity. The *arrows* indicate the $1s \rightarrow t_{2g}$ and e_g transition energies

The absorption spectra of Sr$_3$YCo$_4$O$_{10.5}$ are shown in Fig. 27a. The peaks at about 7.713 and 7.728 keV correspond to the pre-edge and main edge energies, respectively. We searched for the RXS signal near the Co K-edge energy to elucidate Co 3d orbital ordering. Resonating signals at a series of reflections, (h 0 0): $h = 4n \pm 1$, were found. The energy spectrum at (5 0 0) is shown in Fig. 27b. The RXS signal has only the $\sigma \rightarrow \pi'$ scattering component. The azimuthal angle dependence exhibits clear twofold symmetry. The intensity peaks at $\sigma \parallel b$ and becomes zero at $\sigma \parallel c$. A significant h dependence of RXS intensity at (h 0 0) was found as well. On the basis of these results, antiferro-type orbital ordering was proposed, as shown in Fig. 28. Next, the RXS signal at the pre-edge was noted to clarify e_g orbital ordering of the IS state. The absorption spectrum of LaCoO$_3$ has a broad peak (Fig. 27c top) composed of two peaks. The spectrum in the pre-edge region was fitted with two peaks and the tail of the main edge. The high (low)-energy peak corresponds to the $1s \rightarrow e_g$ (t_{2g}) transition. Hence, the t_{2g} and e_g orbital states can be observed separately by tuning the X-ray energy. The energy dependence of the (5 0 0) intensity has a broad peak near the pre-edge energy, and the fitted result clearly indicates the presence of the RXS signal at the $1s \rightarrow e_g$ transition energy. This is direct evidence of not only e_g orbital ordering but also the presence of the IS state because the signal at the $1s \rightarrow e_g$ transition energy reflects the anisotropy of the e_g orbital, and only the IS state has the e_g orbital degrees of freedom. Finally, this RXS study provided us with direct evidence of the existence of the IS state of Co^{3+} and proposed peculiar HS/IS spin-state ordering.

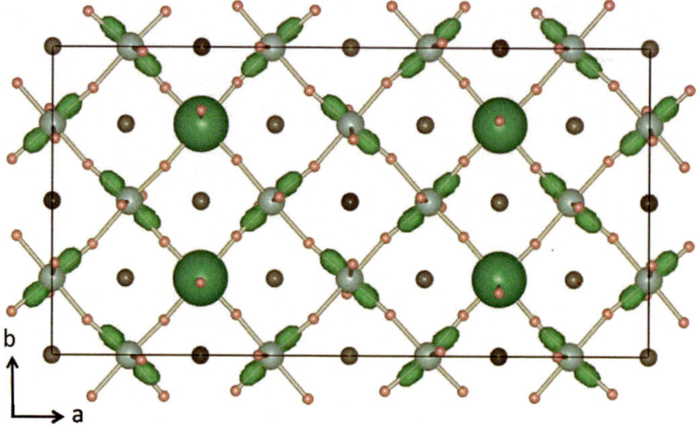

Fig. 28 Orbital and spin ordering structures in CoO_6 layer. The unit cell is indicated by the *solid line*

5   Summary

We reviewed the foundation of the RXS technique and its various applications in $3d$ electron systems. The theoretical framework of the RXS technique was summarized in Sect. 2. In Sect. 3, RXS studies of the perovskite titanate system were presented. It was clarified that the RXS technique can be used to determine orbitally ordered states on the basis of a discussion of the scattering mechanism. In Sect. 4, we presented applications of the RXS technique, namely, charge and orbital ordering in film systems, RXS studies under high pressure, and study of unique spin-state ordering. These indicate the applicability of the technique to resolving typical issues pertaining to the ordering of electronic degrees of freedom. The RXS technique is a powerful technique for investigating charge and orbital ordering in $3d$ transition metal oxides, as described in this chapter. Recently, RXS studies at the $L_{2,3}$ edge (the $2p \rightarrow 3d$ transition energy) have been performed widely. Because the RXS signal at the $L_{2,3}$ edge can directly probe the $3d$ electronic state, a strong magnetic signal of RXS was reported. Furthermore, resonant soft-X-ray scattering was performed globally, as presented in Chap. 5.

References

1. V.E. Dmitrienko, Acta Crystallogr. Sect. A **39**, 29 (1983)
2. D.H. Templeton, L.K. Templeton, Acta Crystallogr. A **41**, 133 (1985)
3. D.H. Templeton, L.K. Templeton, Acta Crystallogr. A **41**, 365 (1985)
4. D.H. Templeton, L.K. Templeton, Acta Crystallogr. A **42**, 478 (1986)
5. Y. Murakami, H. Kawada, H. Kawata, M. Tanaka, T. Arima, Y. Moritomo, Y. Tokura, Phys. Rev. Lett. **80**, 1932 (1998)

6. Y. Murakami, J.P. Hill, D. Gibbs, M. Blume, I. Koyama, M. Tanaka, H. Kawata, T. Arima, Y. Tokura, K. Hirota, Y. Endoh, Phys. Rev. Lett. **81**, 582 (1998)

7. M. Imada, A. Fujimori, Y. Tokura, Rev. Mod. Phys. **70**, 1039 (1998)

8. Y. Tokura, N. Nagaosa, Science **288**, 462 (2000)

9. T.A.W. Beale, G. Beutier, S.R. Bland, A. Bombardi, L. Bouchenoire, O. Bunău, S. Di Matteo, J. Fernández-Rodríguez, J.E. Hamann-Borrero, J. Herrero-Martín, V.L.R. Jacques, R.D. Johnson, A. Juhin, T. Matsumura, C. Mazzoli, A.M. Mulders, H. Nakao, J. Okamoto, S. Partzsch, A.J. Princep, V. Scagnoli, J. Strempfer, C. Vecchini, Y. Wakabayashi, H.C. Walker, D. Wermeille, Y. Yamasaki, Eur. Phys. J. Spec. Top. **208**, 89 (2012)

10. T. Matsumura, H. Nakao, Y. Murakami, J. Phys. Soc. Jpn. **82**, 021007 (2013)

11. Y. Endoh, K. Hirota, S. Ishihara, S. Okamoto, Y. Murakami, A. Nishizawa, T. Fukuda, H. Kimura, H. Nojiri, K. Kaneko, S. Maekawa, Phys. Rev. Lett. **82**, 4328 (1999)

12. K. Nakamura, T. Arima, A. Nakazawa, Y. Wakabayashi, Y. Murakami, Phys. Rev. B **60**, 2425 (1999)

13. M.V. Zimmermann, J.P. Hill, D. Gibbs, M. Blume, D. Casa, B. Keimer, Y. Murakami, Y. Tomioka, Y. Tokura, Phys. Rev. Lett. **83**, 4872 (1999)

14. Y. Wakabayashi, Y. Murakami, Y. Moritomo, I. Koyama, H. Nakao, T. Kiyama, T. Kimura, Y. Tokura, N. Wakabayashi, J. Phys. Soc. Jpn. **70**, 1194 (2001)

15. M.V. Zimmermann, C.S. Nelson, J.P. Hill, D. Gibbs, M. Blume, D. Casa, B. Keimer, Y. Murakami, C.-C. Kao, C. Venkataraman, T. Gog, Y. Tomioka, Y. Tokura, Phys. Rev. B **64**, 195133 (2001)

16. M.V. Zimmermann, C.S. Nelson, Y.-J. Kim, J.P. Hill, D. Gibbs, H. Nakao, Y. Wakabayashi, Y. Murakami, Y. Tokura, Y. Tomioka, T. Arima, C.-C. Kao, D. Casa, C. Venkataraman, T. Gog, Phys. Rev. B **64**, 064411 (2001)

17. J. Voigt, J. Persson, J.W. Kim, G. Bihlmayer, T. Brückel, Phys. Rev. B **76**, 104431 (2007)

18. D. Mannix, D.F. McMorrow, R.A. Ewings, A.T. Boothroyd, D. Prabhakaran, Y. Joly, B. Janousova, C. Mazzoli, L. Paolasini, S.B. Wilkins, Phys. Rev. B **76**, 184420 (2007)

19. R.A. Ewings, A.T. Boothroyd, D.F. McMorrow, D. Mannix, H.C. Walker, B.M.R. Wanklyn, Phys. Rev. B **77**, 104415 (2008)

20. T. Kiyama, Y. Wakabayashi, H. Nakao, H. Ohsumi, Y. Murakami, M. Izumi, M. Kawasaki, Y. Tokura, J. Phys. Soc. Jpn. **72**, 785 (2003)

21. H. Ohsumi, Y. Murakami, T. Kiyama, H. Nakao, M. Kubota, Y. Wakabayashi, Y. Konishi, M. Izumi, M. Kawasaki, Y. Tokura, J. Phys. Soc. Jpn. **72**, 1006 (2003)

22. Y. Wakabayashi, D. Bizen, H. Nakao, Y. Murakami, M. Nakamura, Y. Ogimoto, K. Miyano, H. Sawa, Phys. Rev. Lett. **96**, 017202 (2006)

23. Y. Wakabayashi, D. Bizen, Y. Kubo, H. Nakao, Y. Murakami, M. Nakamura, Y. Ogimoto, K. Miyano, H. Sawa, J. Phys. Soc. Jpn. **77**, 014712 (2008)

24. H. Nakao, J. Nishimura, Y. Murakami, A. Ohtomo, T. Fukumura, M. Kawasaki, T. Koida, Y. Wakabayashi, H. Sawa, J. Phys. Soc. Jpn. **78**, 024602 (2009)

25. H. Nakao, Y. Wakabayashi, T. Kiyama, Y. Murakami, M.V. Zimmermann, J.P. Hill, D. Gibbs, S. Ishihara, Y. Taguchi, Y. Tokura, Phys. Rev. B **66**, 184419 (2002)

26. H. Nakao, M. Tsubota, F. Iga, K. Uchihira, T. Nakano, T. Takabatake, K. Kato, Y. Murakami, J. Phys. Soc. Jpn. **73**, 2620 (2004)

27. M. Kubota, H. Nakao, Y. Murakami, Y. Taguchi, M. Iwama, Y. Tokura, Phys. Rev. B **70**, 245125 (2004)

28. M. Tsubota, F. Iga, K. Uchihira, T. Nakano, S. Kura, T. Takabatake, S. Kodama, H. Nakao, Y. Murakami, J. Phys. Soc. Jpn. **74**, 3259 (2005)

29. H. Nakao, S. Kodama, K. Kiyoto, Y. Murakami, M. Tsubota, F. Iga, K. Uchihira, T. Takabatake, H. Ohsumi, M. Mizumaki, N. Ikeda, J. Phys. Soc. Jpn. **75**, 094706 (2006)

30. L. Paolasini, C. Vettier, F. de Bergevin, F. Yakhou, D. Mannix, A. Stunault, W. Neubeck, M. Altarelli, M. Fabrizio, P.A. Metcalf, J.M. Honig, Phys. Rev. Lett. **82**, 4719 (1999)

31. H. Nakao, K. Ohwada, N. Takesue, Y. Fujii, M. Isobe, Y. Ueda, M.V. Zimmermann, J.P. Hill, D. Gibbs, J.C. Woicik, I. Koyama, Y. Murakami, Phys. Rev. Lett. **85**, 4349 (2000)

32. M. Noguchi, A. Nakazawa, S. Oka, T. Arima, Y. Wakabayashi, H. Nakao, Y. Murakami, Phys. Rev. B **62**, R9271 (2000)
33. K. Ohwada, Y. Fujii, Y. Katsuki, J. Muraoka, H. Nakao, Y. Murakami, H. Sawa, E. Ninomiya, M. Isobe, Y. Ueda, Phys. Rev. Lett. **94**, 106401 (2005)
34. K. Ohwada, Y. Fujii, J. Muraoka, H. Nakao, Y. Murakami, Y. Noda, H. Ohsumi, N. Ikeda, T. Shobu, M. Isobe, Y. Ueda, Phys. Rev. B **76**, 094113 (2007)
35. D. Bizen, N. Shirane, T. Murata, H. Nakao, Y. Murakami, J. Fujioka, T. Yasue, S. Miyasaka, Y. Tokura, J. Phys. Conf. Ser. **150**, 042010 (2009)
36. H. Nakao, D. Bizen, N. Shirane, K. Ikeuchi, Y. Murakami, J. Fujioka, T. Yasue, S. Miyasaka, Y. Tokura, Diam. Light Source Proc. **1**, e115 (2010)
37. R. Fukuta, S. Miyasaka, K. Hemmi, S. Tajima, D. Kawana, K. Ikeuchi, Y. Yamasaki, A. Nakao, H. Nakao, Y. Murakami, K. Iwasa, Phys. Rev. B **84**, 140409 (2011)
38. K. Takubo, T. Kanzaki, Y. Yamasaki, H. Nakao, Y. Murakami, T. Oguchi, T. Katsufuji, Phys. Rev. B **86**, 085141 (2012)
39. T. Akao, Y. Azuma, M. Usuda, Y. Nishihata, J. Mizuki, N. Hamada, N. Hayashi, T. Terashima, M. Takano, Phys. Rev. Lett. **91**, 156405 (2003)
40. N. Ikeda, H. Ohsumi, K. Ohwada, K. Ishii, T. Inami, K. Kakurai, Y. Murakami, K. Yoshii, S. Mori, Y. Horibe, H. Kitô, Nature **436**, 1136 (2005)
41. T. Arima, J.-H. Jung, M. Matsubara, M. Kubota, J.-P. He, Y. Kaneko, Y. Tokura, J. Phys. Soc. Jpn. **74**, 1419 (2005)
42. E. Nazarenko, J.E. Lorenzo, Y. Joly, J.L. Hodeau, D. Mannix, C. Marin, Phys. Rev. Lett. **97**, 056403 (2006)
43. J.E. Lorenzo, C. Mazzoli, N. Jaouen, C. Detlefs, D. Mannix, S. Grenier, Y. Joly, C. Marin, Phys. Rev. Lett. **101**, 226401 (2008)
44. A.M. Mulders, S.M. Lawrence, U. Staub, M. Garcia-Fernandez, V. Scagnoli, C. Mazzoli, E. Pomjakushina, K. Conder, Y. Wang, Phys. Rev. Lett. **103**, 077602 (2009)
45. U. Staub, C. Piamonteze, M. Garganourakis, S.P. Collins, S.M. Koohpayeh, D. Fort, S.W. Lovesey, Phys. Rev. B **85**, 144421 (2012)
46. A. Rodríguez-Fernández, J.A. Blanco, S.W. Lavesey, V. Scgnoli, U. Staub, H.C. Walker, D.K. Shukla, J. Strempfer, Phys. Rev. B **88**, 094437 (2013)
47. K. Horigane, H. Nakao, Y. Kousaka, T. Murata, Y. Noda, Y. Murakami, J. Akimitsu, J. Phys. Soc. Jpn. **77**, 044601 (2008)
48. H. Nakao, T. Murata, D. Bizen, Y. Murakami, K. Ohoyama, K. Yamada, S. Ishiwata, W. Kobayashi, I. Terasaki, J. Phys. Soc. Jpn. **80**, 023711 (2011)
49. J. Fujioka, Y. Yamasaki, H. Nakao, R. Kumai, Y. Murakami, M. Nakamura, M. Kawasaki, Y. Tokura, Phys. Rev. Lett. **111**, 027206 (2013)
50. J. Fujioka, Y. Yamasaki, A. Doi, H. Nakao, R. Kumai, Y. Murakami, M. Nakamura, M. Kawasaki, T. Arima, Y. Tokura, Phys. Rev. B **92**, 195115 (2015)
51. U. Staub, G.I. Meijer, F. Fauth, R. Allenspach, J.G. Bednorz, J. Karpinski, S.M. Kazakov, L. Paolasini, F. d'Acapito, Phys. Rev. Lett. **88**, 126402 (2002)
52. V. Scagnoli, U. Staub, M. Janousch, A.M. Mulders, M. Shi, G.I. Meijer, S. Rosenkranz, S.B. Wilkins, L. Paolasini, J. Karpinski, S.M. Kazakov, S.W. Lovesey, Phys. Rev. B **72**, 155111 (2005)
53. Y. Lu, A. Frano, M. Bluschke, M. Hepting, S. Macke, J. Strempfer, P. Wochner, G. Cristiani, G. Logvenov, H.-U. Habermeier, M.W. Haverkort, B. Keimer, E. Benckiser, Phys. Rev. B **93**, 165121 (2016)
54. L. Paolasini, R. Caciuffo, A. Sollier, P. Ghigna, M. Altarelli, Phys. Rev. Lett. **88**, 106403 (2002)
55. R. Caciuffo, L. Paolasini, A. Sollier, P. Ghigna, E. Pavarini, J. van den Brink, M. Altarelli, Phys. Rev. B **65**, 174425 (2002)
56. N. Tatami, Y. Ando, S. Niioka, H. Kira, M. Onodera, H. Nakao, K. Iwasa, Y. Murakami, T. Kakiuchi, Y. Wakabayashi, H. Sawa, S. Itoh, J. Magn. Magn. Mater. **310**, 787 (2007)
57. D.T. Cromer, D. Liberman, J. Chem. Phys. **53**, 1891 (1970)
58. D.T. Cromer, D. Liberman, Acta Crystallogr. Sect. A **37**, 267 (1981)

59. S. Sasaki, KEK Peport No. 88-14 (1989)
60. J.O. Cross, M. Newville, J.J. Rehr, L.B. Sorensen, C.E. Bouldin, G. Watson, T. Gouder, G.H. Lander, M.I. Bell, Phys. Rev. B **58**, 11215 (1998), http://cars.uchicago.edu/~newville/dafs/diffkk/
61. O. Šipr, A. Šimůnek, S. Bocharov, T. Kirchner, G. Dräger, Phys. Rev. B **60**, 14115 (1999)
62. J.P. Hill, D.F. Mcmorrw, Acta Crystallogr. Sect. A **52**, 236 (1996)
63. S. Ishihara, S. Maekawa, Phys. Rev. Lett. **80**, 3799 (1998)
64. S. Ishihara, S. Maekawa, Phys. Rev. B **58**, 13442 (1998)
65. S. Ishihara, S. Maekawa, Phys. Rev. B **62**, 5690 (2000)
66. I.S. Elfimov, V.I. Anisimov, G.A. Sawatzky, Phys. Rev. Lett. **82**, 4264 (1999)
67. M. Benfatto, Y. Joly, C.R. Natoli, Phys. Rev. Lett. **83**, 636 (1999)
68. M. Takahashi, J. Igarashi, P. Fulide, J. Phys. Soc. Jpn. **68**, 2530 (1999)
69. Y. Murakami, H. Nakao, T. Matsumura, H. Ohsumi, J. Magn. Magn. Mater. **310**, 895 (2007)
70. T. Mizokawa, A. Fujimori, Phys. Rev. B **54**, 5368 (1996)
71. T. Mizokawa, D.I. Khomskii, G.A. Sawatzky, Phys. Rev. B **60**, 7309 (1999)
72. H. Sawada, N. Hamada, K. Terakura, Phys. B **237–238**, 46 (1997)
73. H. Ichikawa, J. Akimitsu, M. Nishi, K. Kakurai, Phys. B **281–282**, 428 (2000)
74. J. Akimitsu, H. Ichikawa, N. Eguchi, T. Miyano, M. Nishi, K. Kakurai, J. Phys. Soc. Jpn. **70**, 3475 (2001)
75. M. Ito, M. Tsuchiya, H. Tanaka, K. Motoya, J. Phys. Soc. Jpn. **68**, 2783 (1999)
76. D.A. Maclean, H.-N. Ng, J.E. Greedan, J. Solid State Chem. **30**, 35 (1979)
77. S. Ishihara, T. Hatakeyama, S. Maekawa, Phys. Rev. B **65**, 064442 (2002)
78. M. Takahashi, J. Igarashi, Phys. Rev. B **64**, 075110 (2001)
79. M. Fabrizio, M. Altarelli, M. Benfatto, Phys. Rev. Lett. **80**, 3400 (1998)
80. Y. Taguchi, Y. Tokura, T. Arima, F. Inaba, Phys. Rev. B **48**, 511 (1993)
81. K. Kumagai, T. Suzuki, Y. Taguchi, Y. Okada, Y. Fujishima, Y. Tokura, Phys. Rev. B **48**, 7636 (1993)
82. Y. Tokura, Y. Taguchi, Y. Moritomo, K. Kumagai, T. Suzuki, Y. Iye, Phys. Rev. B **48**, 14063 (1993)
83. F. Iga, T. Nishiguchi, Y. Nishihara, Phys. B **206–207**, 859 (1995)
84. F. Iga, T. Naka, T. Matsumoto, N. Shirakawa, K. Murata, Y. Nishihara, Phys. B **223–224**, 526 (1996)
85. P. Benedetti, J. Brink, E. Pavarini, A. Vigliante, P. Wochner, Phys. Rev. B **63**, 060408(R) (2001)
86. J. Rodríguez-Carvajal, M. Hennion, F. Moussa, A.H. Moudden, Phys. Rev. B **57**, 3189(R) (1998)
87. T. Koida, M. Lippmaa, T. Fukumura, K. Itaka, Y. Matsumoto, M. Kawasaki, H. Koinuma, Phys. Rev. B **66**, 144418 (2002)
88. H. Yamada, P.-H. Xiang, A. Sawa, Phys. Rev. B **81**, 014410 (2010)
89. M. Izumi, T. Manako, Y. Konishi, M. Kawasaki, Y. Tokura, Phys. Rev. B **61**, 12187 (2000)
90. K. Ohwada, K. Ishii, T. Inami, Y. Murakami, T. Shobu, H. Ohsumi, N. Ikeda, Y. Ohishi, Phys. Rev. B **72**, 014123 (2005)
91. K. Ohwada, Y. Fujii, N. Takesuke, M. Isobe, Y. Ueda, H. Nakao, Y. Wakabayashi, Y. Murakami, K. Ito, Y. Amemiya, H. Fujihisa, K. Aoki, T. Shoubu, Y. Noda, N. Ikeda, Phys. Rev. Lett. **87**, 086402 (2001)
92. K. Asai, O. Yokokura, N. Nishimori, H. Chou, J.M. Tranquada, G. Shirane, S. Higuchi, Y. Okajima, K. Kohn, Phys. Rev. B **50**, 3025 (1994)
93. T. Saitoh, T. Mizokawa, A. Fujimori, M. Abbate, Y. Takeda, M. Takano, Phys. Rev. B **55**, 4257 (1997)
94. G. Maris, Y. Ren, V. Volotchaev, C. Zobel, T. Lorenz, T.T.M. Palstra, Phys. Rev. B **67**, 224423 (2003)
95. M. Itoh, M. Sugihara, I. Natori, K. Motoyama, J. Phys. Soc. Jpn. **64**, 3967 (1995)
96. S. Noguchi, S. Kawamata, K. Okuda, H. Nogiri, M. Motokawa, Phys. Rev. B **66**, 094404 (2002)

97. M.W. Haverkort, Z. Hu, J.C. Cezar, T. Burnus, H. Hartmann, M. Reuther, C. Zobel, T. Lorenz, A. Tanaka, N.B. Brookes, H.H. Hsieh, H.-J. Lin, C.T. Chen, L.H. Tjeng, Phys. Rev. Lett. **97**, 176405 (2006)
98. W. Kobayashi, S. Ishiwata, I. Terasaki, M. Takano, I. Grigoraviciute, H. Yamaguchi, M. Karppinen, Phys. Rev. B **72**, 104408 (2005)
99. W. Kobayashi, S. Yoshida, I. Terasaki, J. Phys. Soc. Jpn. **75**, 103702 (2006)
100. W. Kobayashi, I. Terasaki, AIP Conf. Proc. **850**, 1223 (2006)
101. S. Ishiwata, W. Kobayashi, I. Terasaki, K. Kato, M. Takata, Phys. Rev. B **75**, 220406 (2007)

Observation of Multipole Orderings in f-Electron Systems by Resonant X-ray Diffraction

T. Matsumura

1 Multipole Moment

1.1 Basic Concept

Electrons in a solid have their own spatial characters, i.e., how a wave function $\psi(\mathbf{r})$, an orbital, is extended in space. The anisotropy of the orbital, i.e., to which direction the probability density is extended, play important roles in the physical properties of a material, especially when the electron is relatively localized in a specific atom. The main target of this chapter is on f-electron compounds, which exhibit various kinds of curious ordering phenomena originating from anisotropic charge and magnetization densities. These anisotropies of localized orbitals can be studied systematically by introducing the concept of multipolar degrees of freedom. In f-electron compounds, because of its strong spin-orbit interaction, spin and orbital degrees of freedom are coupled to form a J-multiplet ground state. The electronic state of an f-orbital in an atom should be described by the total angular momentum, $\mathbf{J} = \mathbf{L} + \mathbf{S}$. This J-multiplet is further split into some levels by an additional electric fields from surrounding ions, which is called a crystalline electric field (CEF) [1]. The number and the types of electronic degrees of freedom depends on the CEF eigenstates, which leads to a variety of physical properties. For example, in a cubic CEF, the Γ_7-doublet of a Ce^{3+} ion is a Kramers doublet with magnetic dipole degrees of freedom. There are three components (J_x, J_y, and J_z). By contrast, the Γ_3-doublet of a Pr^{3+} ion is a non-Kramers doublet with two electric quadrupole (O_{20} and O_{22}) and one magnetic octupole (T_{xyz}) components. There is no magnetic dipole degree of freedom. The Γ_8 state of a Ce^{3+} ion is a quartet with three magnetic dipole, five

T. Matsumura (✉)
Graduate School of Advanced Sciences of Matter, Hiroshima University,
Higashi-Hiroshima, Japan
e-mail: tmatsu@hiroshima-u.ac.jp

© Springer-Verlag Berlin Heidelberg 2017
Y. Murakami and S. Ishihara (eds.), *Resonant X-Ray Scattering
in Correlated Systems*, Springer Tracts in Modern Physics 269,
DOI 10.1007/978-3-662-53227-0_3

electric quadrupole, and seven magnetic octupole components. In this chapter, we study how the orderings of multipolar moments in f-electron compounds can be observed by using resonant X-ray diffraction (RXD).

Let us assume a number of electrons are localized in an atom. The electrons will have some probability density, $\rho(\mathbf{r})$. In a central field potential, the charge density $\rho(\mathbf{r})$ will be spherically symmetric. In solids, however, due to CEF and interactions with electrons of surrounding ions, the resultant shape of $\rho(\mathbf{r})$ is somehow modified. It is no more spherically symmetric. To describe the modified charge density, it is useful to adopt multipole expansion. Let us expand the electric potential $\varphi(\mathbf{r})$, produced by the charge density $\rho(\mathbf{r})$, using the spherical harmonics $Y_{Kq}(\theta, \phi)$:

$$
\begin{aligned}
\varphi(\mathbf{r}) &= \int \frac{\rho(\mathbf{r}')}{|\mathbf{r} - \mathbf{r}'|} d\mathbf{r}' \\
&= \sum_{K=0}^{\infty} \sum_{q=-K}^{K} \int \rho(\mathbf{r}') \frac{r'^{K}}{r^{K+1}} \frac{4\pi}{2K+1} Y_{Kq}^{*}(\theta, \phi) Y_{Kq}(\theta', \phi') d\mathbf{r}' \\
&= \sum_{K=0}^{\infty} \frac{1}{r^{K+1}} \sum_{q=-K}^{K} Q_{q}^{(K)} \sqrt{\frac{4\pi}{2K+1}} Y_{Kq}^{*}(\theta, \phi),
\end{aligned}
\tag{1}
$$

where we can define *electric* multipole moments of rank-K:

$$
Q_{q}^{(K)} = \int \rho(\mathbf{r}') r'^{K} \sqrt{\frac{4\pi}{2K+1}} Y_{Kq}^{*}(\theta', \phi') d\mathbf{r}'.
\tag{2}
$$

The rank-K moment has $2K+1$ components. It is noted, if $\rho(\mathbf{r})$ has an inversion symmetry, which is the case when the atom is located at sites with inversion symmetry, $Q_{q}^{(K)}$ with odd K vanishes and only the even rank electric multipole moments remain.

Magnetic multipole moments of the localized electrons can also be defined by expanding the vector potential produced by a current density, i.e., a magnetization density [2]. In case of magnetic multipole moments, when $\rho(\mathbf{r})$ has an inversion symmetry, only the odd rank magnetic multipole remains and the even rank multipole vanishes. The odd rank electric multipoles and even rank magnetic multipoles, which arise when the atom is located at sites without inversion symmetry, are called *odd parity* moments, which we do not deal with in this chapter.

To describe the multipole moments in crystals, it is convenient to use the cubic representation, which is obtained by the transformation from the spherical representation. The angular dependences of the multipole moments up to rank-4 are summarized in Table 1 and are visualised in Fig. 1. The rank-0 moment represents a spherical part of the charge distribution. This is the 0th order approximation. The rank-2 components represent deviations from the spherical charge distribution in the lowest order expansion, which are called *electric quadrupole* moments. Examples of actual charge distribution, obtained by adding the rank-2 components on to the spherical charge distribution of rank-0, are shown in Fig. 2a. If there is an O_{20}-quadrupole, the

Table 1 Multipole moments in cubic representation up to rank 4

Rank	Irrep.	Notation		Operator equivalent	Angular dependence
0	$\Gamma_{1g}\ (A_{1g})$	Q	$z_0^{(0)}$	$C_0^{(0)}=1$	1
1	$\Gamma_{4u}\ (T_{1u})$	J_x	$z_1^{(1)}$	$(-C_1^{(1)}+C_{-1}^{(1)})/\sqrt{2}=J_x$	x
		J_y	$z_2^{(1)}$	$i(C_1^{(1)}+C_{-1}^{(1)})/\sqrt{2}=J_y$	y
		J_z	$z_3^{(1)}$	$C_0^{(1)}=J_z$	z
2	$\Gamma_{3g}\ (E_g)$	O_{20}	$z_1^{(2)}$	$C_0^{(2)}=\{3J_z^2-J(J+1)\}/2$	$(3z^2-r^2)/2$
		O_{22}	$z_2^{(2)}$	$(C_2^{(2)}+C_{-2}^{(2)})/\sqrt{2}=\sqrt{3}(J_x^2-J_y^2)/2$	$\sqrt{3}(x^2-y^2)/2$
	$\Gamma_{5g}\ (T_{2g})$	O_{yz}	$z_3^{(2)}$	$i(C_1^{(2)}+C_{-1}^{(2)})/\sqrt{2}=\sqrt{3}(J_yJ_z+J_zJ_y)/2$	$\sqrt{3}yz$
		O_{zx}	$z_4^{(2)}$	$(-C_1^{(2)}+C_{-1}^{(2)})/\sqrt{2}=\sqrt{3}(J_zJ_x+J_xJ_z)/2$	$\sqrt{3}zx$
		O_{xy}	$z_5^{(2)}$	$i(-C_2^{(2)}+C_{-2}^{(2)})/\sqrt{2}=\sqrt{3}(J_xJ_y+J_yJ_x)/2$	$\sqrt{3}xy$
3	$\Gamma_{2u}\ (A_{2u})$	T_{xyz}	$z_1^{(3)}$	$i(-C_2^{(3)}+C_{-2}^{(3)})/\sqrt{2}$	$\sqrt{15}xyz$
	$\Gamma_{4u}\ (T_{1u})$	T_x^α	$z_2^{(3)}$	$(-\sqrt{5}C_3^{(3)}+\sqrt{3}C_1^{(3)}-\sqrt{3}C_{-1}^{(3)}+\sqrt{5}C_{-3}^{(3)})/4$	$x(5x^2-3r^2)/2$
		T_y^α	$z_3^{(3)}$	$-i(\sqrt{5}C_3^{(3)}+\sqrt{3}C_1^{(3)}+\sqrt{3}C_{-1}^{(3)}+\sqrt{5}C_{-3}^{(3)})/4$	$y(5y^2-3r^2)/2$
		T_z^α	$z_4^{(3)}$	$C_0^{(3)}$	$z(5z^2-3r^2)/2$
	$\Gamma_{5u}\ (T_{2u})$	T_x^β	$z_5^{(3)}$	$(\sqrt{3}C_3^{(3)}+\sqrt{5}C_1^{(3)}-\sqrt{5}C_{-1}^{(3)}-\sqrt{3}C_{-3}^{(3)})/4$	$\sqrt{15}x(y^2-z^2)/2$
		T_y^β	$z_6^{(3)}$	$-i(\sqrt{3}C_3^{(3)}-\sqrt{5}C_1^{(3)}-\sqrt{5}C_{-1}^{(3)}+\sqrt{3}C_{-3}^{(3)})/4$	$\sqrt{15}y(z^2-x^2)/2$
		T_z^β	$z_7^{(3)}$	$(C_2^{(3)}+C_{-2}^{(3)})/\sqrt{2}$	$\sqrt{15}z(x^2-zy^2)/2$
4	$\Gamma_{1g}\ (A_{1g})$	H^0	$z_1^{(4)}$	$(\sqrt{30}C_4^{(4)}+2\sqrt{21}C_0^{(4)}+\sqrt{30}C_{-4}^{(4)})/12$	$5\sqrt{21}(x^4+y^4+z^4-3r^4/5)/12$
	$\Gamma_{3g}\ (E_g)$	H_1^2	$z_2^{(4)}$	$-(\sqrt{42}C_4^{(4)}-2\sqrt{15}C_0^{(4)}+\sqrt{42}C_{-4}^{(4)})/12$	$\sqrt{15}\{7(2z^4-x^4-y^4)-6(3z^2-r^2)r^2\}/12$
		H_2^2	$z_3^{(4)}$	$-(C_2^{(4)}+C_{-2}^{(4)})/\sqrt{2}$	$\sqrt{15}(x^2-y^2)(r^2-7z^2)/4$
	$\Gamma_{4g}\ (T_{1g})$	H_x^α	$z_4^{(4)}$	$-i(C_3^{(4)}+\sqrt{7}C_1^{(4)}+\sqrt{7}C_{-1}^{(4)}+C_{-3}^{(4)})/4$	$\sqrt{35}yz(y^2-z^2)/2$
		H_y^α	$z_5^{(4)}$	$(C_3^{(4)}-\sqrt{7}C_1^{(4)}+\sqrt{7}C_{-1}^{(4)}-C_{-3}^{(4)})/4$	$\sqrt{35}zx(z^2-x^2)/2$
		H_z^α	$z_6^{(4)}$	$i(-C_4^{(4)}+C_{-4}^{(4)})/\sqrt{2}$	$\sqrt{35}xy(x^2-y^2)/2$
	$\Gamma_{5g}\ (T_{2g})$	H_x^β	$z_7^{(4)}$	$i(\sqrt{7}C_3^{(4)}-C_1^{(4)}-C_{-1}^{(4)}+\sqrt{7}C_{-3}^{(4)})/4$	$\sqrt{5}yz(7x^2-r^2)/2$
		H_y^β	$z_8^{(4)}$	$(\sqrt{7}C_3^{(4)}+C_1^{(4)}-C_{-1}^{(4)}-\sqrt{7}C_{-3}^{(4)})/4$	$\sqrt{5}zx(7y^2-r^2)/2$
		H_z^β	$z_9^{(4)}$	$i(-C_2^{(4)}+C_{-2}^{(4)})/\sqrt{2}$	$\sqrt{5}xy(7z^2-r^2)/2$

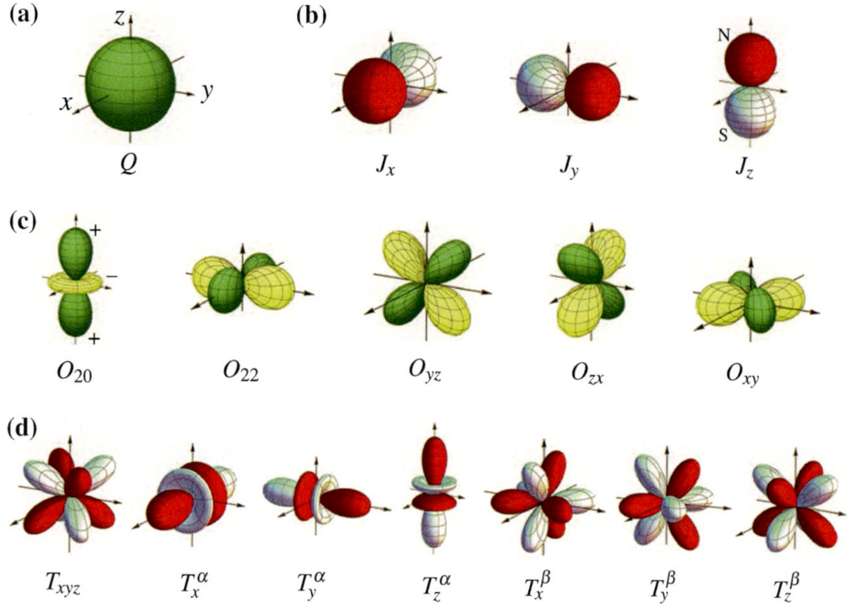

Fig. 1 Angular dependences of the multipole moments up to rank-3 listed in Table 1. **a** Spherical electric charge distribution. **b** Magnetic dipole moments. **c** Electric quadrupole moments. **d** Magnetic octupole moments. N and S represent the distribution of magnetic charges. $+$ and $-$ represent the distribution of electric charges

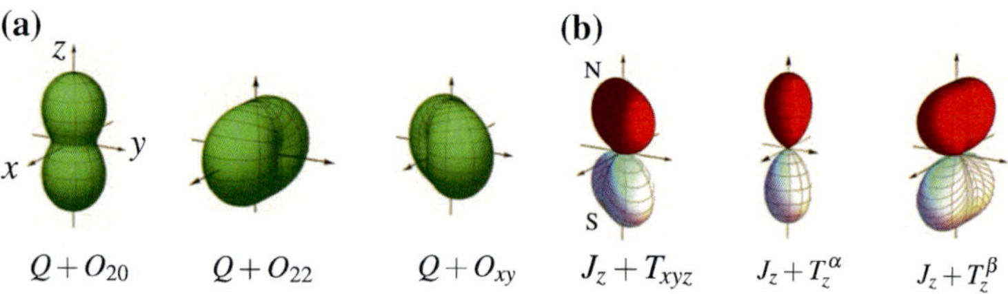

Fig. 2 **a** Examples of actual charge distributions with a quadrupole moment. **b** Examples of actual magnetization densities with dipole and octupole moments

charge distribution is elongated, or compressed, along the z direction, depending on the sign of the O_{20} quadrupole. If there is an O_{22}-quadrupole, it is elongated along the x, or y, direction, depending on the sign. Any type of anisotropic charge distribution can be expressed, in the lowest order approximation, by combining the five types of quadrupole moments.

The magnetic dipole moments in Fig. 1 are normally represented by an arrow to show the direction of a magnetic moment. In Fig. 1, we use magnetic charges of N and S poles to illustrate the angular dependence of a magnetization density. If an isotropic charge distribution of rank-0 is magnetized to the z direction, the magnetized state can be described by putting a pair of N and S poles along the z direction as illustrated in Fig. 1b. This state has only the magnetic dipole moment.

Magnetic octupole moment describes a deviation of the magnetization density from that of the magnetic dipole state. Examples are shown in Fig. 2b. If we add a T_{xyz}-octupole on a J_z-dipole, the resultant magnetization density will be elongated along the [110] direction. It is noted that the anisotropic magnetization density of $J_z + T_{xyz}$ in Fig. 2b is compatible with the anisotropic charge density of $Q + O_{xy}$ in Fig. 2a. This means that, if a magnetic field is applied along the z direction to an electron system with an O_{xy}-type charge distribution, a T_{xyz}-octupole is necessarily induced in addition to the magnetic dipole of J_z. In reverse, when there is only a magnetic octupole of T_{xyz} without a magnetic dipole moment, anisotropic charge distribution of O_{xy} is necessarily induced by applying a magnetic field along the z direction and thereby inducing the J_z-dipole. This type of coupling between electric and magnetic degrees of freedom in magnetic fields is essentially important in f-electron systems. Another point to be noted is that the T_z^α-octupole and the J_z-dipole have the same symmetry, as can be visually understood from the figures; both magnetic state have the rotational symmetry around the z axis. This means that if T_z^α-octupole is induced, J_z-dipole is necessarily induced together, or vice versa.

These basic concepts on higher rank multipole moments are very important especially in localized f-electron systems. The multipole moments of atoms at a distance interact with each other, for example, via conduction electrons in metals, which is called the Ruderman–Kittel–Kasuya–Yosida (RKKY) interaction. It is emphasized that intersite multipole interactions of RKKY type between f-orbitals are rank independent [3]. This means that electric quadrupole or magnetic octupole moments can order by themselves as a primary order parameter as well as normal magnetic dipole orders, which are, of course, frequently realized. There are many cases where the higher rank multipole moments are ordered, or induced, and play important roles in the physical properties.

To experimentally observe the periodic arrangement of multipole moments, we need to rely on diffraction methods using the Bragg's law of $\lambda = 2d \sin \theta$, where d represents the inter-planar spacing of the periodic arrangement. To observe the orderings of magnetic dipole moments, neutron diffraction is the most powerful and effective method because the neutron spin interact directly with the magnetic moments of the atoms. However, it is impossible for neutrons to directly observe the orderings of quadrupole moments because they have no direct interaction. Observation of magnetic octupole orderings is also hard for neutrons, though it is not impossible, because we need to detect a weak signal and to prove that the signal is indeed due to the magnetic octupole and not from the magnetic dipole moment. This is normally performed by measuring the magnetic form factor. By contrast, RXD has a sensitivity to detect multipole moments up to rank-4, i.e., the hexadecapole moments, as it will be explained in the following. It is therefore very valuable that we can directly observe the orderings of multipole moments by RXD.

1.2 Operator Equivalent of Multipole Moments

To calculate the actual multipole moment values in an f-orbital, it is useful to adopt the operator equivalent method. We define a spherical tensor operator $C_q^{(K)}$, using the total angular momentum operator $\mathbf{J} = (J_x, J_y, J_z)$. $C_q^{(K)}$ is defined by

$$C_K^{(K)} = (-1)^K \sqrt{\frac{(2K-1)!!}{(2K)!!}}\,(J_+)^K \tag{3}$$

and

$$[J_-, C_q^{(K)}] = \sqrt{(K+q)(K-q+1)}\, C_{q-1}^{(K)}, \tag{4}$$

where $J_\pm = J_x \pm i J_y$. Using the Wigner–Eckert theorem, the matrix element of $C_q^{(K)}$ between the J-multiplet states can be calculated:

$$\langle JM|C_q^{(K)}|JM'\rangle = \frac{1}{2^K \sqrt{2J+1}} \sqrt{\frac{(2J+K+1)!}{(2J-K)!}}\langle JM|JM'Kq\rangle. \tag{5}$$

The spherical tensor operator can be transformed to the cubic tensor operator using the relations tabulated in Table 1. Using the relations above, we can calculate the matrix elements of the multipole moment operators.

2 Resonant X-ray Diffraction

2.1 Atomic Scattering Factor

When the target atom is located at a center of space inversion, there arises $E1$ and $E2$ atomic scattering factors, which are expressed as

$$f_{E1}(E) = \sum_{\alpha,\beta=1}^{3} \varepsilon_\beta' \varepsilon_\alpha \sum_\Lambda \frac{\langle \psi|R_\beta|\Lambda\rangle\langle\Lambda|R_\alpha|\psi\rangle}{E-(E_\Lambda-E_0)+i\Gamma}, \tag{6}$$

$$f_{E2}(E) = \sum_{\alpha,\beta=1}^{5} K_\beta(\mathbf{k}',\boldsymbol{\varepsilon}')K_\alpha(\mathbf{k},\boldsymbol{\varepsilon})\cdot \sum_\Lambda \frac{\langle \psi|Q_\beta|\Lambda\rangle\langle\Lambda|Q_\alpha|\psi\rangle}{E-(E_\Lambda-E_0)+i\Gamma}, \tag{7}$$

where $|\psi\rangle$ and E_0 are the eigenfunction and the energy of the ground state, respectively [4]. $|\Lambda\rangle$ represents the intermediate state of the resonant process with a core hole and Γ the lifetime broadening. When the transition from $|\psi\rangle$ to $|\Lambda\rangle$ takes place, the quantum number of the orbital angular momentum changes by 1 and 2 for the $E1$ (dipole) and $E2$ (quadrupole) transition processes, respectively. For $L_{2,3}$-edges of a rare-earth element, the $E1$ and $E2$ processes correspond to the $2p \leftrightarrow 5d$ and $2p \leftrightarrow 4f$ transitions, respectively. These absorption edges are widely used because the resonance energies lie in the region from 5 to 10 keV, which is appropriate for fulfilling the Bragg condition in many crystals.

R in the $E1$ term is the dipole operator $(R_1, R_2, R_3) = (x, y, z)$. The five component K-factors in the $E2$ term are written as

$$K_1(\mathbf{a}, \mathbf{b}) = \frac{1}{2}(3a_z b_z - \mathbf{a} \cdot \mathbf{b}),$$

$$K_2(\mathbf{a}, \mathbf{b}) = \frac{\sqrt{3}}{2}(a_x b_x - a_y b_y),$$

$$K_3(\mathbf{a}, \mathbf{b}) = \frac{\sqrt{3}}{2}(a_y b_z + a_z b_y), \tag{8}$$

$$K_4(\mathbf{a}, \mathbf{b}) = \frac{\sqrt{3}}{2}(a_z b_x + a_x b_z),$$

$$K_5(\mathbf{a}, \mathbf{b}) = \frac{\sqrt{3}}{2}(a_x b_y + a_y b_x).$$

The quadrupole operators Q can also be expressed using the K-factors as $Q_\alpha = K_\alpha(\mathbf{r}, \mathbf{r})$.

In most cases, the target atom is at an inversion center. In such cases, as shown in Eqs. (6) and (7), the transitions from $|\psi\rangle$ to $|\Lambda\rangle$ and from $|\Lambda\rangle$ to $|\psi\rangle$ take place through the same rank operator R or Q, i.e., $E1$–$E1$ or $E2$–$E2$, respectively. However, in systems without a centrosymmetry, the $E1$–$E2$ process arises, where the resonant scattering factor also involves $\langle\psi|R|\Lambda\rangle\langle\Lambda|Q|\psi\rangle$. This term, although not written explicitly here, will play an important role in such cases as detecting the odd parity moments.

One difficulty in understanding the RXS process lies in deal with the intermediate state $|\Lambda\rangle$ exactly. By putting the $1/(E - E_\Lambda + E_0 + i\Gamma)$ term outside the summation and treating the resonance as a single oscillator, a convenient formalism using spherical tensors is obtained, which is useful in data analysis [5]. However, we are sometimes confronted with a limitation caused by not treating the energy dependence properly, particularly when dealing with interference effects among several origins of scattering. A more appropriate formalism has been proposed by Nagao and Igarashi [4, 6, 7]. Using the multipole operator equivalents, the resonant atomic scattering factors can be written as

$$f_{E1}(E) = \sum_{v=0}^{2} \alpha_{E1}^{(v)}(E) \sum_{\mu=1}^{2v+1} P_{E1,\mu}^{(v)}(\boldsymbol{\varepsilon}, \boldsymbol{\varepsilon}')\langle z_{E1,\mu}^{(v)}\rangle, \tag{9}$$

$$f_{E2}(E) = \sum_{v=0}^{4} \alpha_{E2}^{(v)}(E) \sum_{\mu=1}^{2v+1} P_{E2,\mu}^{(v)}(\boldsymbol{\varepsilon}, \boldsymbol{\varepsilon}', \mathbf{k}, \mathbf{k}')\langle z_{E2,\mu}^{(v)}\rangle, \tag{10}$$

where $\langle z^{(v)}_{E1,\mu}\rangle$ and $\langle z^{(v)}_{E2,\mu}\rangle$ represent the expectation values of the multipole operator equivalent $z^{(v)}_{\mu}$ for rank-v and the component number μ, which are associated with the $E1$ and $E2$ resonances, respectively. For example, $z^{(1)}_{1} = J_x$ and $z^{(2)}_{5} = \sqrt{3}(J_x J_y + J_y J_x)/2$ as tabulated in Table 1. The $E1$ and $E2$ resonance have sensitivities up to rank 2 and rank 4 moments, respectively. The sensitivity to even rank orbital degrees of freedom is a distinctive characteristic of RXD.

$P^{(v)}_{E1,\mu}$ and $P^{(v)}_{E2,\mu}$ are the geometrical factors for $z^{(v)}_{\mu}$ for $E1$ and $E2$ resonances, respectively. For $E1$,

$$P^{(0)}_{E1,1} = (\boldsymbol{\varepsilon}' \cdot \boldsymbol{\varepsilon}), \tag{11}$$

$$P^{(1)}_{E1,\mu} = -i(\boldsymbol{\varepsilon}' \times \boldsymbol{\varepsilon})_\mu, \tag{12}$$

$$P^{(2)}_{E1,\mu} = K_\mu(\boldsymbol{\varepsilon}', \boldsymbol{\varepsilon}) . \tag{13}$$

The factors for $E2$ are summarized in the "Appendix". These geometrical factors are equivalent to those given by Lovesey et al. [5]. Since the geometrical factors are unambiguously determined by the scattering configuration, $\langle z^{(v)}_{E1,\mu}\rangle$ and $\langle z^{(v)}_{E2,\mu}\rangle$ can be treated as experimental parameters in data analyses.

Let us consider the case of magnetic scattering as an example by assuming a magnetic ion with a finite magnetic dipole moment (m_x, m_y, m_z) in the coordinate system shown in Fig. 3. Using the polarization vectors of $\boldsymbol{\varepsilon}_\sigma = \boldsymbol{\varepsilon}_{\sigma'} = (1, 0, 0)$, $\boldsymbol{\varepsilon}_\pi = (0, \sin\theta, \cos\theta)$, and $\boldsymbol{\varepsilon}_{\pi'} = (0, -\sin\theta, \cos\theta)$, we obtain the rank-1 resonant atomic scattering factors for the σ–σ', σ–π', π–σ', and π–π' channels as the following: $f^{\sigma\sigma'}_{E1} = 0$, $f^{\sigma\pi'}_{E1} \propto (m_y \cos\theta + m_z \sin\theta)$, $f^{\pi\sigma'}_{E1} \propto (-m_y \cos\theta + m_z \sin\theta)$, and

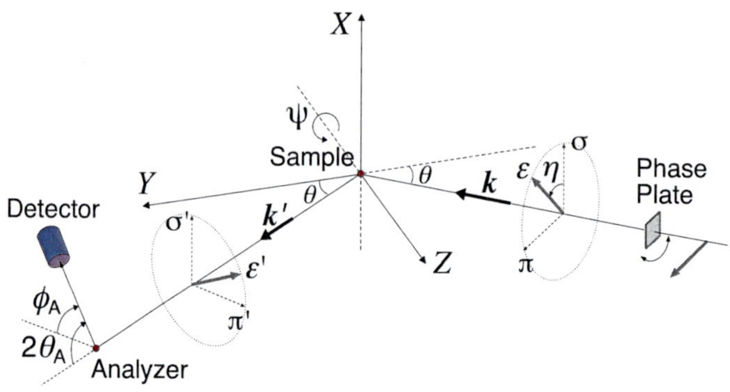

Fig. 3 Scattering geometry of the experiment. σ and π represent the polarization states of the X-ray beam, which is perpendicular (σ) or parallel (π) to the scattering plane spanned by $\mathbf{k}$ and $\mathbf{k}'$. The rotation angle ψ of the sample around the scattering vector $\boldsymbol{\kappa} = \mathbf{k} - \mathbf{k}'$ is called the azimuthal angle. The polarization state of the scattered beam is analyzed by rotating the analyzer crystal by angle ϕ_A. An appropriate crystal is selected so that the scattering angle $2\theta_A$ is close to $90°$. When $\phi_A = 0°$, only the σ-polarized beam is diffracted by the analyzer, whereas only the π-polarized beam is diffracted at $\phi_A = \pm 90°$. It is also possible to manipulate the incident polarization state by using phase plates, allowing full polarization analysis [8–11]

$f_{E1}^{\pi\pi'} \propto m_x \sin 2\theta$, respectively. An important result is that σ–σ' signal vanishes. Non-zero σ–π' and π–σ' signals arise if the magnetic moment has a component parallel to the scattering plane; otherwise, they also disappear. The π–π' signal reflects the magnetic dipole component perpendicular to the scattering plane. Using this relationship, we can investigate the magnetic structure by RXD.

$\alpha_{E1}^{(v)}(E)$ and $\alpha_{E2}^{(v)}(E)$ are the spectral functions representing the energy dependences of the rank-v term for $E1$ and $E2$, respectively. These are complex functions, which are generally written as $\alpha'(E) + i\alpha''(E)$. When scattering processes of different rank coexist, we need to seriously consider the interference effect among them. In such cases, it is essential to consider that the energy profile depends on the rank of the scattering process.

It is remarked that $\langle z_{E1,\mu}^{(v)}\rangle$ and $\langle z_{E2,\mu}^{(v)}\rangle$, in a strict sense, do not represent the same thing because the intermediate states are different. In the case of the L-edge resonances in rare-earth elements, $\langle z_{E1,\mu}^{(v)}\rangle$ represent the multipole moment of the $5d$ orbital, whereas $\langle z_{E2,\mu}^{(v)}\rangle$ represent that of the $4f$ orbital. However, the appearance of $\langle z_{E1,\mu}^{(v)}\rangle$ in the $5d$ orbital, in any case, can be ascribed to the original multipole moment in the $4f$ orbital, which possesses the fundamental degree of freedom. That is to say, they are proportional to each other; in other words, the Wigner–Eckart theorem should be valid. $\langle z_{E1,\mu}^{(v)}\rangle$ and $\langle z_{E2,\mu}^{(v)}\rangle$ can be interpreted as parameters that is *proportional to* the real multipole moment induced in the $4f$ orbital.

Usually, the multipole operator $z_\mu^{(v)}$ is defined with respect to the crystal axes, x, y, and z, which do not coincide with X, Y, and Z in Fig. 3. We need to transform $\boldsymbol{\varepsilon}$, $\boldsymbol{\varepsilon}'$, $\mathbf{k}$, and $\mathbf{k}'$, using Euler rotation, from the XYZ-coordinate to the xyz-coordinate before calculating the geometrical factors of $P_{E1,\mu}^{(v)}$ and $P_{E2,\mu}^{(v)}$.

2.2 *Intensity of X-ray Diffraction*

The intensity of X-ray diffraction is proportional to $|F|^2$, where F is the structure factor of the Bragg diffraction. The energy dependent structure factor $F(E)$, at energies around the absorption edge of the target atom, consists of nonresonant and resonant terms:

$$\begin{aligned}
F(E) &= F_{\mathrm{nr}} + F_{E2}(E) + F_{E1}(E) \\
&= \sum_j \{f_{0,j} + f_{E2,j}(E) + f_{E1,j}(E)\}e^{i\boldsymbol{\kappa}\cdot\mathbf{r}_j},
\end{aligned} \tag{14}$$

where $\boldsymbol{\kappa} = \mathbf{k} - \mathbf{k}'$ is the scattering vector, $\mathbf{r}_j$ the position of the j th atom in the unit cell, and f_0 the atomic scattering factor of nonresonant scattering, which include normal Thomson scattering as well as the nonresonant magnetic scattering. We can express the resonant structure factor as

$$F_{E2}(E) + F_{E1}(E) = \sum_{v=0}^{4} \alpha_{E2}^{(v)}(E) \sum_{\mu=1}^{2v+1} P_{E2,\mu}^{(v)}(\boldsymbol{\varepsilon}, \boldsymbol{\varepsilon}', \mathbf{k}, \mathbf{k}') Z_{E2,\mu}^{(v)}$$

$$+ \sum_{v=0}^{2} \alpha_{E1}^{(v)}(E) \sum_{\mu=1}^{2v+1} P_{E1,\mu}^{(v)}(\boldsymbol{\varepsilon}, \boldsymbol{\varepsilon}') Z_{E1,\mu}^{(v)}, \tag{15}$$

where $Z_{E1,\mu}^{(v)}$ and $Z_{E2,\mu}^{(v)}$ represent the multipole structure factors:

$$Z_{E1,\mu}^{(v)} = \sum_{j} \langle z_{E1,\mu}^{(v)} \rangle_j e^{i\boldsymbol{\kappa} \cdot \mathbf{r}_j}, \tag{16}$$

$$Z_{E2,\mu}^{(v)} = \sum_{j} \langle z_{E2,\mu}^{(v)} \rangle_j e^{i\boldsymbol{\kappa} \cdot \mathbf{r}_j}. \tag{17}$$

All the information with respect to the target sample can be summarized in a 2×2 matrix $\hat{F}$:

$$\hat{F} = \begin{pmatrix} F_{\sigma\sigma'} & F_{\pi\sigma'} \\ F_{\sigma\pi'} & F_{\pi\pi'} \end{pmatrix}, \tag{18}$$

where each element represents the structure factor (scattering amplitude) for σ–σ', σ–π', π–σ', and π–π' processes. The usage of this matrix to calculate the scattering intensity is summarized in the "Appendix".

3 Observation of Multipole Orderings in f-Electron Compounds

3.1 *Spiral Magnetic Order in Ho*

To demonstrate the ability of RXD to detect up to rank-4 moments, the spiral magnetic order of Ho is a good example. Ho metal crystallizes in a hexagonal close packed structure and exhibits a magnetic order below $T_{\mathrm{N}} \sim 130$ K into a spiral structure with an incommensurate propagation vector $\mathbf{q} \sim 0.28\mathbf{c}^*$. The modulation period increases ($\mathbf{q}$ decreases) with decreasing temperature, and at 20 K, the spiral modulation is locked in to a commensurate structure with $\mathbf{q} = \mathbf{c}^*/6$ [12]. The magnetic moments are ferromagnetically aligned within each basal c-plane and rotate from plane to plane. The magnetic structure is schematically illustrated in Fig. 4, simply assuming $\mathbf{q} = \mathbf{c}^*/6$ with equal turn angles from plane to plane. In actuality, the turn angle is not equal due to the bunched structure. The moments are tilted to the c-axis to form a conical structure below 20 K. Since these modifications, however, are not our central issue and does not affect the following considerations, we simply proceed our discussion using the model structure of Fig. 4.

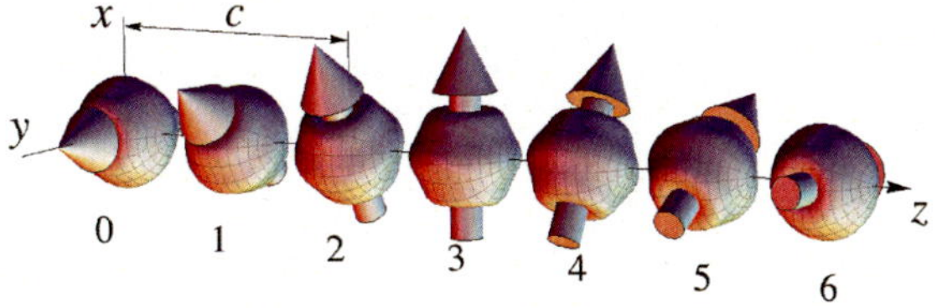

Fig. 4 Schematic illustration of the spiral magnetic order in Ho metal assuming $q = c^*/6$, confined magnetic moments within the c-plane, and equal turn angle of 30° from plane to plane. Anisotropic charge distribution is calculated by assuming a fully saturated magnetic moment

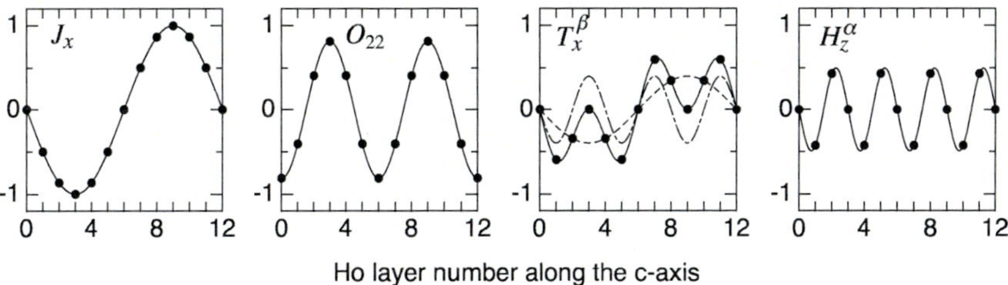

Fig. 5 Oscillations of typical multipole operator equivalents of the Ho ions in Fig. 4. The oscillation of T_x^β is decomposed into q (*dashed line*) and $3q$ (*dot-dashed line*) components

As shown in Fig. 4, the magnetic dipole moment is reversed every six layers. This gives diffraction peaks of magnetic dipole order at $\kappa = (0, 0, \tau \pm \mathbf{q})$ positions, where τ represent a reciprocal lattice vector of the fundamental lattice. By contrast, the period of the charge distribution is half that of the magnetic dipole. As a result, the diffraction peaks of normal Thomson scattering from this aspherical charge distribution is expected to appear at $\kappa = (0, 0, \tau \pm 2\mathbf{q})$ positions [13]. This situation is more quantitatively shown in Fig. 5 as the oscillation of the operator equivalents J_x and O_{22}. The oscillations of the T^β-octupole and H^α-hexadecapole moments are also shown in Fig. 5. We see that they oscillate with a period of 4 layers ($3q$) and 3 layers ($4q$), respectively.

The top panel of Fig. 6 shows the resonance spectrum at $\tau = (0, 0, 4 + q)$, which measures only the q-component of the ordered moments [14, 15]. From Fig. 5 we see that the magnetic dipole moment $\langle \mathbf{J} \rangle$ has a periodicity of 12 layers. Then, the $E1$ resonance at 8.071 keV at $\tau = (0, 0, 4 + q)$ arises from the rank-1 term of the resonance structure factor $\alpha_{E1}^{(1)}(\boldsymbol{\varepsilon} \times \boldsymbol{\varepsilon}') \cdot \langle \mathbf{J} \rangle$. The σ–σ' intensity is prohibited in the $E1$ resonance, whereas it appears in $E2$ due to the non-zero geometrical factor $P_{E2}^{(1)}$ for σ–σ'. It is noted that the the octupole (rank-3 scattering process) also has the q-component as shown in Fig. 5, which also contributes to the σ–σ' intensity at $E2$, though it is difficult to separate.

The results for $2q$ can also be understood in terms of quadrupole (rank-2 scattering process). The strong σ–σ' intensity at $E1$ is consistent with the geometrical factor $P_{E1}^{(2)}$, and can be associated with the non-zero expectation value of the quadrupole operators with Γ_{3g} and Γ_{5g} symmetries, which oscillate with the $2q$ periodicity.

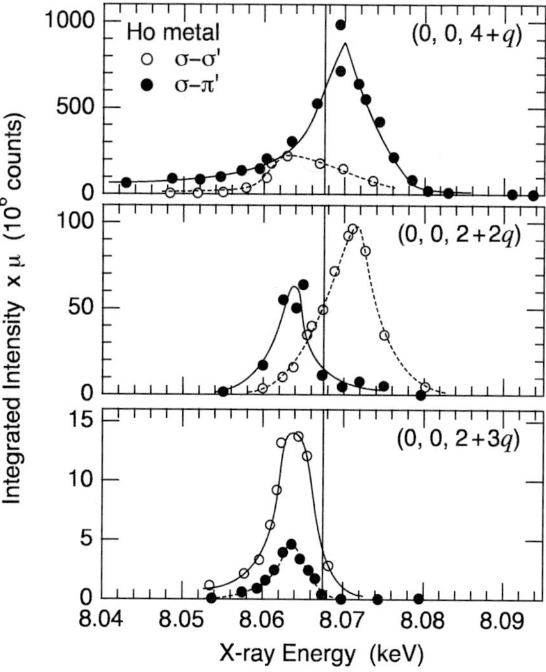

Fig. 6 Energy dependences of the diffraction intensities from the spiral magnetic order in Ho [14, 15]

$P_{E2}^{(2)}$ shows that the σ–π' intensity appears at $E2$. The σ–π' intensity should also exist at $E1$, but is estimated to be much weaker than σ–σ'. It is noted that the hexadecapole (rank-4 scattering process) with the same Γ_{3g} and Γ_{5g} symmetry (not shown) also have the $2q$-component, and contributes to this $E2$ resonance.

The resonance at the $3q$ position can be associated with the magnetic octupole. Since only the $E2$ resonance has a sensitivity to the octupole moments, no signal is expected at $E1$. The $4q$ signal arises from the hexadecapole moments. The data without polarization analysis is shown in [14]. The calculated energy spectra for q, $2q$, $3q$, and $4q$ peaks are presented in [16].

3.2 Antiferroquadrupole and Antiferromagnetic Order in DyB$_2$C$_2$

In most of magnetic materials, as in Ho metal, the primary order parameter is a magnetic dipole moment. The higher rank multipoles are just necessarily induced together. In DyB$_2$C$_2$, by contrast, an electric quadrupole moment becomes the primary order parameter. As shown in Fig. 7a, b, there are two successive phase transitions at $T_Q = 25$ K and $T_N = 15$ K [17]. Since there is no anomaly in magnetic susceptibility at 25 K and no magnetic order is detected between 15 and 25 K, the

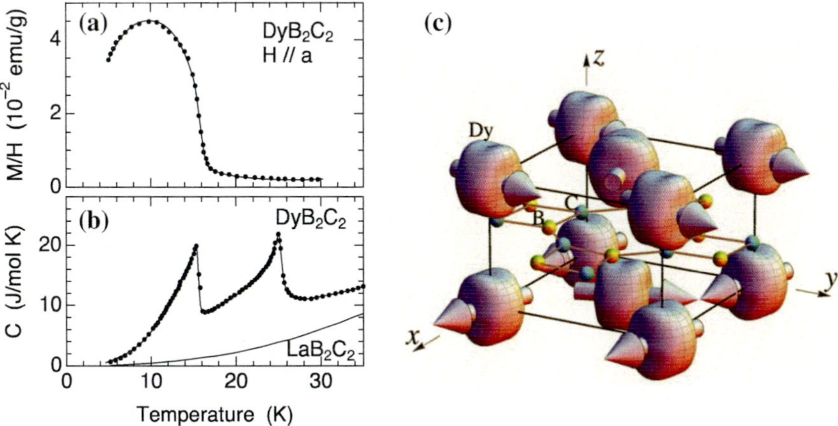

Fig. 7 Temperature dependence of **a** magnetic susceptibility and **b** specific heat of DyB$_2$C$_2$. Specific heat of LaB$_2$C$_2$ is shown as a reference [17]. **c** Schematic illustration of AFM- and AFQ-orderings in DyB$_2$C$_2$, showing the structures of magnetic dipole moments and anisotropic charge distributions in a chemical unit cell. The unit cell of the AFQ and AFM structure is double the chemical unit cell along the z axis

transition at 25 K is concluded as nonmagnetic. From the specific heat, the released entropy is $R \ln 2$ at 15 K and $R \ln 4$ at 25 K, which shows that the CEF ground state consists of slightly separated pairs of Kramers doublets, forming a quasi-quartet. A quartet has $4 \times 4 - 1 = 15$ components in total, three dipoles, five quadrupoles, and seven octupoles. The nonmagnetic transition at 25 K in DyB$_2$C$_2$ strongly suggests that the quadrupolar degree of freedom is frozen, leaving a Kramers doublet with magnetic dipolar degrees of freedom, which is frozen below 15 K by the magnetic ordering.

Schematic illustration of the quadrupole and magnetic ordered state is shown in Fig. 7c. Since the quadrupole moments are ordered in an alternating way, this state is called an antiferroquadrupole (AFQ) order. Because of this AFQ order, the direction of the magnetic moments are confined along the principal axis of the quadrupole moment. Then, the resultant structure of the antiferromagnetic (AFM) order is strongly affected by the underlying AFQ order.

The magnetic and quadrupolar unit cell of the ordered structure is double the chemical unit cell, $a \times a \times 2c$. Then, we expect non-zero structure factor at $\kappa = 2\pi(0, 0, l/2)$ with $l = $ odd. Let us calculate the multipole structure factors. There are four Dy ions in the unit cell of the ordered phase: $\mathbf{r}_1 = (0, 0, 0)$, $\mathbf{r}_2 = (1/2, 1/2, 0)$, $\mathbf{r}_3 = (0, 0, 1)$, and $\mathbf{r}_4 = (1/2, 1/2, 1)$. Then, the structure factor for the $(0, 0, l/2)$ reflection is

$$Z_\mu^{(v)} = \sum_{j=1}^{4} \langle z_\mu^{(v)} \rangle_j e^{i\boldsymbol{\kappa} \cdot \mathbf{r}_j}$$

$$= \langle z_\mu^{(v)} \rangle_1 + \langle z_\mu^{(v)} \rangle_2 - \langle z_\mu^{(v)} \rangle_3 - \langle z_\mu^{(v)} \rangle_4. \tag{19}$$

The magnetic dipole moments in the AFM phase in Fig. 7 are expressed as

$$
\begin{aligned}
\langle \mathbf{z}^{(1)} \rangle_1 &= (\cos\phi, -\sin\phi, 0), \\
\langle \mathbf{z}^{(1)} \rangle_2 &= (-\sin\phi, \cos\phi, 0), \\
\langle \mathbf{z}^{(1)} \rangle_3 &= (\sin\phi, \cos\phi, 0), \\
\langle \mathbf{z}^{(1)} \rangle_4 &= (-\cos\phi, -\sin\phi, 0),
\end{aligned}
\tag{20}
$$

where ϕ represents a cant angle. We used notations of the rank-1 tensor as summarized in Table 1. We see that $Z^{(1)}_{1,3}$ vanishes and only $Z^{(1)}_2 = 2(\cos\phi - \sin\phi)$ remains non-zero if $\phi \neq \pi/4$. Then, the final structure factor for σ–π' is

$$
F^{(1)}_{E1,\sigma\pi'} = -2i\cos\theta(\cos\phi - \sin\phi)\sin\psi.
\tag{21}
$$

The structure factor of the other domain, where all the moments are rotated by 90° about the z-axis, is

$$
F^{(1)}_{E1,\sigma\pi'} = -2i\cos\theta(\cos\phi - \sin\phi)\cos\psi.
\tag{22}
$$

By adding $|F^{(1)}_{E1,\sigma\pi'}|^2$ from these two domains, we have a constant intensity, which is independent of the azimuthal angle ψ.

The expectation values of the quadrupole moments are $\langle \mathbf{z}^{(2)} \rangle_1 = (0, -\cos 2\phi, 0, 0, -\sin 2\phi)$, $\langle \mathbf{z}^{(2)} \rangle_2 = (0, \cos 2\phi, 0, 0, -\sin 2\phi)$, $\langle \mathbf{z}^{(2)} \rangle_3 = (0, \cos 2\phi, 0, 0, \sin 2\phi)$, and $\langle \mathbf{z}^{(2)} \rangle_4 = (0, -\cos 2\phi, 0, 0, \sin 2\phi)$. Therefore, only $Z^{(2)}_5 = -2\sin 2\phi$ is non-zero. The resonant structure factor for the rank-2 order parameter is

$$
F^{(2)}_{E1,\sigma\sigma'} = 2\sqrt{3}\sin 2\phi \sin 2\psi,
\tag{23}
$$

$$
F^{(2)}_{E1,\sigma\pi'} = 2\sqrt{3}\sin 2\phi \sin\theta \cos 2\psi,
\tag{24}
$$

for $E1$ and

$$
F^{(2)}_{E2,\sigma\sigma'} = \frac{3}{2}\sqrt{\frac{6}{7}}\sin 2\phi \sin^2\theta \sin 2\psi,
\tag{25}
$$

$$
F^{(2)}_{E2,\sigma\pi'} = -3\sqrt{\frac{3}{14}}\sin 2\phi \sin 3\theta \cos 2\psi,
\tag{26}
$$

for $E2$. These are the same for the other domain.

The experimental results shown in Fig. 8 can be explained by these structure factors. The σ–σ' intensity at $\psi = 45°$ and the σ–π' intensity at $\psi = 0°$ increase below $T_Q = 25$ K. This clearly shows that the AFQ order grows up below 25 K. Furthermore, the azimuthal angle dependences of the $E1$ resonance intensity for σ–σ' ($\propto \sin^2 2\psi$) and σ–π' ($\propto \cos^2 2\psi$), as shown in Fig. 9, are also consistent between theory and experiment. The signal from the AFM order is also reflected in the σ–π' intensity at $\psi = 45°$, where the rank-2 structure factor is zero. The intensity increases

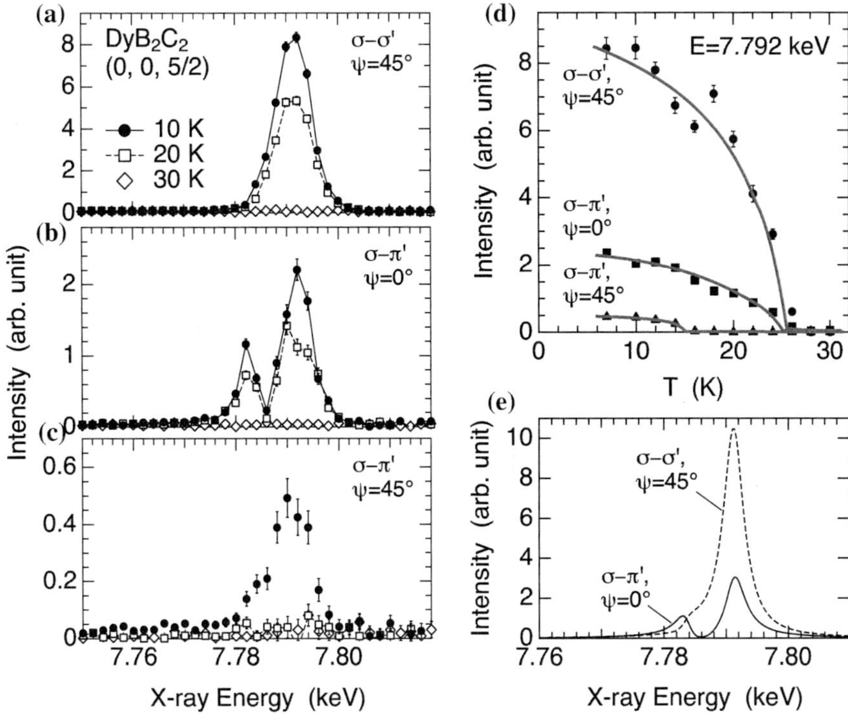

Fig. 8 **a–c** Energy dependences of the $(0, 0, 5/2)$ reflection of DyB_2C_2 in the AFM (10 K), AFQ (20 K), and paramagnetic (30 K) phases. **d** Temperature dependences of the $E1$ resonance intensity. The data are from [18–20]. **e** Calculated energy dependence for $\sigma-\sigma'$ at $\psi = 45°$ and $\sigma-\pi'$ at $\psi = 0°$ by taking into account the interference between $E1$ and $E2$ resonances

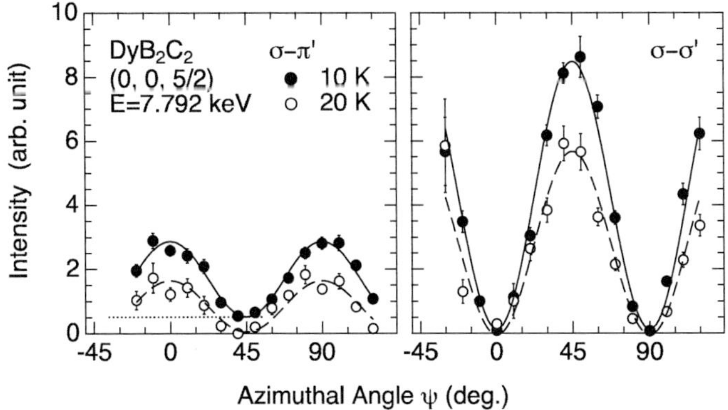

Fig. 9 Azimuthal angle dependences of the $E1$ resonant intensities for $\sigma-\sigma'$ and $\sigma-\pi'$ [18, 19]

below 15 K due to the appearance of the AFM order, which gives the ψ independent intensity of $|F^{(1)}_{E1,\sigma\pi'}|^2$.

We notice that the spectral shapes of the σ–σ' intensity at $\psi = 45°$ and the σ–π' intensity at $\psi = 0°$ are different; there seems no $E2$ resonance in the former, whereas is is clearly observed in the latter. This is somehow strange because the amplitudes of Eqs. (25) and (26) are not so much different. This is due to the interference effect between the $E1$ and $E2$ resonances. Let us express the total structure factor in the following way:

$$F_{\varepsilon\varepsilon'}(E) = \frac{A_2 e^{i\gamma_2}}{E - \Delta_2 + i\Gamma_2} F^{(2)}_{E2,\varepsilon\varepsilon'} + \frac{A_1 e^{i\gamma_1}}{E - \Delta_1 + i\Gamma_1} F^{(2)}_{E1,\varepsilon\varepsilon'}, \qquad (27)$$

where γ_2 and γ_1 represent the phase parameters of the Lorentzian-type spectral function. Although an accurate form of the spectral function should be calculated by more realistic theoretical calculation, we use this form by introducing the phase parameter on the normal Lorentzian-type spectral function since it is technically very difficult at the present to experimentally deduce the rank separated spectral functions. Figure 8e shows the calculated energy dependences with $\gamma_1 - \gamma_2 = 2.53$. Note that only the phase difference affects the intensity $|F|^2$. This type of interference effect can be utilized in separating scatterings of different origins, which will be dealt with in a later section.

3.3 Antiferromagnetic Octupole Order

Antiferromagnetic octupole order is an unusual ordering of magnetic moments, where the time reversal symmetry is broken but there appears no magnetic dipole moment. There appears only a distribution of magnetization density (a distribution of non-uniform current density). Even in such an ordering, the ground state degeneracy is lifted, and there appears a clear anomaly in specific heat. Magnetic susceptibility also exhibit some anomaly. However, no magnetic dipole order is detected by neutron diffraction. $Ce_xLa_{1-x}B_6$ for $x < 0.8$ and NpO_2 are the two typical examples of such an unusual ordering.

3.3.1 $Ce_{0.7}La_{0.3}B_6$

By doping La into a typical AFQ compound CeB_6, which will be described afterwards in detail, a mysterious ordered phase appears. The magnetic phase diagram of $Ce_{0.7}La_{0.3}B_6$ is shown in Fig. 10, which is slightly different from the phase diagram of CeB_6 [21, 22]. Phase I is a paramagnetic phase. Phase II and III are AFQ and AFM-dipole ordered phases, respectively. These phases are commonly observed in a wide concentration range of $Ce_xLa_{1-x}B_6$. For $x < 0.8$, there appears a mysterious ordered phase at zero field, which has been called phase IV. Although there is a clear

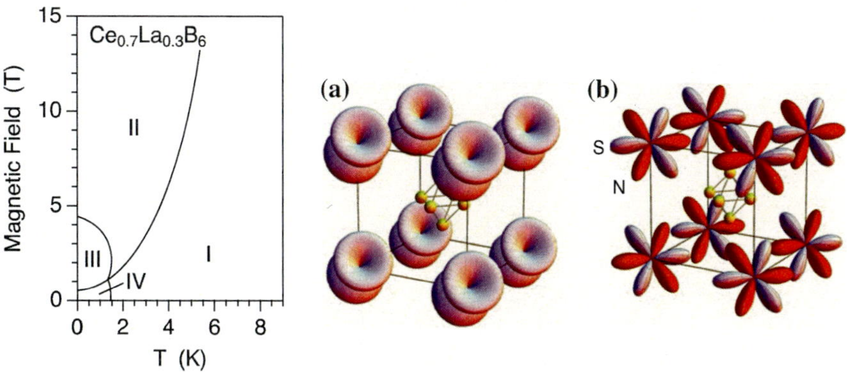

Fig. 10 Magnetic phase diagram of $Ce_{0.7}La_{0.3}B_6$ and schematic illustrations of **a** anisotropic charge distributions and **b** magnetization densities in the $(T_x^\beta + T_y^\beta + T_z^\beta)$-type AF-octupole ordered state

anomaly in specific heat and in magnetic susceptibility, no magnetic dipole order has been observed. What was proposed from various experimental studies was a T^β-type antiferromagnetic octupole ordering [23].

It is well established that the ground state of a Ce ion ($J = 5/2$) in the cubic CEF is the Γ_8-quartet, which are expressed as

$$|\alpha\rangle = \left|\frac{1}{2}\right\rangle, \tag{28}$$

$$|\beta\rangle = \left|-\frac{1}{2}\right\rangle, \tag{29}$$

$$|\gamma\rangle = \sqrt{\frac{5}{6}}\left|\frac{5}{2}\right\rangle + \sqrt{\frac{1}{6}}\left|-\frac{3}{2}\right\rangle, \tag{30}$$

$$|\delta\rangle = \sqrt{\frac{1}{6}}\left|\frac{3}{2}\right\rangle + \sqrt{\frac{5}{6}}\left|-\frac{5}{2}\right\rangle. \tag{31}$$

In a normal AFM-dipole order, this quartet splits into four singlets. In an AFQ order, the quartet splits into two Kramers doublets. For example, a doublet of $(|\beta\rangle - i|\gamma\rangle)/\sqrt{2}$ and $(|\delta\rangle - i|\alpha\rangle)/\sqrt{2}$ are the eigenfunctions of O_{xy} with an eigenvalue of $+1$. Another doublet of $(|\beta\rangle + i|\gamma\rangle)/\sqrt{2}$ and $(|\delta\rangle + i|\alpha\rangle)/\sqrt{2}$ are also the eigenfunctions of O_{xy}, but with an eigenvalue of -1. When these two doublets are ordered alternately, an O_{xy}-AFQ order is formed. This state is actually realized in CeB_6 at zero field. We can also make another type of linear combinations, with singlet–doublet–singlet levels, where the ground state is an eigenfunction of $(T_x^\beta + T_y^\beta + T_z^\beta)$ [23]. When the neighbouring ion has an opposite eigenvalue, we call it a T^β-type AF-octupole (AFO) order. Since these are also the eigenstates of $(O_{yz} + O_{zx} + O_{xy})$ with the same eigenvalue, a ferro-type quadrupole (FQ) order

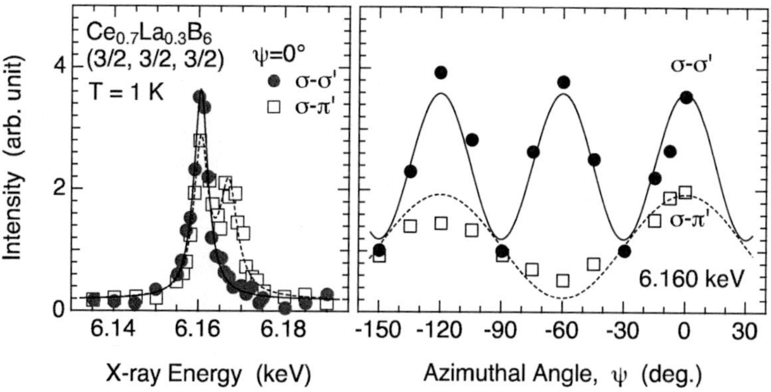

Fig. 11 Energy dependence and azimuthal angle dependence of the resonant intensity of the (3/2, 3/2, 3/2) superlattice reflection in Ce$_{0.7}$La$_{0.3}$B$_6$ [25]

is simultaneously realized, which has been detected by a dilatometric measurement [24]. This situation, where the T^β-type AFO order and the FQ order coexist simultaneously, is schematically illustrated in Fig. 10a, b.

The AFO ordered state was directly detected by RXD [25]. The data is shown in Fig. 11. The resonance signal, which appear below 1.5 K, is dominated by the $E2$ resonance. In particular, no $E1$ resonance is observed for σ–σ', which strongly suggests that the order parameter is not rank-2, but is a higher order multipole.

The AFO ordered structure in Fig. 10b is described by a propagation vector $\kappa = (1/2, 1/2, 1/2)$. The lattice parameter of the cubic unit cell is doubled. Then, the structure factor should be calculated by including the eight Ce ions shown in Fig. 10b, which are divided into two sublattices, A and B, with opposite values of the octupole moment. The $e^{i\kappa \cdot \mathbf{r}}$ factor is $+1$ for Ce ions at $(0, 0, 0)$, $(1, 1, 0)$, $(1, 0, 1)$, and $(0, 1, 1)$ on the A sublattice, whereas it is -1 at $(1, 0, 0)$, $(0, 1, 0)$, $(0, 0, 1)$, and $(1, 1, 1)$ on the B sublattice. Therefore, the octupole structure factor $Z^{(3)}$ is proportional to the difference of the octupole moment between the A and B sublattices. The next thing we need to take into account is the domain distribution. There are four domains, where the order parameter is expressed by $(T_x^\beta + T_y^\beta + T_z^\beta)$, $(-T_x^\beta + T_y^\beta + T_z^\beta)$, $(T_x^\beta - T_y^\beta + T_z^\beta)$, and $(T_x^\beta + T_y^\beta - T_z^\beta)$. By calculating the structure factors for these four domains and by summing up the squares of them with the ratio of domain populations, we have a calculated scattering intensity [26]. The lines in Fig. 11 show calculated azimuthal angle dependences by assuming a domain ratio of 3:1:1:1 [4]. The agreement between the experiment and calculation is satisfactory. The reason for the three times larger population for the $(T_x^\beta + T_y^\beta + T_z^\beta)$ domain, although it is unclear yet, might somehow be associated with the (1 1 1) surface of the sample used in the experiment. Another experiment by studying the incident and final polarization dependences also supports the T^β-AFO order [11]. An uniform lattice distortion along the [111] axis, as expected from the FQ order shown in Fig. 10a, has also been detected by X-ray diffraction [27]. The lattice is elongated along the [111] axis, and the crystal symmetry in the AFO phase is determined to be rhombohedral.

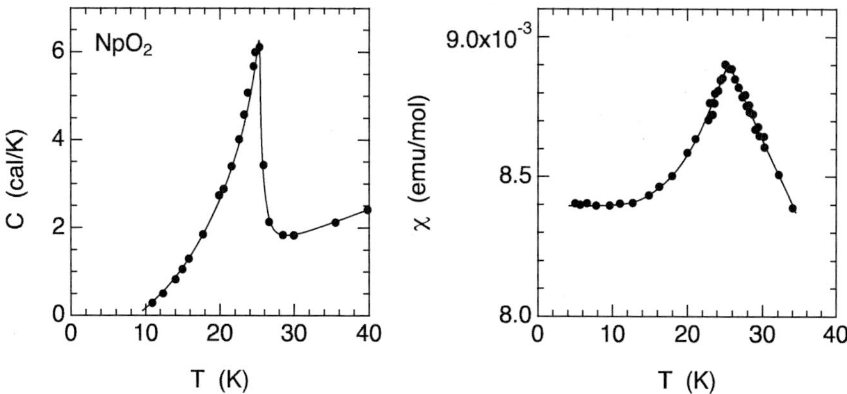

Fig. 12 Temperature dependence of magnetic susceptibility and specific heat of NpO_2 [29, 30]

3.3.2 Triple-q Magnetic Octupole Order in NpO_2

The phase transition in NpO_2 has also been a long mystery [28]. Specific heat and magnetic susceptibility clearly shows an anomaly as shown in Fig. 12, which strongly suggests an AFM order, no magnetic order has been detected by neutron diffraction [29, 30]. There are some studies reporting that the time reversal symmetry is broken. This has been another case of hidden order.

Figure 13 shows the results of RXD experiment, performed at the M_4-edge of Np, using the $3d - 5f$ transition [31]. Since the resonance is $E1$, the signal contains information up to rank-2. The azimuthal angle dependence is well explained by assuming an triple-q AFQ order as shown in Fig. 13d. However, this result does not mean that the primary order parameter is the AFQ. As in CeB_6, the CEF ground state of NpO_2 is also a Γ_8-quartet, which allows 15 degrees of freedom from rank-1 to rank-3. If AFQ-order were the case, the Γ_8-quartet ground state will be split into two doublets with Kramers's degeneracy. The time reversal symmetry will be preserved, which is contradictory to the experiment. In the case of NpO_2, the observed AFQ order can be understood by assuming an triple-q AFO order of T^β-octupole. If we write the T^β-type octupole of Np as $\langle T^\beta \rangle = \langle \mu_x T_x^\beta + \mu_y T_y^\beta + \mu_z T_z^\beta \rangle$, the octupole moment at the j-th Np site is described by using

$$\mu_j = \sum_{i=1}^{3} \mathbf{m}_i e^{i\mathbf{q}_i \cdot \mathbf{r}_j}, \tag{32}$$

where $\mathbf{q}_1 = 2\pi(1, 0, 0)$, $\mathbf{q}_2 = 2\pi(0, 1, 0)$, $\mathbf{q}_3 = 2\pi(0, 0, 1)$, $\mathbf{m}_1 = (1, 0, 0)$, $\mathbf{m}_2 = (0, 1, 0)$, and $\mathbf{m}_3 = (0, 0, 1)$. Since $\mathbf{q}_i$ is parallel to the Fourier component $\mathbf{m}_i$, this structure is called longitudinal. The resultant structure of octupole moments are shown in Fig. 13e.

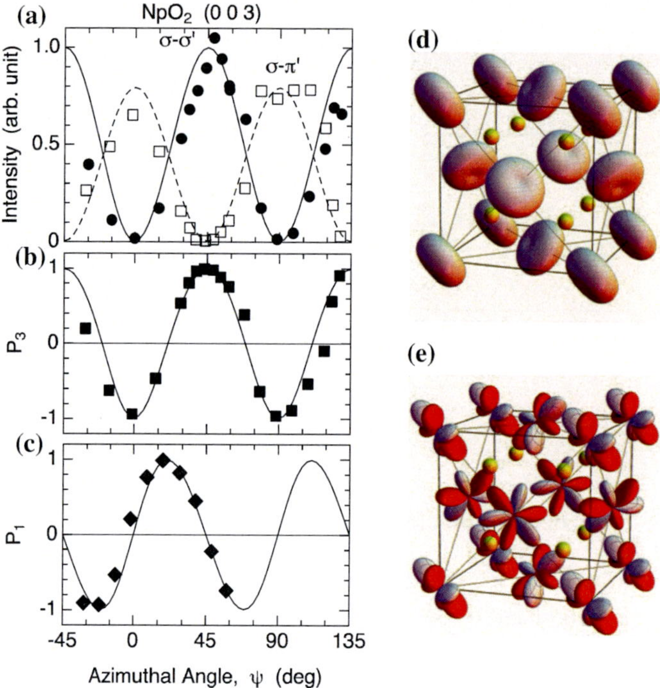

Fig. 13 **a** Azimuthal angle dependences of the $E1$ resonant intensity of the (0 0 3) reflection at the M_4-edge of NpO_2 (3.846 keV) [31]. **b, c** Azimuthal angle dependences of the Stokes parameters P_3 and P_1 obtained from **a**. **d** Schematic illustration of the ordered structure of the quadrupole moments which is coupled with the AFO order. **e** Ordered structure of magnetization densities in the triple-**q** T^β-AFO phase of NpO_2

As in the case of $Ce_{0.7}La_{0.3}B_6$, this octupole order accompanies an quadrupole moment $\langle O_5 \rangle = \langle \mu_x O_{yz} + \mu_y O_{zx} + \mu_z O_{xy} \rangle$, which is also shown in Fig. 13d. The AFO order accompanies an triple-**q** AFQ order. The experiment detects this AFQ order. The azimuthal angle dependence for the triple-**q** AFQ order is calculated in the same way as before. By summing up the scattering amplitudes from the four Np ions in the unit cell, we obtain the following $E1$ structure factor:

$$\hat{F}_{E1} = \langle O_{xy} \rangle \frac{\sqrt{3}}{2} \begin{pmatrix} -\sin 2\psi & \sin\theta\cos 2\psi \\ -\sin\theta\cos 2\psi & -\sin^2\theta\sin 2\psi \end{pmatrix}. \tag{33}$$

Only $\langle O_{xy} \rangle$ remains in the structure factor. The σ–σ' and σ–π' intensities vary with $\sin^2 2\psi$ and $\cos^2 2\psi$. This result is in good agreement with the experimental result. There is no domain effect in this structure.

Another point to be noted here is it is valuable to analyze not only σ' or π' polarizations, but also measure the $\pm 45°$ linear polarizations. From the scattered intensities for σ' and π', we know the Stokes parameter P_3'.

$$P_3' = \frac{I_{\sigma'} - I_{\pi'}}{I_{\sigma'} + I_{\pi'}}. \tag{34}$$

At $\psi = 0$ the scattered beam is π-polarized ($P_3' = -1$), since $F_{\sigma\sigma'}$ vanishes and $F_{\sigma\pi'}$ is non-zero. At $\psi = 45$ the scattered beam is σ-polarized ($P_3' = 1$).

By contrast, the data show that there is no difference between the intensities of σ' and π' at $\psi = 22.5°$ and $\psi = -22.5°$. However, this does not mean $\psi = 22.5°$ and $\psi = -22.5°$ are equivalent. They are different, which is reflected in the Stokes parameter P_1'. P_1 can be obtained by measuring the intensities at $\phi_A = \pm45°$ and taking the difference:

$$P_1' = \frac{I_{+45} - I_{-45}}{I_{+45} + I_{-45}}. \tag{35}$$

The difference between $\psi = 22.5°$ and $\psi = -22.5°$ arises from the difference in sign of $F_{\sigma\sigma'}$ for $+\psi$ and $-\psi$, which determines to which direction the incident linear polarization rotates. This information can be obtained by measuring the intensities at $\phi_A = \pm45°$. As shown in Fig. 13c, the data further supports our model of the triple-q AFQ ordering.

3.4 Interference Effects

In some cases, the scattering amplitudes of different origins coincide at the same Bragg condition. For example, when there are two different origins and F is written by $F_1 + F_2$, the observed intensity is proportional to $|F_1 + F_2|^2$, and is not the simple sum of $|F_1|^2 + |F_2|^2$. This effect provides us with a valuable chance to investigate the relationship between F_1 and F_2 in detail. Let us see three experimental cases of such situation.

3.4.1 Ferromagnetic Order in EuS

Normally, the resonant scatterings are much smaller than the normal Thomson (charge) scattering from the lattice. Ferromagnetic states are difficult to detect, since the Bragg scattering from the magnetic order ($\propto |F_M|^2$) overlaps with that of the charge scattering from the lattice ($\propto |F_C|^2$). In other words, $|F_C|^2$ is much larger than $|F_M|^2$. However, F_M could be extracted from the data if we use the interference between F_C and F_M. When F_C and F_M coexist, the structure factor is expressed by

$$F(E) = F_{C,0} + F_{C,r}\frac{e^{i\gamma_1}}{E - \Delta + i\Gamma_1} + F_M\frac{e^{i\gamma_2}}{E - \Delta + i\Gamma_2}, \tag{36}$$

where $F_{C,0}$ represents the nonresonant term in the total structure factor of the lattice, $F_{C,r}$ and F_M the resonant structure factor of the magnetic ion for the rank-0

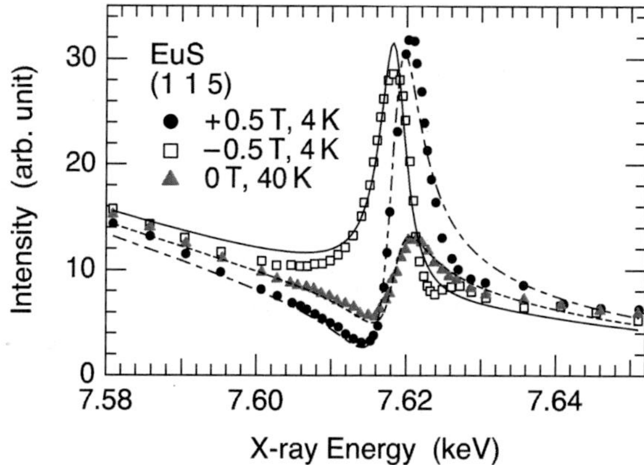

Fig. 14 Energy dependence of the 115 reflection intensity of EuS [32]. The fitting curve for the 40 K data, in the paramagnetic state, is obtained by assuming $\Delta = 7.6185$ keV, $\Gamma_1 = 2.6$ eV, $\gamma_1 = -1.75$, and $F_{C,\mathrm{r}} = 1.8$. Those for the data at 4 K in the ferromagnetic state are calculated by assuming $\Gamma_2 = 2.2$ eV, $\gamma_2 = 0.72$, and $F_M = \pm 3.6$ depending on the field direction, in addition to the same $F_{C,\mathrm{r}}$ term as that at 40 K. The nonresonant term is assumed to be $F_{C,0} = 2.9 - 31(E - \Delta)$, where the energy dependence is due to extrinsic absorption effects

(charge) and rank-1 (magnetic dipole) moment, respectively. The first and second term constitute F_C, where the second term is usually referred to as an anomalous scattering term. In $|F|^2$, there arises an interference term, which is proportional to $\alpha_{E1}^{(1)}(E) F_C^* F_M$. The ratio of the interference term to $|F_C|^2$ is $\sim |F_M/F_C|$, whereas that of pure magnetic signal to $|F_C|^2$ is $\sim |F_M/F_C|^2$. If $|F_C|^2$ is small enough so that the interference term is detectable, we can extract the ferromagnetic signal from the total intensity in which the charge scattering is superimposed.

EuS, which crystallizes in a NaCl-type structure, is a ferromagnet with $T_C = 16.5$ K. At the L_2-edge of Eu, the 2θ angle of the 115 reflection is very close to $90°$. With $a = 5.96$ Å, the 2θ angle of the 115 reflection at 7.62 keV (L_2-edge) is $90.35°$, with which the charge scattering intensity for $\pi-\pi'$ is reduced by a factor of $\cos^2 2\theta = 3.7 \times 10^{-5}$. When we measure the Bragg peak at the L_2-edge in the ferromagnetic state by applying a magnetic field along the direction perpendicular to the scattering plane, $\pi-\pi'$ scattering occurs in the $E1$ process. These two scatterings interfere with each other as shown in Fig. 14. When the field direction is reversed, the direction of the ferromagnetic moment is also reversed, and F_M changes its sign. As a result, the energy spectrum changes drastically. The basic structure of the spectrum can be understood by Eq. (36). The fitting parameters are written in the caption of Fig. 14.

3.4.2 Magnetic Field Induced Multipoles in CeB$_6$

CeB$_6$ is one of the most well known f electron compounds that has attracted researchers for more than 30 years because of its mysterious ordered phases. Now, the

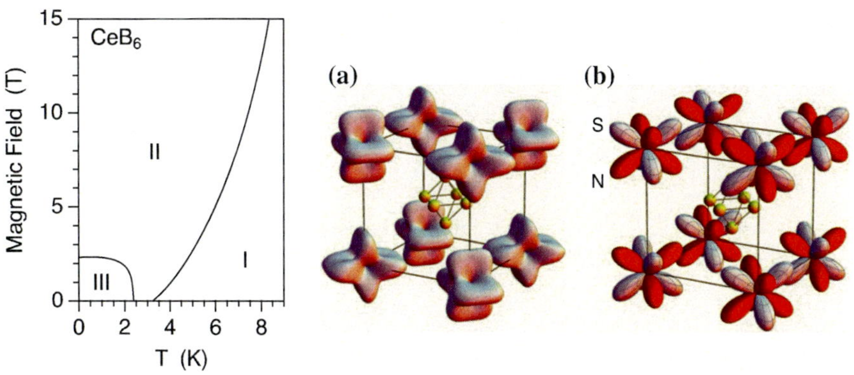

Fig. 15 Magnetic phase diagram of CeB$_6$ [33] and schematic illustrations of **a** anisotropic charge distributions in the O_{xy}-type AFQ ordered phase and **b** magnetization densities of the T_{xyz}-AFO induced in magnetic fields on the O_{xy}-AFQ state

AFQ order is established by various experimental and theoretical studies [22]. At zero field, CeB$_6$ exhibits successive phase transitions at $T_Q = 3.3$ K and $T_N = 2.3$ K. Below $T_Q = 3.3$ K, an AFQ order, where the O_{xy}-quadrupole is an order parameter, occurs as shown by the anisotropic charge distributions in Fig. 15.

If we apply a magnetic field along the z-axis in the AFQ ordered state, the T_{xyz}-type magnetic octupole is induced [34]. This situation is also explained in Fig. 2. One of the mysteries in CeB$_6$ has been why T_Q increases with increasing field. Magnetic field normally splits the degeneracy of the system by the Zeeman effect. In case of CeB$_6$ the Γ_8 CEF ground state is split into four singlets by the field. Then, the ordered phase will be suppressed, and usually the transition temperature decreases. This has been a serious problem for a long time. One promising scenario for explaining this phenomenon is to take into account the AFO interaction between the field-induced T_{xyz}-octupole, which stabilizes the ordered phase.

In Fig. 16, we show the energy spectra of the $(3/2, 3/2, 1/2)$ reflection, corresponding to $\kappa = (1/2, 1/2, 1/2)$, of CeB$_6$ at the L_3 edge. The field is applied along the $[\bar{1}10]$ direction, which is paralell to the X axis in Fig. 3. In this field direction, the AFQ order parameter changes to $O_{yz} - O_{zx}$ in magnetic fields, and the J_z-AFM is induced in addition to T_{xyz}-AFO and uniform magnetization along the field direction [35]. Thus, various kinds of multipole moments are induced and interference among resonant scatterings from these moments takes place. In Fig. 16, the resonant signal at 5.726 keV at zero field is purely due to the AFQ order, which disappears above T_Q. With increasing the field, the spectrum becomes complex and exhibits a drastic difference when the field direction is reversed. To discuss the field reversal asymmetry in the spectrum, we show in Fig. 16c, d the average and difference intensity, $I_{ave} = (I_+ + I_-)/2$ and $\Delta I = (I_+ - I_-)/2$, respectively, at ± 5 T. To analyse these spectra, we need $\alpha_{E1}^{(v)}(E)$ and $\alpha_{E2}^{(v)}(E)$ for different multipole rank v. At the present stage, however, we need to rely on theoretical calculation. The calculated spectral functions are shown in Fig. 17.

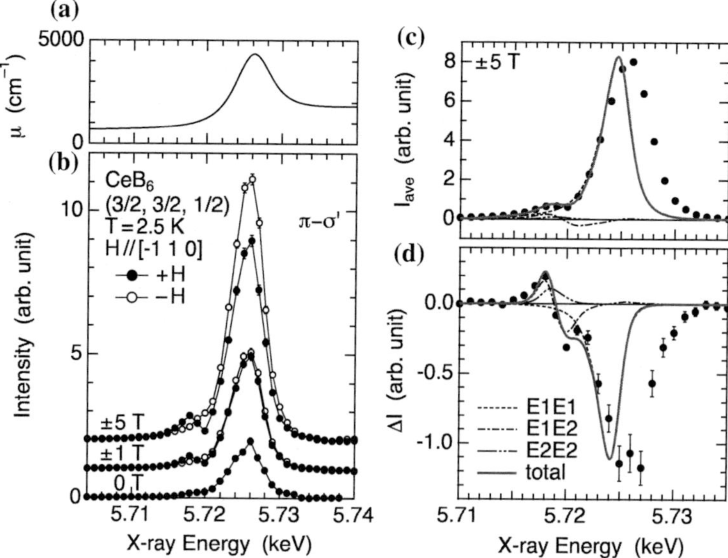

Fig. 16 **a** Energy dependence of the X-ray absorption coefficient of CeB$_6$. **b** Energy dependence of the (3/2, 3/2, 1/2) reflection in magnetic fields along [$\bar{1}$10] direction. **c**, **d** Averaged and difference intensity, respectively, for ±5 T with calculated spectra. The *curves* indicated by E1E1, E1E2, and E2E2 show contributions from $F_{E1}^{*}F_{E1}$, $F_{E1}^{*}F_{E2} + F_{E2}^{*}F_{E1}$, and $F_{E2}^{*}F_{E2}$, respectively. Data are taken from [35]

We see that the spectral functions shown in Fig. 17 have different line shapes. Of course, none of these cannot be expressed by a single Lorentzian function $e^{i\gamma}/(E - \Delta + i\Gamma)$. The single Lorentzian function we used in the analyses in the present chapter is a first step data treatment when we do not know more reliable spectral function. By introducing the phase parameter γ, we may use this function as an empirical fitting function.

By changing the external conditions of temperature, magnetic field, and field direction, the multipole moment values $\langle z_\mu^{(v)} \rangle$ change. Another important point in our analysis is that the odd rank tensor changes its sign when the field direction is reversed. This is because, in the AFQ ordered phase, the quadrupole moments are fixed and the magnetic moments are paramagnetic. The even rank tensors do not change signs with the field reversal. Therefore, the intensity at $+H$ and $-H$ are expressed as

$$I(E, \pm H) = \left| F_{E1}^{(2)}(E, H) + F_{E2}^{(2)}(E, H) + F_{E2}^{(4)}(E, H) \right.$$

$$\left. \pm i\{F_{E1}^{(1)}(E, H) + F_{E2}^{(1)}(E, H) + F_{E2}^{(3)}(E, H)\} \right|^2. \quad (37)$$

The difference and average intensities can be calculated from this expression. We see from the analysis, after a minor approximation, that $I_{\text{ave}}(\Delta_1) \propto (F_{E1}^{(2)})^2$,

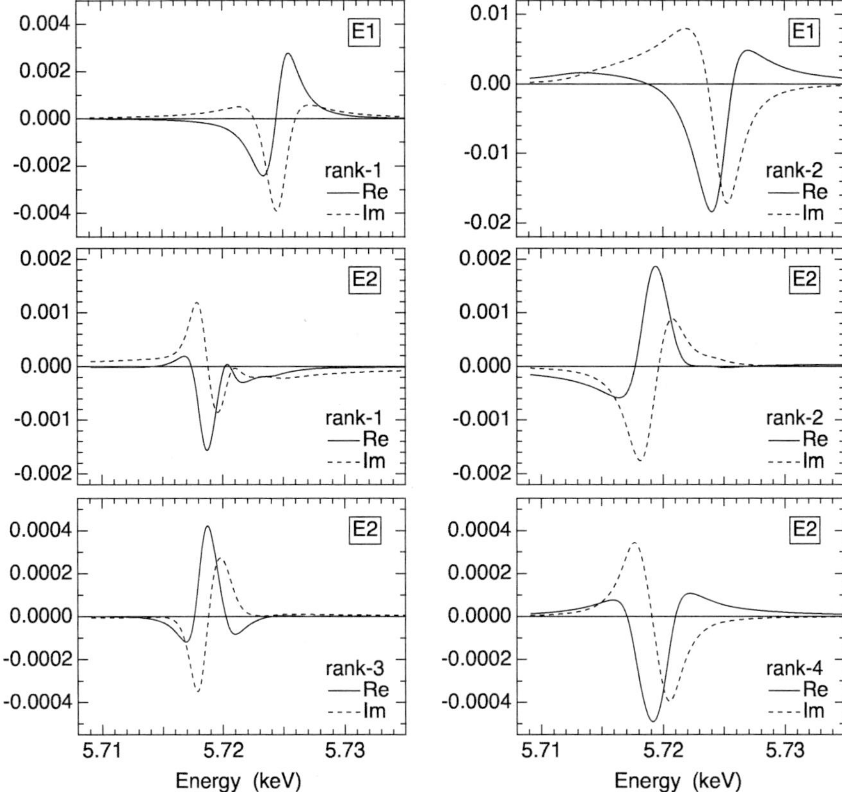

Fig. 17 Calculated spectral functions $\alpha_{E1}^{(\nu)}(E)$ and $\alpha_{E2}^{(\nu)}(E)$ for CeB$_6$ [7]

$\Delta I(\Delta_1) \propto F_{E1}^{(1)} F_{E1}^{(2)}$, and $\Delta I(\Delta_2) \propto F_{E1}^{(2)} F_{E2}^{(3)}$. Then, it is possible to extract the field dependences of $F_{E1}^{(2)}$, $F_{E1}^{(1)}$, and $F_{E2}^{(3)}$, which are proportional to $\langle O_{yz} - O_{zx} \rangle$, $\langle J_z \rangle$, and $\langle T_{xyz} \rangle$, respectively. Finally, we can plot the field dependence of these multipole order parameters as shown in Fig. 18 [35]. Although such a plot for $\langle J_z \rangle$ is normally performed by neutron diffraction, those for $\langle O_{yz} - O_{zx} \rangle$ and $\langle T_{xyz} \rangle$ are extremely difficult. This point is a great advantage of RXD.

3.4.3 Field Induced Charge Order in SmRu$_4$P$_{12}$

The filled skutterudite compounds, which crystallizes in a body centred cubic structure with a space group $Im\bar{3}$, as shown in Fig. 19a, exhibit various kinds of interesting ordered phases. They are strongly associated with the characteristic of Fermi surface nesting with $\mathbf{q} = (1, 0, 0)$ in this series of compounds [36]. An interesting point is that the f-electron's degree of freedom is coupled with the Fermi surface. Typical examples can be seen in PrRu$_4$P$_{12}$ and PrFe$_4$P$_{12}$. In these compounds, the ordered

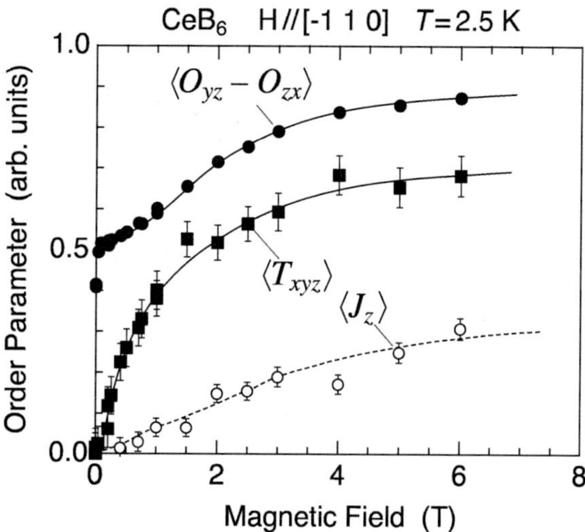

Fig. 18 Magnetic field dependences of the $\langle O_{yz} - O_{zx} \rangle$-AFQ, $\langle J_z \rangle$-AFM, and $\langle T_{xyz} \rangle$-AFO moments in the AFQ phase of CeB$_6$ [35]

phases are nonmagnetic [37]. SmRu$_4$P$_{12}$, by contrast, exhibits an AFM dipole order with $T_N = 16.5$ K, which is established by neutron diffraction [38]. Below T_N, there appear a mysterious ordered phase, which is more enhanced with increasing magnetic field. The magnetic phase diagram is shown in Fig. 19b [39]. This field induced ordered phase in SmRu$_4$P$_{12}$, named phase II, has been a long mystery.

In phase II, it was discovered that lattice distortion with $\mathbf{q} = (1, 0, 0)$ is induced in magnetic fields, which can be observed as an appearance of energy independent Thomson (charge) scattering as shown in Fig. 19c [40]. Superimposed on the Thomson scattering is a resonant scattering, especially at the $E2$ energy. The response is reversed by reversing the field direction. This shows that the Thomson scattering, with the π–π' process, interfere with the π–π' resonant scattering from magnetic order. Let us analyze the energy spectrum around the $E2$ resonance using the following structure factor:

$$F_{E2,\pi\pi'}(E) = F_{C,\pi\pi'} + \alpha_{E2}(E)F_{M,\pi\pi'}, \tag{38}$$

where we assume that $\alpha_{E2}(E)$ is expressed as $e^{i\gamma}/(E - \Delta + i\Gamma)$. In the present case, F_C changes its sign by the field reversal, whereas the sign of $F_{M,\pi\pi'}$ does not change. We can explain the interference effect satisfactorily as shown by the calculated curves in Fig. 19c.

The average and difference intensities are enhanced simultaneously in phase II as shown in Fig. 19d. The average intensity is mostly due to $|F_{C,\pi\pi'}|^2$. The difference intensity reflects $(F_{C,\pi\pi'}^* F_{M,\pi\pi'} + F_{M,\pi\pi'}^* F_{C,\pi\pi'})$. Then, we can deduce the fact that $F_{M,\pi\pi'}$ is induced by applying a field in phase II. Since $F_{M,\pi\pi'}$ reflects an AFM component perpendicular to the scattering plane, this result shows that the AFM component in phase II is parallel to the magnetic field. This is quite an anomalous state because normal AFM component prefers to be perpendicular to the field direction. The AFM component parallel to the field direction shows that there appear a long

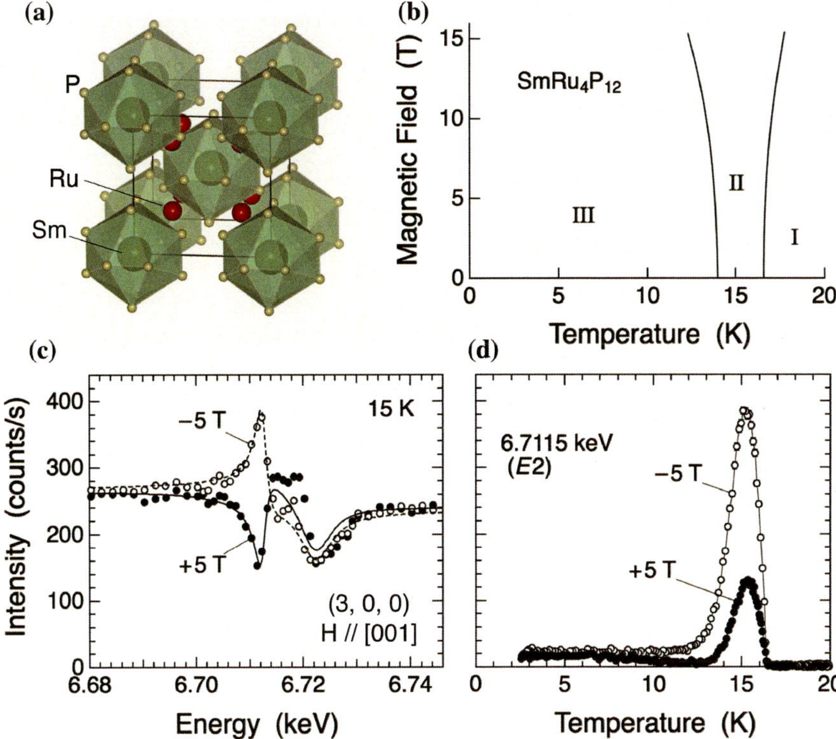

Fig. 19 a Crystal structure of SmRu$_4$P$_{12}$. **b** Magnetic phase diagram of SmRu$_4$P$_{12}$, consisting of the paramagnetic phase I, intermediate AFM phase II, and the low-temperature AFM phase III [39]. **c** Energy spectrum of the 300 reflection in magnetic fields of ±5 T. **d** Temperature dependence of the 300 peak intensity at 6.7115 keV (E2) in magnetic fields of ±5 T [40]

and short magnetic moments along the field direction. A plausible explanation for this state is that the CEF state of the f-orbital, which is linked with the moment value, is coupled with the nesting property of the conduction electrons, as has been theoretically predicted [41]. This state can be viewed as a hexadecapole order of f electrons without losing the cubic symmetry. This induces a density modulation of the conduction electron and the lattice distortion with $\mathbf{q} = (1, 0, 0)$.

4 Summary

- We reviewed the concept of multipole moments in localized f electron systems. Each of the multipole moments can be a primary order parameter independently. In magnetic fields, however, through spin-orbit interaction, they are coupled with each other and a number of multipole moments can be induced simultaneously.

- We described how the multipole order parameters can be detected by resonant X-ray diffraction. Using $E1$ and $E2$ resonances, multipole moments up to rank-2 (quadrupole) and rank-4 (hexadecapole), respectively, can be detected. It is emphasized that we need to use rank dependent spectral functions. Since it is difficult to know experimentally, we introduced a experimental fitting function of Lorentzian oscillators with phase factors.
- We described some experimental cases of observing multipole orderings: magnetic dipole order in Ho metal, antiferro-quadrupole order in DyB_2C_2, and antiferromagnetic octupole order in $Ce_{0.7}La_{0.3}B_6$ and NpO_2. By utilizing the interference effect among coexisting scattering amplitudes at a Bragg spot, e.g., by measuring the field reversal effect, we can extract individual scattering amplitudes as in the case of ferromagnetic order in EuS. Extraction of the field dependence of antiferro-dipole, -qudrupole, and -octupole moments in CeB_6 is a more interesting example. In $SmRu_4P_{12}$, the interference between Thomson scattering from lattice distortion and resonant scattering from the field-induced antiferromagnetic dipole order provided a valuable information on the magnetic structure in magnetic fields.
- On the other hand, with RXD, it is still difficult to determine the absolute values of the order parameters. We do not know how large the multipole moments are. The unit of vertical axis of Fig. 18 is arbitraly. This will be a problem to be studied in future.

Appendix

Geometrical Factors

We summarize here the geometrical factors $P_{E2,\mu}^{(v)}$ for $z_{E2,\mu}^{(v)}$ at $E2$ resonance [4, 6, 7]. $\varepsilon_{j,j',j''}$ is the Levi–Civita tensor, where $\varepsilon_{j,j',j''} = 1$ when $(j, j', j'') = (x, y, z)$, (y, z, x), or (z, x, y), and $\varepsilon_{j,j',j''} = -1$ when $(j, j', j'') = (x, z, y)$, (y, x, z), or (z, y, x).

$$P_{E2,1}^{(0)} = \frac{3}{4\sqrt{5}}\left\{(\mathbf{k}' \cdot \mathbf{k})(\boldsymbol{\varepsilon}' \cdot \boldsymbol{\varepsilon}) + (\mathbf{k}' \cdot \boldsymbol{\varepsilon})(\boldsymbol{\varepsilon}' \cdot \mathbf{k})\right\} \tag{39}$$

$$P_{E2,j}^{(1)} = -i\frac{3}{4\sqrt{10}}\left\{(\boldsymbol{\varepsilon}' \cdot \boldsymbol{\varepsilon})(\mathbf{k}' \times \mathbf{k}) + (\mathbf{k}' \cdot \mathbf{k})(\boldsymbol{\varepsilon}' \times \boldsymbol{\varepsilon})\right.$$
$$\left. +(\mathbf{k}' \cdot \boldsymbol{\varepsilon})(\boldsymbol{\varepsilon}' \times \mathbf{k}) + (\boldsymbol{\varepsilon}' \cdot \mathbf{k})(\mathbf{k}' \times \boldsymbol{\varepsilon})\right\}_j ; \quad (j = 1, 2, 3) \tag{40}$$

$$P_{E2,j}^{(2)} = -\frac{3}{2\sqrt{14}}\left\{(\boldsymbol{\varepsilon}' \cdot \boldsymbol{\varepsilon})K_j(\mathbf{k}, \mathbf{k}') + (\mathbf{k}' \cdot \mathbf{k})K_j(\boldsymbol{\varepsilon}, \boldsymbol{\varepsilon}')\right.$$
$$\left. +K_j(\mathbf{k}' \times \mathbf{k}, \boldsymbol{\varepsilon}' \times \boldsymbol{\varepsilon})\right\} ; \quad (j = 1, 2, 3, 4, 5) \tag{41}$$

$$P_{E2,1}^{(3)} = i\frac{1}{4\sqrt{2}}\sum_{j=1}^{3}\left\{(\mathbf{k}'\times\mathbf{k})_j K_{j+2}(\boldsymbol{\varepsilon}',\boldsymbol{\varepsilon}) + (\boldsymbol{\varepsilon}'\times\boldsymbol{\varepsilon})_j K_{j+2}(\mathbf{k}',\mathbf{k})\right.$$

$$\left.+(\mathbf{k}'\times\boldsymbol{\varepsilon})_j K_{j+2}(\boldsymbol{\varepsilon}',\mathbf{k}) + (\boldsymbol{\varepsilon}'\times\mathbf{k})_j K_{j+2}(\mathbf{k}',\boldsymbol{\varepsilon})\right\} \tag{42}$$

$$P_{E2,j+1}^{(3)} = i\frac{3}{8}\sqrt{\frac{5}{2}}\left\{(\mathbf{k}'\times\mathbf{k})_j\varepsilon_j'\varepsilon_j + (\boldsymbol{\varepsilon}'\times\boldsymbol{\varepsilon})_j k_j'k_j\right.$$

$$\left.+(\mathbf{k}'\times\boldsymbol{\varepsilon})_j\varepsilon_j'k_j + (\boldsymbol{\varepsilon}'\times\mathbf{k})_j k_j'\varepsilon_j\right\} + \frac{1}{2}P_{E2,j}^{(1)} \; ; \; (j = 1,2,3) \tag{43}$$

$$P_{E2,j+4}^{(3)} = i\frac{3}{16}\sqrt{\frac{3}{2}}\left\{(\mathbf{k}'\times\mathbf{k})_j\sum_{j',j''}\varepsilon_{j,j',j''}(\varepsilon_{j'}'\varepsilon_{j'} - \varepsilon_{j''}'\varepsilon_{j''})\right.$$

$$+(\boldsymbol{\varepsilon}'\times\boldsymbol{\varepsilon})_j\sum_{j',j''}\varepsilon_{j,j',j''}(k_{j'}'k_{j'} - k_{j''}'k_{j''})$$

$$+(\mathbf{k}'\times\boldsymbol{\varepsilon})_j\sum_{j',j''}\varepsilon_{j,j',j''}(\varepsilon_{j'}'k_{j'} - \varepsilon_{j''}'k_{j''})$$

$$\left.+(\boldsymbol{\varepsilon}'\times\mathbf{k})_j\sum_{j',j''}\varepsilon_{j,j',j''}(k_{j'}'\varepsilon_{j'} - k_{j''}'\varepsilon_{j''})\right\} \; ; \; (j = 1,2,3) \tag{44}$$

$$P_{E2,1}^{(4)} = \sqrt{\frac{2}{15}}\left\{\frac{15}{4}(k_x'k_x\varepsilon_x'\varepsilon_x + k_y'k_y\varepsilon_y'\varepsilon_y + k_z'k_z\varepsilon_z'\varepsilon_z) - \sqrt{5}P_{E2,1}^{(0)}\right\} \tag{45}$$

$$P_{E2,2}^{(4)} = -\frac{\sqrt{42}}{4}(k_x'k_x\varepsilon_x'\varepsilon_x + k_y'k_y\varepsilon_y'\varepsilon_y - 2k_z'k_z\varepsilon_z'\varepsilon_z) - \frac{1}{2}\sqrt{\frac{6}{7}}\left\{(\mathbf{k}'\cdot\mathbf{k})K_1(\boldsymbol{\varepsilon}',\boldsymbol{\varepsilon})\right.$$

$$\left.+(\boldsymbol{\varepsilon}'\cdot\boldsymbol{\varepsilon})K_1(\mathbf{k}',\mathbf{k}) + (\mathbf{k}'\cdot\boldsymbol{\varepsilon})K_1(\boldsymbol{\varepsilon}',\mathbf{k}) + (\boldsymbol{\varepsilon}'\cdot\mathbf{k})K_1(\mathbf{k}',\boldsymbol{\varepsilon})\right\} \tag{46}$$

$$P_{E2,3}^{(4)} = \frac{3\sqrt{14}}{4}(k_x'k_x\varepsilon_x'\varepsilon_x - k_y'k_y\varepsilon_y'\varepsilon_y) - \frac{3}{2}\sqrt{\frac{2}{21}}\left\{(\mathbf{k}'\cdot\mathbf{k})K_2(\boldsymbol{\varepsilon}',\boldsymbol{\varepsilon})\right.$$

$$\left.+(\boldsymbol{\varepsilon}'\cdot\boldsymbol{\varepsilon})K_2(\mathbf{k}',\mathbf{k}) + (\mathbf{k}'\cdot\boldsymbol{\varepsilon})K_2(\boldsymbol{\varepsilon}',\mathbf{k}) + (\boldsymbol{\varepsilon}'\cdot\mathbf{k})K_2(\mathbf{k}',\boldsymbol{\varepsilon})\right\} \tag{47}$$

$$P_{E2,j+3}^{(4)} = \frac{2}{8\sqrt{6}}\left\{K_{j+2}(\mathbf{k}',\mathbf{k})\sum_{j',j''}\varepsilon_{j,j',j''}(\varepsilon_{j'}'\varepsilon_{j'} - \varepsilon_{j''}'\varepsilon_{j''})\right.$$

$$+K_{j+2}(\boldsymbol{\varepsilon}',\boldsymbol{\varepsilon})\sum_{j',j''}\varepsilon_{j,j',j''}(k_{j'}'k_{j'} - k_{j''}'k_{j''})$$

$$+K_{j+2}(\mathbf{k}',\boldsymbol{\varepsilon})\sum_{j',j''}\varepsilon_{j,j',j''}(\varepsilon_{j'}'k_{j'} - \varepsilon_{j''}'k_{j''})$$

$$\left.+K_{j+2}(\boldsymbol{\varepsilon}',\mathbf{k})\sum_{j',j''}\varepsilon_{j,j',j''}(k_{j'}'\varepsilon_{j'} - k_{j''}'\varepsilon_{j''})\right\} \; ; \; (j = 1,2,3) \tag{48}$$

$$P_{E2,j+6}^{(4)} = \frac{3}{4\sqrt{42}}\left[\{7k_j'k_j - 3(\mathbf{k}'\cdot\mathbf{k})\}K_{j+2}(\boldsymbol{\varepsilon}',\boldsymbol{\varepsilon})\right.$$

$$+\{7\varepsilon_\mu'\varepsilon_\mu - 3(\boldsymbol{\varepsilon}'\cdot\boldsymbol{\varepsilon})\}K_{j+2}(\mathbf{k}',\mathbf{k})$$

$$+\{7\varepsilon_j'k_j - 3(\boldsymbol{\varepsilon}'\cdot\mathbf{k})\}K_{j+2}(\mathbf{k}',\boldsymbol{\varepsilon})$$

$$\left.+\{7k_j'\varepsilon_j - 3(\mathbf{k}'\cdot\boldsymbol{\varepsilon})\}K_{j+2}(\boldsymbol{\varepsilon}',\mathbf{k})\right] \; ; \; (j = 1,2,3) \tag{49}$$

Scattering Intensity and Polarization Analysis

We use the scattering-amplitude-operator method to analyze the experimental results including polarization analysis [42]. This method is useful for describing the observed intensity at the detector in a general scattering geometry shown in Fig. 3, where the incident linear polarization is rotated to an arbitrary angle η using phase plates and a crystal analyzer system is inserted to analyze the polarization state of the scattered X-ray.

All the information with respect to the target sample is summarized in a 2×2 matrix $\hat{F}$ consisting of four elements of the scattering amplitude for $\sigma-\sigma'$, $\sigma-\pi'$, $\pi-\sigma'$, and $\pi-\pi'$ processes:

$$\hat{F} = \begin{pmatrix} F_{\sigma\sigma'} & F_{\pi\sigma'} \\ F_{\sigma\pi'} & F_{\pi\pi'} \end{pmatrix}. \tag{50}$$

This determines the state of the target system. By using the identity matrix $\hat{I}$ and the Pauli matrix $\hat{\sigma}$, $\hat{F}$ can generally be expressed as

$$\begin{aligned} \hat{F} &= \beta\hat{I} + \boldsymbol{\alpha} \cdot \hat{\sigma} \\ &= \begin{pmatrix} \beta + \alpha_3 & \alpha_1 - i\alpha_2 \\ \alpha_1 + i\alpha_2 & \beta - \alpha_3 \end{pmatrix}, \end{aligned} \tag{51}$$

where the parameters β and $\boldsymbol{\alpha} = (\alpha_1, \alpha_2, \alpha_3)$ are

$$\begin{aligned} \beta &= (F_{\sigma\sigma'} + F_{\pi\pi'})/2, \\ \alpha_1 &= (F_{\pi\sigma'} + F_{\sigma\pi'})/2, \\ \alpha_2 &= i(F_{\pi\sigma'} - F_{\sigma\pi'})/2, \\ \alpha_3 &= (F_{\sigma\sigma'} - F_{\pi\pi'})/2. \end{aligned} \tag{52}$$

Next, to calculate the scattering cross-section, information on the incident photon state is necessary. This is described by the density matrix

$$\hat{\mu} = (\hat{I} + \mathbf{P} \cdot \hat{\sigma})/2, \tag{53}$$

where the Stokes vector $\mathbf{P} = (P_1, P_2, P_3)$ represents the polarization state of the incident photon. P_1, P_2, and P_3 represent $\pm 45°$, left or right handed circular, and σ or π polarization state, respectively [42]. For example, $\mathbf{P} = (0, 0, 1)$ and $(0, 0, -1)$ mean the perfectly σ and π polarized state, i.e., $\eta = 0°$ and $90°$ in Fig. 3, respectively. In general, since the beam is not perfectly polarized, we need to consider a situation with $P_1^2 + P_2^2 + P_3^2 < 1$.

Once we know the matrix $\hat{F}$, the scattering cross-section $(d\sigma/d\Omega)$ can be calculated by

$$\left(\frac{d\sigma}{d\Omega}\right) = \mathrm{Tr}\{\hat{\mu}\hat{F}^\dagger \hat{F}\}$$
$$= \beta^\dagger\beta + \boldsymbol{\alpha}^\dagger \cdot \boldsymbol{\alpha} + \beta^\dagger(\mathbf{P}\cdot\boldsymbol{\alpha}) + (\mathbf{P}\cdot\boldsymbol{\alpha}^\dagger)\beta + i\mathbf{P}\cdot(\boldsymbol{\alpha}^\dagger\times\boldsymbol{\alpha}). \tag{54}$$

The Stokes vector of the scattered X-ray, $\mathbf{P}'$, can be obtained from

$$\left(\frac{d\sigma}{d\Omega}\right)\mathbf{P}' = \mathrm{Tr}\{\hat{\mu}\hat{F}^\dagger\hat{\sigma}\hat{F}\}$$
$$= \beta^\dagger\boldsymbol{\alpha} + \boldsymbol{\alpha}^\dagger\beta - i(\boldsymbol{\alpha}^\dagger\times\boldsymbol{\alpha}) + \beta^\dagger\beta\mathbf{P} - i\beta^\dagger(\mathbf{P}\times\boldsymbol{\alpha})$$
$$+ i(\mathbf{P}\times\boldsymbol{\alpha}^\dagger)\beta + \boldsymbol{\alpha}^\dagger(\mathbf{P}\cdot\boldsymbol{\alpha}) - \boldsymbol{\alpha}^\dagger\times(\mathbf{P}\times\boldsymbol{\alpha}). \tag{55}$$

As an example, we show a case of fundamental Bragg reflection from the crystal lattice by non-resonant Thomson scattering. In this case, the scattering amplitude is written as

$$\hat{F} = -F_c\begin{pmatrix} 1 & 0 \\ 0 & \cos 2\theta \end{pmatrix}, \tag{56}$$

where F_c represents the structure factor of the reflection. From Eq. (52), $\beta = -F_c\cos^2\theta$, $\alpha_1 = \alpha_2 = 0$, and $\alpha_3 = -F_c\sin^2\theta$ are obtained. Then, from Eq. (54), the scattering cross-section becomes

$$\left(\frac{d\sigma}{d\Omega}\right) = |F_c|^2\left\{1 - \frac{1}{2}(1 - P_3)\sin^2 2\theta\right\}, \tag{57}$$

and from Eq. (55), $\mathbf{P}'$ satisfies

$$\left(\frac{d\sigma}{d\Omega}\right)P_1' = |F_c|^2 P_1 \cos 2\theta,$$

$$\left(\frac{d\sigma}{d\Omega}\right)P_2' = |F_c|^2 P_2 \cos 2\theta,$$

$$\left(\frac{d\sigma}{d\Omega}\right)P_3' = |F_c|^2\left\{P_3 + \frac{1}{2}(1 - P_3)\sin^2 2\theta\right\}. \tag{58}$$

The Stokes vector of the incident X-ray, with the polarization angle η in Fig. 3, is written as

$$\mathbf{P} = P_L(\sin 2\eta,\ 0,\ \cos 2\eta), \tag{59}$$

where P_L represents the degree of linear polarization of the incident X-ray.

The intensity after diffracted by the analyzer crystal is also described by the Thomson scattering, and Eq. (57) is applied. It is noted, however, that P_3 must be transformed to the value for the diffraction at the analyzer, which we write as P_{3A}:

$$P_{3A} = P_1' \sin 2\phi_A + P_3' \cos 2\phi_A. \tag{60}$$

Finally, the intensity at the detector is expressed as

$$I = K \left(\frac{d\sigma}{d\Omega} \right) \left\{ 1 - \frac{1}{2}(1 - P_{3A}) \sin^2 2\theta_A \right\}, \tag{61}$$

where $(d\sigma/d\Omega)$ is the scattering cross-section of the sample expressed by Eq. (57) and K represents a constant factor.

References

1. K.R. Lea, M.J.M. Leask, W.P. Wolf, J. Phys. Chem. Solids **23**, 1381 (1962)
2. H. Kusunose, J. Phys. Soc. Jpn. **77**, 064710 (2008)
3. H. Shiba, O. Sakai, R. Shiina, J. Phys. Soc. Jpn. **68**, 1988 (1999)
4. T. Nagao, J.-I. Igarashi, Phys. Rev. B **74**, 104404 (2006)
5. S.W. Lovesey, E. Balcar, K.S. Knight, J. Fernandez-Rodriguez, Phys. Rep. **411**, 233 (2005)
6. T. Nagao, J.-I. Igarashi, J. Phys. Soc. Jpn. **77**, 084710 (2008)
7. T. Nagao, J.-I. Igarashi, Phys. Rev. B **82**, 024402 (2010)
8. C. Mazzoli, S.B. Wilkins, S. Di Matteo, B. Detlefs, C. Detlefs, V. Scagnoli, L. Paolasini, P. Ghigna, Phys. Rev. B **76**, 195118 (2007)
9. B. Detlefs, S.B. Wilkins, R. Caciuffo, J.A. Paixao, K. Kaneko, F. Honda, N. Metoki, N. Bernhoeft, J. Rebizant, G.H. Lander, Phys. Rev. B **77**, 024425 (2008)
10. T. Inami, S. Michimura, T. Matsumura, J. Phys.: Conf. Ser. **425**, 132011 (2013)
11. T. Matsumura, S. Michimura, T. Inami, T. Otsubo, H. Tanida, F. Iga, M. Sera, Phys. Rev. B **89**, 014422 (2014)
12. G.P. Felcher, G.H. Lander, T. Arai, S.K. Shinha, F.H. Spedding, Phys. Rev. B **13**, 3034 (1976)
13. D.T. Keating, Phys. Rev. **178**, 732 (1969)
14. D. Gibbs, G. Grubel, D.R. Harshman, E.D. Isaacs, D.B. McWhan, D. Mills, C. Vettier, Phys. Rev. B **43**, 5663 (1991)
15. D. Gibbs, D.R. Harshman, E.D. Isaacs, D.B. McWhan, D. Mills, C. Vettier, Phys. Rev. Lett. **61**, 1241 (1988)
16. J.P. Hannon, G.T. Trammell, M. Blume, D. Gibbs, Phys. Rev. Lett. **61**, 1245 (1988)
17. H. Yamauchi, H. Onodera, K. Ohoyama, T. Onimaru, M. Kosaka, M. Ohashi, Y. Yamaguchi, J. Phys. Soc. Jpn. **68**, 2057 (1999)
18. K. Hirota, N. Oumi, T. Matsumura, H. Nakao, Y. Wakabayashi, Y. Murakami, Y. Endoh, Phys. Rev. Lett. **84**, 2706 (2000)
19. T. Matsumura, N. Oumi, K. Hirota, H. Nakao, Y. Murakami, Y. Wakabayashi, T. Arima, S. Ishihara, Y. Endoh, Phys. Rev. B **65**, 094420 (2002)
20. T. Matsumura, D. Okuyama, N. Oumi, K. Hirota, H. Nakao, Y. Murakami, Y. Wakabayashi, Phys. Rev. B **71**, 012405 (2005)
21. T. Tayama, T. Sakakibara, K. Tenya, H. Amitsuka, S. Kunii, J. Phys. Soc. Jpn. **66**, 2268 (1997)
22. Y. Kuramoto, H. Kusunose, A. Kiss, J. Phys. Soc. Jpn. **78**, 072001 (2009)
23. K. Kubo, Y. Kuramoto, J. Phys. Soc. Jpn. **73**, 216 (2004)

24. M. Akatsu, T. Goto, Y. Nemoto, O. Suzuki, S. Nakamura, S. Kunii, J. Phys. Soc. Jpn. **72**, 205 (2003)
25. D. Mannix, Y. Tanaka, D. Carbone, N. Bernhoeft, S. Kunii, Phys. Rev. Lett. **95**, 117206 (2005)
26. H. Kusunose, Y. Kuramoto, J. Phys. Soc. Jpn. **74**, 3139 (2005)
27. T. Inami, S. Michimura, Y. Hayashi, T. Matsumura, M. Sera, F. Iga, Phys. Rev. B **90**, 041108(R) (2014)
28. P. Santini, S. Carretta, G. Amoretti, R. Caciuffo, N. Magnani, G.H. Lander, Rev. Mod. Phys. **81**, 807 (2009)
29. D.W. Osborne, E.F. Westrum Jr., J. Chem. Phys. **21**, 1884 (1953)
30. P. Erdös, G. Solt, Z. Żolnierek, A. Blaise, J.M. Fournier, Physica **102B**, 164 (1980)
31. J.A. Paixão, C. Detlefs, M.J. Longfield, R. Caciuffo, P. Santini, N. Bernhoeft, J. Rebizant, G.H. Lander, Phys. Rev. Lett. **89**, 187202 (2002)
32. D. Hepfeld, O.H. Seeck, J. Voigt, J. Bos, K. Fischer, T. Brückel, Europhys. Lett. **59**, 284 (2002)
33. R.G. Goodrich, D.P. Young, D. Hall, L. Balicas, Z. Fisk, N. Harrison, J.B. Betts, A. Migliori, F.M. Woodward, J.W. Lynn, Phys. Rev. B **69**, 054415 (2004)
34. R. Shiina, H. Shiba, P. Thalmeier, J. Phys. Soc. Jpn. **66**, 1741 (1997)
35. T. Matsumura, T. Yonemura, K. Kunimori, M. Sera, F. Iga, T. Nagao, J.I. Igarashi, Phys. Rev. B **85**, 174417 (2012)
36. H. Harima, J. Phys. Soc. Jpn. **77**(Suppl. A), 114 (2008)
37. Y. Aoki, H. Sugawara, H. Harima, H. Sato, J. Phys. Soc. Jpn. **74**, 209 (2005)
38. C.-H. Lee, S. Tsutsui, K. Kihou, H. Sugawara, H. Yoshizawa, J. Phys. Soc. Jpn. **81**, 063702 (2012)
39. K. Matsuhira, Y. Doi, M. Wakeshima, Y. Hinatsu, H. Amitsuka, Y. Shimaya, R. Giri, C. Sekine, I. Shirotani, J. Phys. Soc. Jpn. **74**, 1030 (2005)
40. T. Matsumura, S. Michimura, T. Inami, Y. Hayashi, K. Fushiya, T.D. Matsuda, R. Higashinaka, Y. Aoki, H. Sugawara, Phys. Rev. B **89**, 161116(R) (2014)
41. R. Shiina, J. Phys. Soc. Jpn. **82**, 083713 (2013)
42. S.W. Lovesey, S.P. Collins, *X-ray Scattering and Absorption by Magnetic Materials* (Clarendon, Oxford, 1996)

Hard X-ray Resonant Scattering for Studying Magnetism

Taka-hisa Arima

1 Basics of Resonant X-ray Magnetic Scattering

Solid matter such as alkaline metals, noble metals, and group VI semiconductors, where the correlation energy between electrons is relatively weak, does not show any magnetic order. On the other hand, when the electron correlation becomes significant, magnetically ordered states can be stabilized, because the electron-electron interaction is dependent on their spin states through the Pauli exclusion principle. Evidence of magnetic order in condensed matter may be provided by several techniques. Neutron diffraction has widely been applied to studying magnetic structure, which denotes the spatial arrangement of magnetic moments in condensed matter, for more than half a century [1]. Because a neutron carries no electric charge but has a tiny magnetic moment, it cannot feel the electric field but does feel the magnetic field in matter. The neutron scattering length of a magnetic moment is 2.695 fm/μ_B, which is comparable to that of a nucleus. Furthermore, the quantum wavelength of a neutron can be a few angstroms, if its kinetic energy is comparable to its thermal energy at room temperature. As a result, neutrons can exhibit Bragg reflections (magnetic reflections) originating from the periodic arrangement of magnetic moments. The magnetic structure in a magnetically ordered phase can be determined by analyzing the neutron diffraction data, just as the atomic arrangement in a crystal can be determined by analyzing X-ray diffraction data. Elastic scattering of neutrons is thus very powerful for determining the magnetic structure in matter.

T. Arima (✉)
Department of Advanced Materials Science, University of Tokyo,
Kashiwa 277-8561, Japan
e-mail: arima@k.u-tokyo.ac.jp; takahisa.arima@riken.jp

T. Arima
RIKEN Center for Emergent Matter Science, Wako 351-0198, Japan

© Springer-Verlag Berlin Heidelberg 2017
Y. Murakami and S. Ishihara (eds.), *Resonant X-Ray Scattering in Correlated Systems*, Springer Tracts in Modern Physics 269,
DOI 10.1007/978-3-662-53227-0_4

Any beam of wavelength in the angstrom range could also be used to determine magnetic structure, if it interacts with the electron spin and orbital moments. X-ray is such a good candidate, since it is a short-wavelength electromagnetic wave. The interaction of X-ray photons with magnetic matter has in fact been studied for a long time, and applied to X-ray magnetic circular dichroism (XMCD) and X-ray magnetic scattering. X-ray magnetic scattering was first observed in antiferromagnetic NiO by Bergevin and Brunel in 1972 [2]. This was almost a quarter century after Shull and a coworker reported neutron magnetic scattering [1], though the discovery of the X-ray [3] was much earlier than that of the neutron [4]. This was due mainly to the fact that the interaction of X-rays with the magnetic moment of an electron is much weaker than that with the electron charge. Bergevin and Brunel observed that magnetic reflections of (1/2 1/2 1/2) and (3/2 3/2 3/2) in NiO were weaker by eight orders of magnitude than crystallographic (1 1 1) reflection for CuKα radiation. The negligibly small magnetic contribution to X-ray scattering made this technique almost useless for studying magnetic structure except for very special cases.

A possible resonant effect on X-ray magnetic scattering was pointed out by Blume in the 1980s [5]. He predicted that the magnetic signal could be enhanced at absorption edges of magnetic elements. After a while, Namikawa and coworkers observed resonant X-ray magnetic scattering in a ferromagnetic nickel single crystal at the Ni K absorption edge [6]. They measured the change in intensity of (220) Bragg reflection with switching magnetization direction and obtained the asymmetry ratio R_a of the intensity, defined as

$$R_{\mathrm{a}} \equiv \frac{I^{\uparrow} - I^{\downarrow}}{I^{\uparrow} + I^{\downarrow}}. \tag{1}$$

Here, $I^{\uparrow}$ ($I^{\downarrow}$) denotes the intensity of the Bragg reflection in a magnetic field applied antiparallel (parallel) to $\boldsymbol{k} \times \boldsymbol{k}'$. $\boldsymbol{k}$ and $\boldsymbol{k}'$ are the propagation vectors of incident and scattered X-ray beams, respectively. In other words, $I^{\uparrow}$ ($I^{\downarrow}$) is the X-ray scattering intensity when the majority spin is parallel (antiparallel) to $\boldsymbol{k} \times \boldsymbol{k}'$. As shown by open symbols in Fig. 1, R_a shows a dip of approximately -1×10^{-3} at the Ni K edge. The dominant process of resonant scattering at the K absorption edge of Ni is the intratomic transition of an electron between 1 s and 4p. The exchange splitting of the Ni 4p band may cause an additional term in the amplitude of nonresonant magnetic scattering. The observed dip of the asymmetry ratio at the K edge is explained by interference between the magnetic and charge scattering.

This interference can also induce XMCD in ferromagnets. Schütz and coworkers observed XMCD of the order of 10^{-4} in iron at the Fe K edge [7]. The observed resonant enhancements of the magnetic signal at the K edge of transition metals were still not strong enough for practical use. As discussed later, much stronger resonant effects were predicted at the $L_{2,3}$ edges of transition metal atoms. In fact, it was found that the magnitude of XMCD at the $L_{2,3}$ edges of Pt in an Fe-Pt alloy exceeded $10\,\%$ [8]. These days, XMCD has become useful for obtaining element-selective information on magnetization with high spatial resolution.

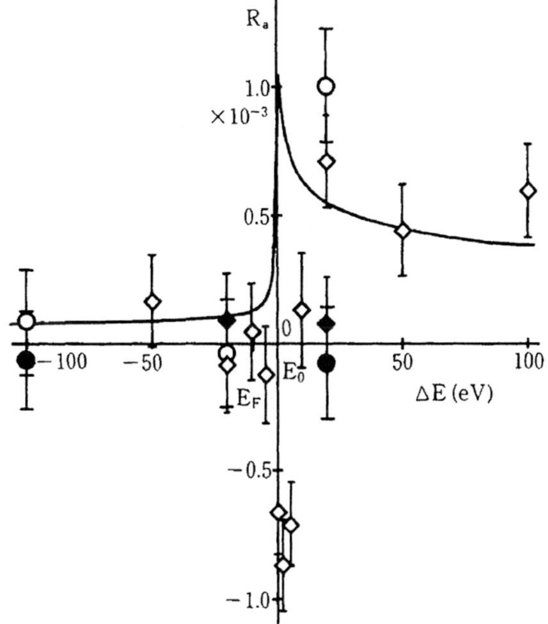

Fig. 1 *Symbols*: experimental data of asymmetry ratio (see text) of 220 Bragg reflection of a ferromagnetic Ni crystal. *Solid line*: calculated asymmetric ratio of 220 Bragg reflection without considering the resonant effect. Excerpt from Ref. [6] by courtesy of J. Phys. Soc. Jpn

X-ray scattering and X-ray absorption are explained by using the Hamiltonian $\mathscr{H}$ for electrons in an oscillating electromagnetic field,

$$\mathscr{H} = \frac{1}{2m} \sum_j \left(\boldsymbol{p}_j + e\boldsymbol{A}(\boldsymbol{r}_j, t) \right)^2 + \sum_j V(\boldsymbol{r}_j) + 2\mu_B \sum_j \boldsymbol{s}_j \cdot \nabla \times \boldsymbol{A}(\boldsymbol{r}_j, t)$$
$$- \frac{2\mu_B}{mc^2} \sum_j \boldsymbol{s}_j \cdot \left(\nabla V(\boldsymbol{r}_j) + \frac{\partial \boldsymbol{A}(\boldsymbol{r}_j, t)}{\partial t} \right) \times \left(\boldsymbol{p}_j + e\boldsymbol{A}(\boldsymbol{r}_j, t) \right). \tag{2}$$

Here, the scalar potential $V(\boldsymbol{r}_j)$ represents the Coulomb interaction. The incident and scattered X-ray beams are expressed by the vector potential $\boldsymbol{A}(\boldsymbol{r}, t)$. This Hamiltonian contains $\boldsymbol{A}$-linear terms,

$$\mathscr{H}_1 = \frac{e}{m} \sum_j \boldsymbol{A}(\boldsymbol{r}_j) \cdot \boldsymbol{p}_j + 2\mu_B \sum_j \boldsymbol{s}_j \cdot \left(\nabla \times \boldsymbol{A}(\boldsymbol{r}_j) \right)$$
$$- \frac{2\mu_B}{mc^2} \sum_j \boldsymbol{s}_j \cdot \frac{\partial \boldsymbol{A}(\boldsymbol{r}_j, t)}{\partial t} \times \boldsymbol{p}_j - \frac{2e\mu_B}{mc^2} \sum_j \boldsymbol{s}_j \cdot \nabla V(\boldsymbol{r}_j) \times \boldsymbol{A}(\boldsymbol{r}_j, t) \tag{3}$$
$$\approx \frac{e}{m} \sum_j \boldsymbol{A}(\boldsymbol{r}_j) \cdot \boldsymbol{p}_j + 2\mu_B \sum_j \boldsymbol{s}_j \cdot \left(\nabla \times \boldsymbol{A}(\boldsymbol{r}_j) \right), \tag{4}$$

and A-quadratic terms,

$$\mathcal{H}_2 = \frac{e^2}{2m} \sum_j A^2(r_j, t) - \frac{2e\mu_B}{mc^2} \sum_j s_j \cdot \frac{\partial A(r_j, t)}{\partial t} \times A(r_j, t). \qquad (5)$$

In Eq. (4), the terms of the order of v/c are omitted by following Blume [5]. Because the creation and annihilation operators of a photon are linear in A, X-ray scattering process should be quadratic in A. The matrix element of elastic scattering arising from the first-order perturbation of $\mathcal{H}_2$ is described as

$$< g|\mathcal{H}_2|g >, \qquad (6)$$

and is irrelevant to the resonant effect. The resonant scattering of an electromagnetic wave is, in general, a second-order process. The first term in Eq. (4) is directly related to the orbital motions of electrons, while the second term expresses the Zeeman term. It is of note that the former can also be spin-dependent at a magnetic ion through the spin–orbit coupling. The leading term in the resonant magnetic scattering on a magnetic ion at r_0 is hence expressed [9, 10] as

$$\frac{1}{m} \sum_n \sum_{j'j} \frac{\left\langle g \left| \varepsilon' \cdot p_{j'} e^{-ik' \cdot r_{j'}} \right| n \right\rangle \left\langle n \left| \varepsilon \cdot p_j e^{ik \cdot r_j} \right| g \right\rangle}{E_g - E_n + \hbar\omega}. \qquad (7)$$

Here, ε (ε') and k (k') denote the unit polarization vector and propagation vector of the incident (scattered) X-ray, respectively, as schematically depicted in Fig. 2. Expanding the exponent functions in Eq. (7) with respect to r_j and $r_{j'}$ around r_0, one obtains the lowest-order term represented as

$$\frac{1}{m} \sum_n \sum_{j'j} \frac{\left\langle g \left| \varepsilon' \cdot p_{j'} \right| n \right\rangle \left\langle n \left| \varepsilon \cdot p_j \right| g \right\rangle}{E_g - E_n + \hbar\omega} e^{i Q \cdot r_0}. \qquad (8)$$

Here, Q is the scattering vector defined as

$$Q = k - k'. \qquad (9)$$

The amplitude in Eq. (8) should exhibit a resonant enhancement associated with the electric-dipole (E1) transitions between the initial (g) and intermediate (n) states, because of the decrease in the denominator. This process corresponds to the absorption of a photon due to an electron excitation followed by the emission of a photon due to an electron relaxation through the E1 transitions. Such E1–E1 magnetic scattering is closely associated with XMCD near the absorption edge. According to Ref. [9], the amplitude f_{E1}^{res} of the X-ray scattering at a magnetic atom through the E1–E1 channel can be represented as

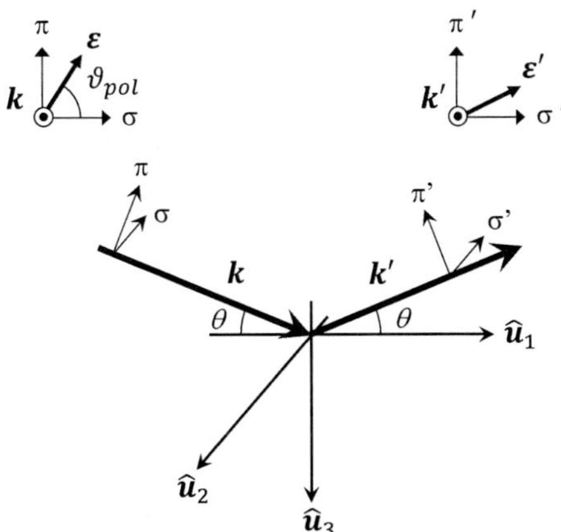

Fig. 2 Definition of coordinate systems. Unit vectors $\hat{k}$ and $\hat{k}'$ represent the propagation vectors of incident and scattered X-ray beams. Spin and orbital angular moments are expressed by using a three-dimensional Cartesian coordinate system defined by $\hat{u}_1 = (\hat{k} + \hat{k}')/2\cos\theta$, $\hat{u}_2 = \hat{k} \times \hat{k}'/\sin 2\theta$, and $\hat{u}_3 = (\hat{k} - \hat{k}')/2\sin\theta$. The direction of the linear polarization ε of the incident beam is expressed by an argument ϑ_{pol} in the coordinate system of σ and π, where π is opposite to $\hat{u}_2$, and $\sigma = \pi \times \hat{k}$ (*top-left inset*). The polarization ε' of the scattered beam is represented by using another coordinate system of σ' and π' (*top-right inset*)

$$f_{E1}^{\text{res}} \propto \varepsilon' \cdot \varepsilon \left[F_{11} + F_{1-1} \right] - i\varepsilon' \times \varepsilon \cdot \hat{z} \left[F_{11} - F_{1-1} \right] \\ + (\varepsilon' \cdot \hat{z})(\varepsilon \cdot \hat{z}) \left[2F_{10} - F_{11} - F_{1-1} \right]. \tag{10}$$

Here, $\hat{z}$ is the unit vector pointing along the magnetic moment. F_{1j} ($j = \pm 1$ or 0) corresponds to the E1 transition between atomic wave functions, where the $\hat{z}$ component of the total orbital angular momentum L_z changes by j. The first term on the right side of Eq. (10) is irrelevant to the magnetic moment. The second term, which is clearly connected with XMCD in a ferromagnet, gives rise to resonant magnetic scattering through the E1–E1 channel. The third term is quadratic in the magnetic moment and is associated with X-ray magnetic linear dichroism (XMLD), which can be regarded as a high-frequency extension of the so-called optical Voigt effect.

Expanding the vector potential in Eq. (4) with respect to r_j, the second term gives the spin contribution to magnetic multipole transitions.

$$2i\mu_{\text{B}} \sum_j s_j \cdot (k \times \varepsilon) \left(1 + ik \cdot r_j + \cdots \right). \tag{11}$$

The first term of Eq. (11) does not produce any significant matrix element between the core and valence states. The second term, classified into a magnetic quadrupole (M2) process, is certainly relevant to the resonant X-ray scattering. Nonetheless, its matrix element between core states and empty valence states is smaller than the E2 matrix element by a factor of v/c. In other words, the Zeeman term [second term of Eq. (4)] would play a major part only in nonresonant magnetic scattering.

2 Usefulness of Resonant X-ray Magnetic Scattering

There are several experimental techniques to study magnetism in matter. Macroscopic magnetization and specific heat provide information about free energy and entropy as a function of magnetic field and temperature. Nuclear magnetic resonance and muon spin rotation have been used as local probes of magnetic order. A static magnetic order should produce a static magnetic field in the matter, which can be probed by Larmor precessions of the spin moments of nuclei or muons. Electron spin resonance probes the effective magnetic field acting on electrons. As mentioned in Sect. 1, neutron elastic scattering (diffraction) is one of the most powerful techniques to determine magnetic structure because the scattering amplitude contains information on the Fourier transformation of the spatial distribution of magnetic moments, $\mu(r)$. Then, one may ask whether or not resonant X-ray scattering has some advantages over neutron diffraction.

There are some essential differences between neutron magnetic diffraction and resonant X-ray magnetic diffraction. The neutron scattering length of a magnetic moment depends only on its magnitude and direction. In contrast, the X-ray scattering amplitude of a magnetic atom depends both on the wave functions and on the photon energy, as indicated by Eq. (7). As a result, some magnetic moments contribute more than others to resonant X-ray reflection. The $|Q|$ dependence of the scattering amplitude is also different between neutron scattering and resonant X-ray scattering. The amplitude of neutron magnetic scattering at a magnetic atom, referred to as magnetic form factor, exhibits a faster decay with the increase in $|Q|$ than the conventional atomic form factor in X-ray Thomson scattering. Only widely spread valence electrons contribute to the magnetic form factor, while the conventional form factor is determined by the spatial distribution of all electrons at the atom. On the other hand, the amplitude of resonant X-ray magnetic scattering in Eq. (7) depends weakly on $|Q|$. The weak $|Q|$ dependence is ascribed to the tightly localized nature of core electrons. As a consequence, the intensity data of neutron magnetic reflections at various Q give information about $\mu(r)$, which cannot, in principle, be obtained by resonant X-ray reflection.

Although resonant X-ray scattering is not suitable for quantitative analysis of $\mu(r)$, the amplitude of resonant X-ray scattering contains useful information about the wave functions of unoccupied valence electrons because the core states are well-approximated by those of an isolated atom. If the transition between a core state and a spin-polarized atomic orbital is allowed in the E1 process, the resonant enhancement of magnetic X-ray scattering can be so large that high-quality data are easily obtained. Moreover, one can also investigate the dependence of scattering amplitude on the directions of ε and ε' as discussed in Sect. 3.4. As a consequence, the spin and orbital states of the valence electrons can be analyzed using Eq. (8) in a rather straightforward way.

In many magnetic materials, magnetic moments are hosted by partially filled d or f orbitals of transition metals, lanthanides, or actinides. The empty d orbitals may be studied through E1 transitions from the 2p core states, as shown in Fig. 3a. Here, 2p

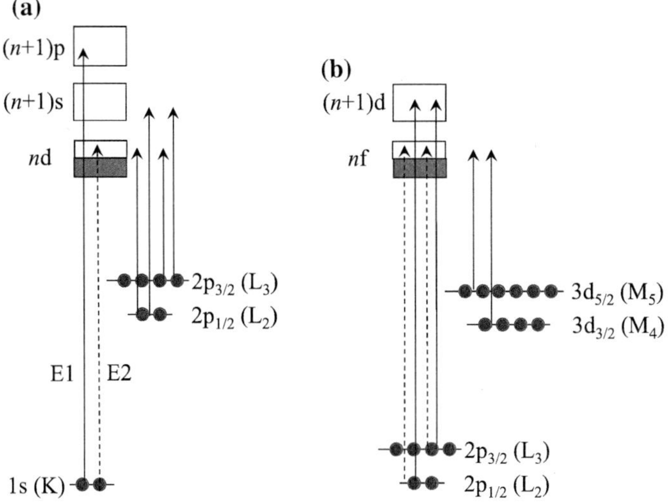

Fig. 3 X-ray induced transitions of an electron from core states to valence states in **a** d- and **b** f-electron systems. *Solid* and *broken arrows* show E1 and E2 processes, respectively

states are split into higher-lying quartet $(J = 3/2)$ and lower-lying doublet $(J = 1/2)$ by a fairly large spin–orbit interaction. Thus the absorption edges from $2p_{3/2}$ (L_3) and $2p_{1/2}$ (L_2) are observed at separate energies. Large resonant enhancements of X-ray magnetic scattering are expected at $L_{2,3}$ absorption edges in the d electron systems. However, L-edge resonant X-ray scattering has not been applied to many 3d/4d transition-metal compounds.

The $L_{2,3}$ edges of 3d transition metals are located between 450 and 960 eV, as shown in Fig. 4. The corresponding wavelengths of 13–28 Å would be too long to satisfy the Bragg condition for many magnetic 3d transition-metal compounds. In magnetic 3d transition-metal compounds, K-edge resonant scattering at much shorter wavelength has been often utilized instead of L-edge resonant scattering. A large rising of the X-ray absorption at K edge is assigned to the E1-allowed intratomic 1s–4p transition. The spin state of 3d electrons affects the 4p energies to some extent through the Coulomb interaction. The consequent tiny spin-splitting in the 4p states may cause a weak resonant enhancement of the amplitude of X-ray magnetic scattering, as already indicated in Fig. 1. A weak peak due to 1s–3d transition might also be discerned at a few eV lower than the main absorption edge in the X-ray absorption spectrum (XAS). If the transition metal site is on an inversion center, the E2 transition is the dominant process for 1s–3d excitation. Thus, one cannot expect a large resonant enhancement of the magnetic scattering. If the local inversion symmetry is broken, the E1 transition matrix between 1 s and 3d becomes nonzero through the 3d–4p hybridization. The resonant magnetic scattering would be caused by both E1 and E2 transitions. As a consequence, some unique X-ray response may show up, as discussed in Sect. 6.

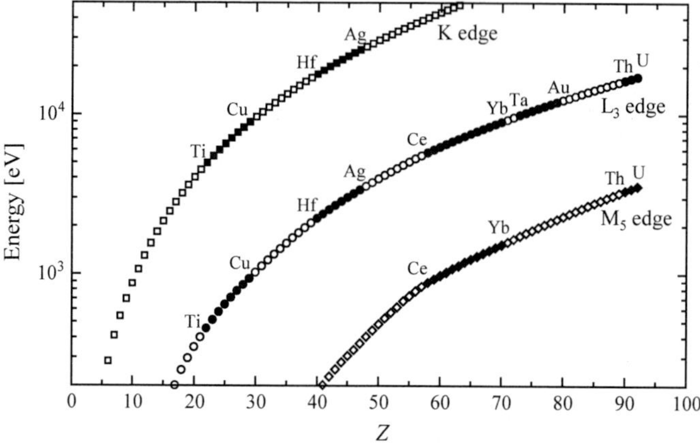

Fig. 4 K, L$_3$, and M$_5$ absorption edges for all elements. *Filled symbols* indicate the data for elements that often host magnetic moments. The data are obtained from Ref. [11]

The L$_3$ edges of 4d transition elements are located between 2.3 and 3.4 keV (see Fig. 4). The corresponding wavelengths are sufficiently short for searching a few low-index magnetic Bragg reflections. Nonetheless, not many successful studies on resonant magnetic X-ray scattering in 4d transition-metal compounds have been reported yet. The L$_3$ absorption edges of 5d transition elements are located at approximately 10 keV. The corresponding wavelengths of approximately 1 Å are suitable for diffraction measurements. Recently, 5d transition oxide compounds have been attracting attention because some theories predict that the strong spin–orbit coupling may induce some exotic electronic states of matter. Resonant X-ray scattering has become a powerful tool for studying the electronic states in magnetic 5d transition metal compounds in concert with a growing interest in this class of materials.

Lanthanide elements are the other key players of magnetism in matter. The magnetic moments are carried by 4f electrons in most cases. The unoccupied 4f states can be investigated by using an E1-allowed electron excitation from the 3p core states, as shown in Fig. 3b. Here 3d states are split into a higher-lying sextet ($J = 5/2$) and lower-lying quartet ($J = 3/2$) by spin–orbit interaction. The absorption edges from 3d$_{5/2}$ (M$_5$) and 3d$_{3/2}$ (M$_4$) are observed at separate energies, as L$_{2,3}$ edges. Large resonant enhancements of X-ray magnetic scattering are expected at the M$_{4,5}$ absorption edges in the f electron systems. The wavelengths of X-ray corresponding to the M$_{4,5}$ edges of lanthanide elements are located between 7.2 and 14 Å, as shown in Fig. 4, which would be too long to satisfy the Bragg condition in many compounds of lanthanides. Therefore, L$_{2,3}$-edge resonant X-ray scattering is often utilized for studying magnetic structure in many lanthanide compounds. At the L$_{2,3}$ edges of lanthanide atoms, 2p–4f and 2p–5d transitions take place through the E2 and E1 processes, respectively, as shown in Fig. 3b. Magnetic scattering exhibits a resonant enhancement at each transition and hence, provides useful information on magnetic order. The details will be discussed in Sect. 4. Some compounds of early actinide elements like uranium, thorium, and neptunium are known to undergo magnetic transitions. The $M_{4,5}$ absorption edges associated with E1-allowed 3d–5f transitions are located at approximately 3.5 keV, near the boundary between soft and hard X-rays, as shown in Fig. 4. Resonant magnetic scattering might be a good probe of magnetic order, although there have not been many examples.

As mentioned above, resonant X-ray scattering can probe the orbital and spin states in d and f electron systems.

Resonant X-ray scattering may also have several advantages over neutron diffraction from a practical point of view.

- The available photon flux density of the X-ray beam from a synchrotron source is much higher than that of neutron beams. Even a tiny crystal of a few tens of micrometers or a thin film can be used for measuring magnetic scattering.
- The X-ray beam can be focused on a spot smaller than 100 nm because of the high brilliance of new-generation X-ray sources and the recent development of X-ray optics. The focused submicron beam can be used to obtain a real-space image of the magnetic domains.
- The short-pulse nature of the synchrotron source may enable us to perform nano or picosecond scale time-resolved measurements of magnetic order after giving an impulsive stimulus.
- Some magnetic ions are not suitable for neutron measurements because of a large neutron absorption coefficient. Typical examples include Sm, Eu, Gd, and Ir. On the other hand, a large resonant enhancement of X-ray scattering is expected at the $L_{2,3}$ edges of these elements. The resonant X-ray scattering measurement is useful for the magnetic structure analysis of compounds containing these elements.

These unique features can make the resonant X-ray scattering a complementary tool to neutron diffraction in studying magnetic structures. Possible evolutions of the synchrotron source may develop this technique to be even more useful in near future.

3 Measurement Techniques

3.1 Methodology for Ferromagnets

In a simple ferromagnet, the contribution of resonant X-ray scattering to the study magnetism may be limited. It is more useful for studying magnetism in a rather complicated ferromagnetic system. For example, the spatial profiles of element-selective magnetization in a ferromagnetic superlattice system could be reproduced by analyzing the Q dependence of the magnetic scattering amplitude, as discussed in Sect. 4.

In ferromagnetic materials, magnetic Bragg reflections must be superimposed on conventional Bragg reflections. The magnetic component is generally so weak in amplitude that some modulation technique must be used to detect the interference term between the charge and magnetic signals. Equation (10) predicts that the resonant magnetic scattering in the E1 approximation should show characteristic polarization dependence. For example, if the magnetic moment is parallel to the $\hat{u}_2$ axis, the resonant scattering is observed only in the $\pi-\pi'$ channel (see Fig. 2). On the other hand, the magnetic moment parallel to the scattering plane ($\hat{u}_1-\hat{u}_3$ plane) gives rise to the scattering in the $\sigma-\pi'$ and $\pi-\sigma'$ channels.

An effective way to detect the magnetic scattering is the measurement of differential signal with switching magnetization direction. If the magnetization is perpendicular to the scattering plane, only the π-polarized incident X-ray can exhibit the resonant magnetic scattering. For the π-polarized incidence, both the charge and magnetic components should be polarized along π'. The resultant interference could be observed as an asymmetry ratio in the intensity of a Bragg reflection with switching magnetization direction by an external magnetic field, similar to Namikawa's measurement (see Fig. 1).

If the magnetization is parallel to the scattering plane, circularly polarized incidence should induce interference between the charge and magnetic terms. The sign of the interference can be reversed by switching the circular polarization of the incident X-ray. In this configuration, a modulation of the incident circular polarization may be useful to observe a tiny magnetic signal. In most cases, the magnetic component is much weaker than the charge component, and hence, proportional to the asymmetry ratio R_a of the intensity with switching the circular polarization of the incident beam, defined as

$$R_a \equiv \frac{I_R - I_L}{I_R + I_L}. \tag{12}$$

The application of a lock-in technique may enable the detection of a tiny signal. One can confirm the result by measuring the circular-polarization modulation signal in the oppositely magnetized case. The modulation signal should be reversed by switching the magnetization direction.

3.2 Methodology for Antiferromagnets

Typical antiferromagnetic order accompanies the formation of a magnetic superlattice. One hence expects that additional pure magnetic reflections of X-rays as well as neutrons should appear in the antiferromagnetic phase. The intensity of a pure magnetic reflection is the square of the amplitude of magnetic scattering. The signal is generally so weak that an improvement in the signal-to-noise ratio is essential for a successful measurement. Energy analysis of the scattered X-ray is an effective way to separate the elastic component from the X-ray fluorescence that is induced by the irradiation of X-rays with a photon energy near the absorption edges.

Polarization analysis of scattered X-rays for several Q points is usually performed to to examine the origin of the X-ray superlattice scattering. If the magnetic moments are parallel to the scattering plane, resonant magnetic scattering may occur in the

$\sigma-\pi'$ and $\pi-\sigma'$ channels, in contrast to Thomson scattering. A rotation of the sample around the scattering vector Q (azimuth rotation) may also rotate the directions of magnetization moments and affect the polarization state of the scattered beam. As a result, the intensity of the π' (σ') polarized component in the scattered X-ray must exhibit unique dependence on the azimuth angle. The photon energy and polarization of X-rays can be analyzed by utilizing a Bragg reflection of a suitable analyzer crystal, as discussed in Sect. 3.4.

The magnetic unit cell is not always different from the crystallographic unit cell. Antiferromagnetic order without a magnetic superlattice can take place if plural magnetic moments on crystallographically identical magnetic sites in a primitive cell are arranged in an antiferromagnetic way. Then, most magnetic Bragg reflections are superimposed on the fundamental reflections, as in the case of ferromagnetic matter. The circular-polarization modulation technique would be useful for detecting the interference between the charge and magnetic signals. Some magnetic Bragg reflections may appear at the Q points where the conventional Bragg reflection is forbidden by glide or screw symmetries. Investigation of resonant X-ray scattering at the forbidden Q points may be useful for studying this type of magnetic order. Here, however, one must be careful that asymmetric distribution of electrons at the magnetic sites can also form Bragg reflections at the same Q positions, which are discernible irrespective of magnetic order. This class of reflections is termed Templeton–Templeton scattering or anomalous-tensor-susceptibility (ATS) scattering [12–14]. The polarization dependence of the ATS scattering is determined by the glide or screw symmetries, and is well investigated. The difference in polarization dependence could distinguish the magnetic signal from nonmagnetic ATS signals.

3.3 Polarization Control

Similar to other resonant X-ray scattering measurements, control of the polarization of incident X-ray beam and analysis of the polarization of scattered X-ray beam are very important for the successful measurement of resonant magnetic X-ray scattering. For example, the polarization dependence in resonant X-ray scattering is key information for examining the origin of the resonant scattering as well as for revealing the directions of magnetic moments.

These days, almost all measurements of X-ray magnetic scattering are carried out using synchrotron radiation or free electron laser. At many experimental stations for hard X-ray scattering with photon energy above 3.5 keV, a horizontally polarized beam is provided. Dynamical scattering theory predicts that X-ray polarization can be controlled using a nearly perfect single crystal [15]. A perfect crystal, which is set close to a Bragg condition, should produce a phase shift between the π and σ polarization components. This phase shift can be applied to a phase plate for hard X-rays [16]. Figure 5 shows the calculated polarization state of 12.385-keV X-ray after passing through a diamond phase plate of a 2.7-mm thickness. The abscissa shows the angular displacement θ_{XPR} from the (220) Bragg reflection condition. One

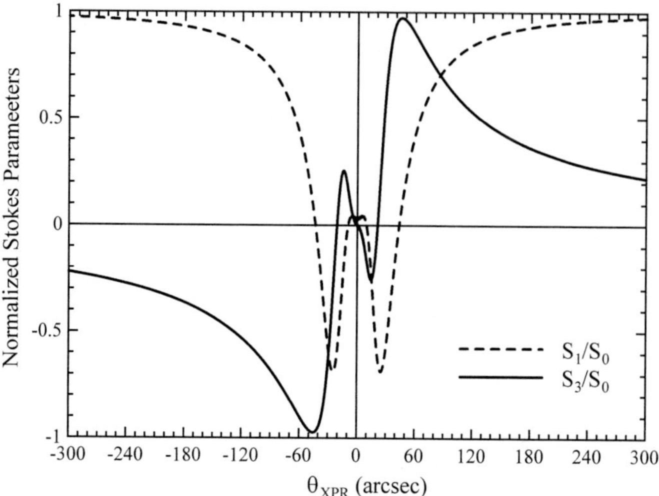

Fig. 5 Calculated polarization state of the X-ray beam of a photon energy 12.385 keV propagating through a 2.7 mm-thick diamond phase plate. The divergence of the beam is assumed to be of Gaussian type with a full width of half maximum of 15 arcsec. θ_{XPR} is defined as the rotational shift of the phase plate from the (220) Bragg condition. The horizontal linear component and circular component are expressed by normalized Stokes parameters S_1/S_0 and S_3/S_0, respectively. The polarization of the incident X-ray beam is horizontal ($S_1/S_0 = 1$). The scattering plane of (220) reflection is slanted by 45 from the *horizontal plane*

should note here that the polarization state is sensitive to the angle θ_{XPR}. The error of θ_{XPR} must be controlled on the order of millidegrees. The purity of the polarization should also be smeared by the divergence of the incident beam near $\theta_{XPR} = 0$. The calculation in Fig. 5 is for a small beam divergence of 15 arcsec. Then, the degree of circular polarization, represented by the normalized Stokes parameter S_3/S_0, reaches almost ± 1 when θ_{XPR} is ± 45 arcseconds. In other words, the diamond plate can operate as a $\pm\lambda/4$ plate, and circularly polarized X-rays with controlled handedness can be obtained. One can also switch the handedness of the circular polarization by oscillating the phase plate. A tiny magnetic signal can be detected as the periodic modulation of the scattering intensity by combining a lock-in technique.

The thickness and material of the phase plate should be well tuned in response to the request of the measurement. The phase retardation of the transmitted beam is proportional to the thickness of the crystal and inverse of the deviation from a Bragg condition. The effect of beam divergence can hence be reduced by using a thicker crystal. On the other hand, the beam intensity after the phase plate is exponentially decreased as the thickness increases. It is also known that a crystal of heavier elements gives a larger phase shift but a stronger absorption. The phase retardation is also dependent on the X-ray wavelength. In general, the phase plate should be thicker for controlling X-rays with the higher photon energy. If a diamond sufficiently thick for high photon energies cannot be prepared, a thin silicon crystal may be another good candidate.

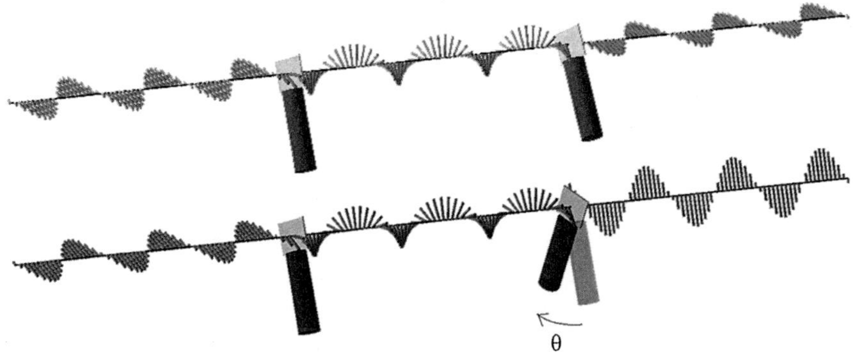

Fig. 6 Polarization control of X-ray by $+\lambda/4$ and $-\lambda/4$ phase plates. Horizontally polarized beam is changed into a circular polarization by the first quarter phase plate. The second quarter phase plate again changes the polarization from circular to linear. The polarization direction after transmitting the second plate is rotated by the same angle as the rotation angle of the second plate about the beam axis

One might think that the phase plate could also operate as a $\pm\lambda/2$ plate. However, the control of linear polarization using a $\lambda/2$ phase plate is not very practical at present, because the beam divergence is significantly large. As shown in Fig. 5, the dip of normalized Stokes parameter S_1/S_0 is ≈ -0.7, indicating that the vertically polarized component can reach only 85 %. The divergence of the X-ray beam should be further improved for the practical use of an X-ray $\lambda/2$ plate. A combination of two $\lambda/4$ plates is more effective to change the linear polarization to an arbitrary direction, as shown in Fig. 6. The horizontally polarized beam is first changed into the circularly polarization by the first $\lambda/4$ phase plate. The second phase plate again changes the polarization from circular to linear. The polarization direction can be controlled by rotating the second plate around the beam.

In principle, one can also change the polarization direction of the incident beam by rotating the sample around k [17]. For this purpose, one must use a special six-circle diffractometer. In many cases, however, similar information has been obtained by performing a measurement of azimuth-angle dependence of the scattering intensity. Here one must be careful that the azimuth rotation of the sample might cause a shift of the area irradiated by the incident beam, which would degrade the data quality.

3.4 Polarization Analysis

The polarization state of the scattered X-ray is analyzed by utilizing the polarization dependence of a Bragg reflection of a crystal. The intensity ratio of Thomson scattering of the $\pi-\pi'$ channel to $\sigma-\sigma'$ channel is $\cos^2 2\theta$ at a scattering angle of 2θ. One can analyze the X-ray polarization within an error of 1 % by using an analyzer crystal satisfying a relation $85° < 2\theta < 95°$. A crystal with a moderate mosaicity and

Fig. 7 Schematic views of
experimental configuration
for extracting (*top*) σ' and
(*bottom*) π' components of a
Bragg reflection by using a
90-degree Bragg reflection
of an analyzer crystal

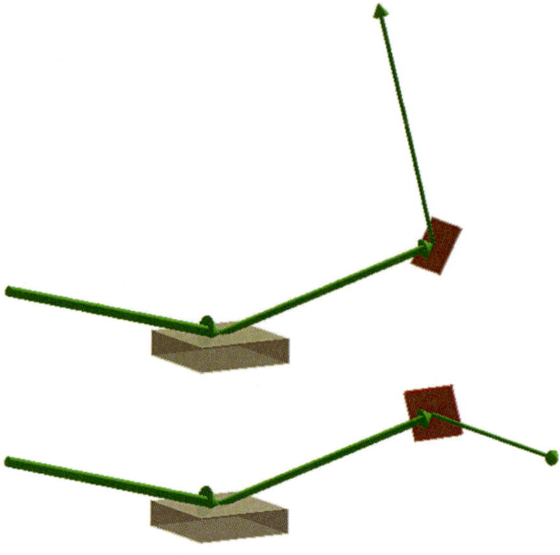

intense Bragg reflections like pyrolytic graphite crystals is suitable for a polarization analyzer. The typical experimental setup is shown in Fig. 7. The selected polarization component of the scattered beam is switched between σ' and π' by rotating the analyzer crystal by 90 degrees around $\boldsymbol{k}'$. The polarization analyzer simultaneously operates as an energy analyzer to reduce inelastic scattering and fluorescence.

One must be aware that the π'-to-σ' ratio in scattering from the sample could not directly be obtained just by the rotation of the analyzer crystal. The observed peak intensity is dependent on the beam divergence. Because the divergence of the X-ray beam from the synchrotron source is generally larger along the horizontal axis than along the vertical axis. This difference more or less affects the peak profile of the reflection of the analyzer crystal. The integrated intensities must be obtained for both the configurations by measuring the peak profiles of the reflection.

3.5   Beam Focusing

An advantage of X-ray microscopy over neutron microscopy is better spatial resolution based on the beam-focusing technology. In principle, the spatial resolution of X-ray microscopy could be much higher than that of optical microscopy, because the wavelength is much shorter. However, the refractive index of matter for X-rays is so close to unity that it is practically impossible to fabricate a refraction-based X-ray lens. Instead, X-ray diffraction could be utilized to obtain X-ray lenses. A typical example is a Fresnel zone plate, which consists of N pairs of opaque and transparent bands bounded by circles of radii,

$$r_n = \sqrt{nf\lambda}; \qquad n = 1, 2, 3, \ldots, N. \tag{13}$$

Here, λ and f are wavelength and focal length, respectively. The spatial resolution (spot size) d of the Fresnel zone plate is estimated by diffraction theory. Rayleigh's criterion tells us that the diffraction-limit resolution is approximately the same as the outermost zone width Δr_N. Cutting-edge technology enables the fabrication of a zone plate with Δr_N of ≈ 10 nm [18]. Nonetheless, the Fresnel zone plates are mainly used for focusing soft X-rays. Because the zone plate for hard X-rays should be much thicker than that for soft X-rays, it is very difficult to fabricate a Fresnel zone with $\Delta r_N \approx 10$ nm.

In general, concave mirrors can be also used to focus the beam. The refractive index of matter for X-rays is a little smaller than unity, which allows the total reflection for grazing-angle incidence. The critical angle θ_c in radian units for the total reflection can be calculated as

$$\theta_c[\mathrm{rad}] \approx 0.016\lambda[\mathrm{nm}]\sqrt{\rho[\mathrm{g/cm^3}]}, \tag{14}$$

using the free-electron approximation. The Kirkpatrick–Baez mirror is composed of a pair of one-dimensional elliptic mirrors, focusing the X-ray beam along the vertical and horizontal axes, respectively, with the same focal points [19]. The focusing size d of the mirror system is limited by two factors. One is given by geometrical optics as

$$d \geq \frac{D\ell_2}{\ell_1}, \tag{15}$$

where D denotes the X-ray source size. ℓ_1 is the distance from the source to the mirror. ℓ_2 is the distance from the mirror to the object. The other factor is based on diffraction and called Rayleigh's criterion [19], expressed as

$$d \geq \frac{\lambda}{2\theta_c}. \tag{16}$$

Equations (14) and (16) indicate that the lower limit of d is only determined by the density of the surface material of the total-reflection mirror. The value is approximately 10 nm. At the present stage, Kirkpatrick–Baez mirror system is more useful to focus hard X-ray beam down to submicrometer order than X-ray lens systems. Moreover, a recent development of focusing mirrors has reduced the spot size down to 7 nm [20], which would make the resonant X-ray scattering technique more useful in the research field of nanomagnetism.

4 Resonant X-ray Magnetic Scattering at Lanthenide L Edges

Partially filled 4f shells in rare-earth atoms often host magnetic moments. Although the 2p–4f transitions ($L_{2,3}$) are not E1-allowed, the E2 transitions at $L_{2,3}$ edges are often utilized for resonant magnetic scattering measurements. Resonant X-ray scattering at the L edges also probes the ordering of electric and magnetic multipolar moments related to 4f electrons in rare-earth compounds. The details are discussed by Matsumura elsewhere in this book.

Gibbs and coworkers first observed a relatively strong resonant enhancement in the magnetic X-ray scattering of Holmium metal [21]. Ten 4f electrons at each Ho atom form total spin and orbital angular momenta of $2\hbar$ and $6\hbar$, respectively. Below the magnetic ordering temperature $T_N = 133$ K, the magnetic dipole momenta lie within the (001) plane of hexagonal lattice and are arranged in a spiral. The pitch of the helix changes from about $3.3c$ just below T_N to $6c$ at low temperatures. They found that the magnetic signal of $(0, 0, 2 + \tau)$, where τ is the magnetic wavenumber, showed a resonant enhancement by a factor of fifty, when the incident X-ray energy was tuned at the L_3 absorption edge of Ho.

They further investigated the resonant spectra of scattering intensity at $(0,0,4 + \tau)$, $(0,0,2 + 2\tau)$, $(0,0,2 + 3\tau)$, and $(0,0,2 + 4\tau)$ in two polarization states π' and σ'. The incident polarization was estimated to be $S_1/S_0 \approx 0.77$. The results are summarized in Fig. 8. The π' component is always stronger in the scattering at τ and peaked above the absorption edge. The weaker σ' component shows a peak below the edge in contrast. The higher harmonics were investigated at 25 K, where the spiral pitch was incommensurate to the lattice ($\tau \sim 0.186$). Second harmonics were also stronger in π' below the edge, while a larger peak was observed above the edge in σ'. Third harmonics only showed a peak below the edge.

Such a rather complicated behavior was explained by considering the electron transition processes. The resonant peaks below the edge should arise from the E2 transition between $2p_{3/2}$ and $4f$, while the peaks above the edge should be assigned to the E1 transition between $2p_{3/2}$ and $5d$ [9, 22]. In general, these transitions are different in photon energy by approximately 5–10 eV, which is large enough for determining the magnetic states in the 4f and 5d orbitals separately. The orbital-selective resonance of the magnetic scattering at slightly different energies makes this technique useful for investigating the magnetism of rare-earth intermetallic compounds.

A large enhancement of the magnetic X-ray scattering at the L edges of rare earths also allows a study of ferromagnetic matter. Sève and coworkers investigated the spatial distribution of the 5d magnetic moments induced by ferromagnetic Fe moments in La/Fe and Ce/Fe multilayers by L_2-resonant X-ray scattering [23]. One can assume that the distribution of the magnetic moments should have the same period as the spatial modulation of the composition. The magnetic signal should be superimposed with the conventional superlattice reflections corresponding to the stacking period of multilayers. Asymmetry ratios R_a at several low-angle superlattice peaks shown in Fig. 9 were obtained in a magnetic field in the $\pm\hat{\boldsymbol{u}}_1$ directions. An

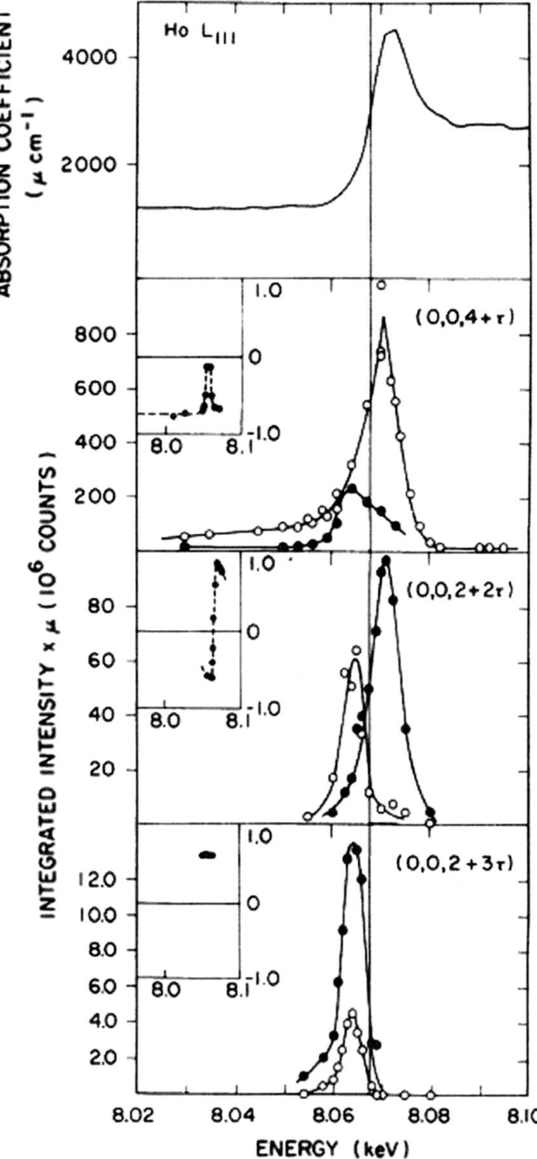

Fig. 8 Absorption and integrated intensities of several superlattice reflections of Ho as a function of the photon energy approximately the L$_3$ edge. The measurement of the spectrum of superlattice $(0\ 0\ 4 + \tau)$ was performed at 20 K, while the $(0\ 0\ 2 + 2\tau)$ and $(0\ 0\ 2 + 3\tau)$ reflections were studied at 25 K. *Filled* and *open circles* correspond to the scattered X-ray in the σ' and π' polarization channels. Excerpt from Ref. [21] by courtesy of American Physical Society

analysis of R_a indicates that Ce moments in Ce/Fe multilayers gradually decrease in absolute magnitude from the interfaces towards the center of the Ce layers and antiferromagnetically oscillate across the Ce layer.

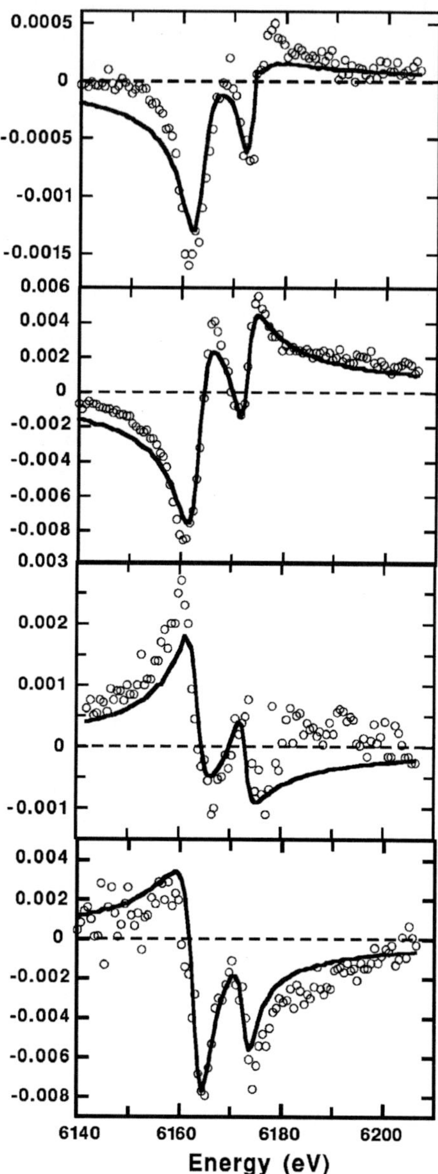

Fig. 9 Spectra of the asymmetry ratios at the Ce L_2 edge at four low-angle superlattice peaks of a Ce_{22}/Fe_{30} multilayer. *Open circles* show the experimental data. *Solid lines* were obtained by fitting the magnetic profile across the Ce layer

5 5d Transition-Metal Oxide Compounds

5d transition elements and their compounds have been attracting a great deal of interest of material scientists, because several unique phenomena like large normal/inverse spin Hall effect and topological Mott insulating states are expected to arise from the strong spin–orbit coupling. Resonant X-ray scattering at the L_2 and

L_3 edges is a useful tool for studying magnetism in 5d transition-metal compounds
for the following reasons.

1. Electron excitations from 2p to 5d are allowed in E1 processes.
2. Energy difference between 2p and 5d levels is approximately 10 keV. Wavelengths at the L edges are of the order of 1 Å, which is suitable for diffraction measurements.
3. The spin–orbit couplings in the valence 5d as well as the core 2p states are so large ($\sim$1 eV) that a large asymmetry in the oscillator strength between X-ray beams of helicities ± 1 is expected.

The above mentioned particularities in the resonant X-ray magnetic scattering in $5d$
transition-metal oxide allow us to explain the data in a simple way. In fact, magnetic
structures in several 5d transition-metal oxide compounds have been determined by
this technique as follows.

5.1 Sr_2IrO_4

Sr_2IrO_4 is a layered oxide compound with a K_2NiF_4-related structure. A unit cell
contains four IrO_2 layers, as depicted in Fig. 10a. The nominal valence of an Ir ion is
$+4$ with a $3d^5$ configuration. Each Ir atom is surrounded by six oxygen atoms. IrO_6
octahedra are linked by sharing corners to form two-dimensional square lattice. The
crystal structure belongs to tetragonal $I4_1/acd$ with the lattice parameters $a \sim 5.5$
Å and $c \sim 26$ Å due to a rotation of the $IrOf_6$ octahedra about the c axis by ± 11
degrees. The transport and magnetic properties of a single crystal were studied by Cao
and coworkers [24]. The temperature dependence of resistivity is semiconducting.
Magnetic susceptibility shows weak ferromagnetism of easy-plane type below 240 K.

Kim and coworkers performed synchrotron X-ray resonant magnetic scattering
to reveal the magnetic structure at low temperatures [25]. Figure 10c–e show the
profiles of resonant X-ray scattering at the Ir L_3 absorption edges. (1 0 $4n+2$),
(0 1 $4n$), and (0 0 $2n+1$) reflections are clearly observed. The intensities of (0
0 $2n+1$) are much weaker than (1 0 $4n+2$). Here, one should note that these
reflections do not satisfy the reflection condition of the body-centered lattice. The
superlattice appears only below 240 K, as shown in Fig. 10f. Moreover, in a magnetic
field of ≈ 0.3 T, (1 0 $4n+2$) reflections disappear and (1 0 $2n+1$) reflections emerge
instead (see Fig. 10e). These behaviors strongly suggest that the resonant reflections
should be ascribed to antiferromagnetic order of Ir moments. Strong (1 0 ℓ) and
(0 1 ℓ) reflections indicate checker-board type magnetic ordering in each Ir square
lattice. The Ir moments are canted to produce the net magnetization in each layer,
indicated by the presence of (0 0 ℓ). The arrangements of Ir moments in magnetic
fields of 0 and 0.3 T are schematically shown in Fig. 10b. The alternation in peak
position is attributable to a change in the stacking of the antiferromagnetically ordered
layers. A metamagnetic transition also accompanies the rearrangement of the net

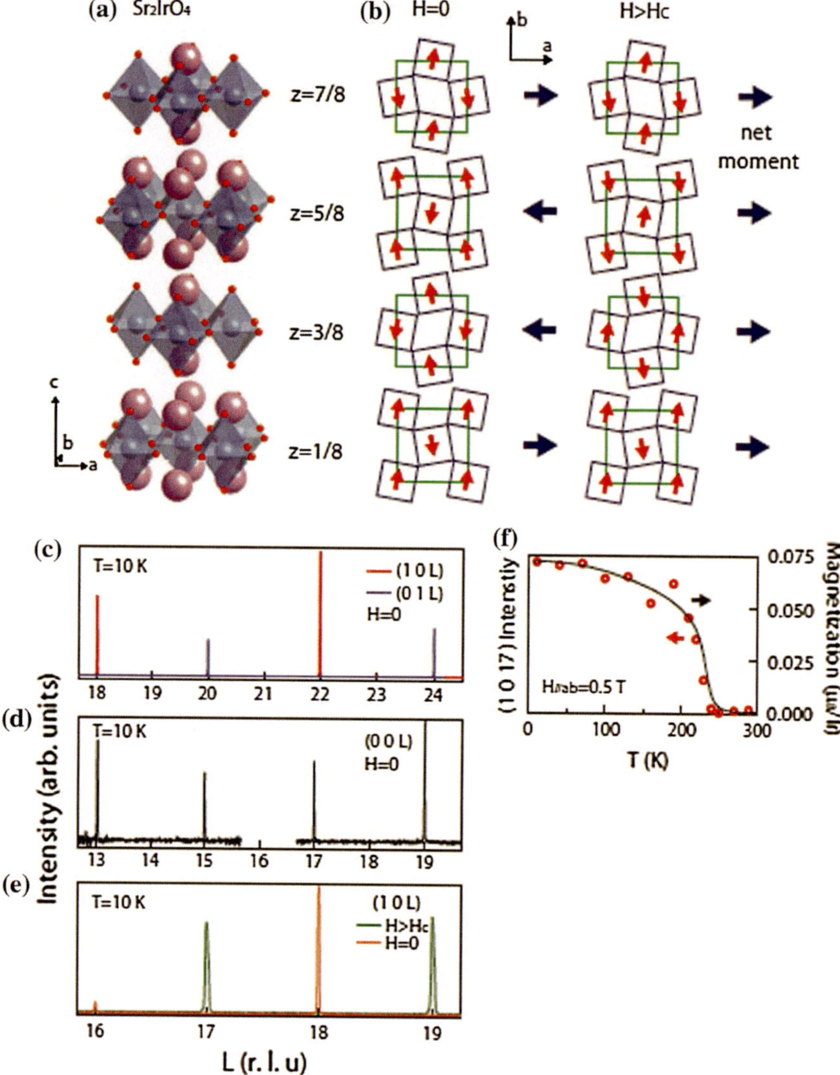

Fig. 10 Crystal and magnetic structures of Sr_2IrO_4. **a** Layered crystal structure consisting of a tetragonal unit cell (space group $I4_1/acd$) with lattice parameters $a \sim 5.5$ Å and $c \sim 26$ Å. The *blue*, *red*, and *purple spheres* represent Ir, O, and Sr atoms, respectively. **b** Canted antiferromagnetic ordering pattern of Ir magnetic moments (*arrows*) within IrO_2 planes perpendicular to the c axis and their stacking pattern along c axis in the zero-field and in the weakly ferromagnetic state, determined from the X-ray data shown in **c**–**e**. **c** and **d** L-scan profile of magnetic X-ray diffraction ($\lambda = 1.1$ Å) along $(1\ 0\ L)$ and $(0\ 1\ L)$ (**c**) and $(0\ 0\ L)$ direction (**d**) at 10 K in zero-field. The huge fundamental Bragg peak at $(0\ 0\ 16)$ was removed in **d**. **e** L-scan of magnetic X-ray diffraction ($\lambda = 1.1$ Å) along $(1\ 0\ L)$ direction at 10 K in the zero field and in the in-plane magnetic field of ≈ 0.3 T. **f** Temperature dependence of intensity of magnetic $(1\ 0\ 19)$ peak (*red circles*) in the in-plane magnetic field of ≈ 0.3 T. The temperature dependent magnetization in the in-plane field of 0.5 T is shown by the *solid line*. The measurements were performed at SPring-8. Excerpt from Ref. [25] by courtesy of Science

magnetization on each layer. Antiferromagnetic stacking with a period of four layers is changed to ferromagnetic stacking, resulting in the weak ferromagnetic moment reported by Cao et al. [24]

Figure 11a shows spectra of X-ray (1 0 22) reflection of Sr_2IrO_4 at 10 K as well as XAS. The magnetic superlattice reflection exhibits a large resonant enhancement just below the Ir L_3 edge of approximately 11.2 keV, while the enhancement is relatively weak at the L_2 edge, approximately 12.8 keV. It is very likely that this contrast should arise from a spin–orbital entangled electron state at the Ir site. The electron state at Ir^{4+} ion in Sr_2IrO_4 is approximated by that in a regular IrO_6 cluster for simplicity. When a transition-metal is coordinated by a ligand octahedron, the d orbitals are split into a higher-lying e_g quartet and a lower-lying t_{2g} sextet. The magnitude of the O_h ligand field, denoted as $10Dq$, is evaluated to be 3–5 eV, which is much larger than the ferromagnetic Hund coupling among the $5d$ electrons. Five $5d$ electrons of the Ir^{4+} ion occupy the lower-lying t_{2g} states and a hole with $s = 1/2$ remains. The sixfold degeneracy in the t_{2g} state may be in part lifted to form a Kramers doublet at low temperatures. In Sr_2IrO_4, IrO_6 octahedra are tetragonally distorted. The elongation of octahedron may lift the sixfold degeneracy in the t_{2g} state and destabilize xy orbital. Another probable origin of level splitting is the spin–orbit coupling. In a regular IrO_6 octahedron, the t_{2g} state is split into a $j_{\text{eff}} = 1/2$ doublet and a $j_{\text{eff}} = 3/2$ quartet in the limit of large $10Dq$. Here, the effective total angular momentum j_{eff} is the difference between the spin angular momentum s and the orbital angular momentum ℓ. The ideal $j_{\text{eff},z} = \pm 1/2$ states are expressed as

$$\left| j_{\text{eff},z} = +1/2 \right\rangle = \frac{1}{\sqrt{3}} \left(|xy, \downarrow\rangle - |yz, \uparrow\rangle + i\,|zx, \uparrow\rangle \right), \tag{17}$$

$$\left| j_{\text{eff},z} = -1/2 \right\rangle = \frac{1}{\sqrt{3}} \left(|xy, \uparrow\rangle + |yz, \downarrow\rangle + i\,|zx, \downarrow\rangle \right). \tag{18}$$

The resonant spectrum of X-ray magnetic scattering at the Ir L edges provides key information to reveal the hole state in t_{2g}. The E1 transition of a hole between the pure $j_{\text{eff}} = 1/2$ state and the $2p_{1/2}$ core state is not allowed.

$$\left\langle 2p_{1/2};\ j_z = \pm 1/2\, |\boldsymbol{r}|\, 5d j_{\text{eff}} = 1/2;\ j_{\text{eff},z} = \pm 1/2 \right\rangle = 0. \tag{19}$$

Substituting Eq. (19) for Eq. (7), the magnetic X-ray scattering should not be enhanced at the L_2 edge if the hole is in the $j_{\text{eff}} = 1/2$ doublet. The resonant enhancement of magnetic scattering is expected only at the Ir L_3 edge. On the other hand, if the tetragonal field would dominate the energy diagram of the t_{2g} state, a hole would occupy the xy orbital. The resonant enhancements at the L_2 and L_3 edges would be comparable. The large contrast in resonant enhancement between the L_2 and L_3 edges shown in Fig. 11a strongly suggests that the hole state should be approximated by the $j_{\text{eff}} = 1/2$ doublet.

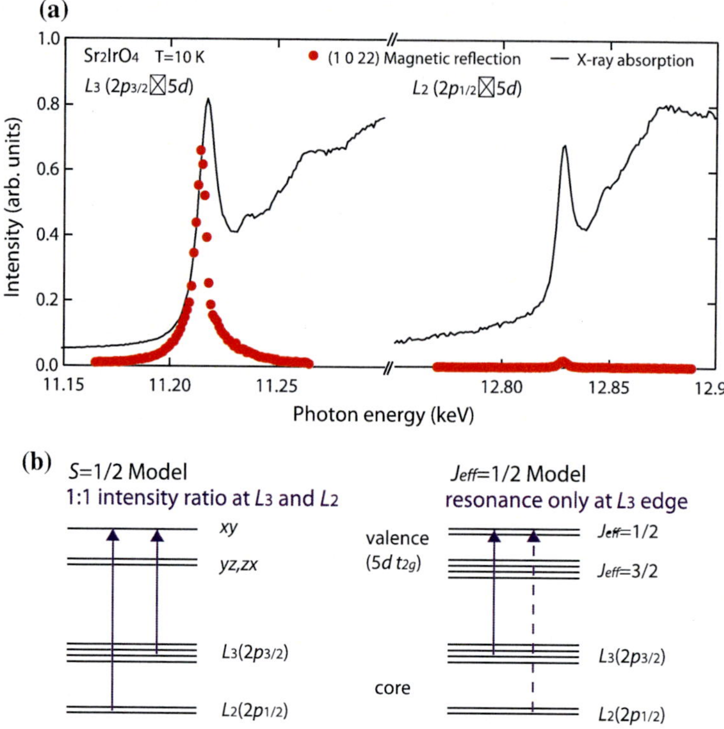

Fig. 11 Resonant enhancement of the magnetic reflection (1 0 22) at the Ir L edge. **a** *Solid lines* indicate X-ray absorption spectra indicating the presence of Ir L_3 (2p$_{3/2}$) and L_2 (2p$_{1/2}$) edges approximately 11.22 and 12.83 keV, respectively. *Dotted lines* represent the intensity of magnetic (1 0 22) reflection. **b** Calculation of X-ray scattering matrix elements expects equal resonant scattering intensities at L_3 and L_2 for the $S = 1/2$ model. For $J_{\text{eff}} = 1/2$ model, in contrast, the resonant enhancement occurs only for L_3 edge and zero enhancement is expected at the L_2 edge. Excerpt from Ref. [25] by courtesy of Science

The ideal $j_{\text{eff}} = 1/2$ state should have a magnetic moment $m_z = \pm 1\,\mu_B$ consisting of $\langle s_z \rangle = \mp 1/6$ and $\langle \ell_z \rangle = \mp 2/3$. The ratio of $\langle \ell_z \rangle$ to $\langle s_z \rangle$ is hence expected to be 4. The ratio in the real crystal can be evaluated by the polarization dependence of *nonresonant* magnetic scattering. Blume and Gibbs calculated the susceptibility tensor $F(\boldsymbol{Q})$ for magnetic scattering in the off-resonant condition [26] as

$$
\begin{aligned}
F_{\sigma \to \sigma'}(\boldsymbol{Q}) &= S_2(\boldsymbol{Q}) \sin 2\theta, \\
F_{\pi \to \sigma'}(\boldsymbol{Q}) &= -2 \sin^2 \theta \left[(L_1(\boldsymbol{Q}) + S_1(\boldsymbol{Q})) \cos \theta - S_3(\boldsymbol{Q}) \sin \theta \right], \\
F_{\sigma \to \pi'}(\boldsymbol{Q}) &= 2 \sin^2 \theta \left[(L_1(\boldsymbol{Q}) + S_1(\boldsymbol{Q})) \cos \theta + S_3(\boldsymbol{Q}) \sin \theta \right], \\
F_{\pi \to \pi'}(\boldsymbol{Q}) &= \sin 2\theta \left[2 L_2(\boldsymbol{Q}) \sin^2 \theta + S_2(\boldsymbol{Q}) \right].
\end{aligned}
\tag{20}
$$

Here, $S_i(\boldsymbol{Q})$ and $L_i(\boldsymbol{Q})$ $(i = 1, 2, 3)$ denote Fourier coefficients of the i-th direction cosines of spin and orbital angular momenta with wavenumber $\boldsymbol{Q}$, respectively, in the coordinate system $\{\hat{\boldsymbol{u}}_1, \hat{\boldsymbol{u}}_2, \hat{\boldsymbol{u}}_3\}$ shown in Fig. 2. 2θ is the scattering angle. If the ordered moments are collinearly arranged along $\hat{\boldsymbol{u}}_2$, the intensity $I(\boldsymbol{Q})$ of a purely magnetic reflection should depend on the polarization direction ϑ_{pol} of the incident beam, as

$$I(\boldsymbol{Q}) \propto S_2^2(\boldsymbol{Q}) \cos^2 \vartheta_{\mathrm{pol}} + \left[2L_2(\boldsymbol{Q}) \sin^2 \theta + S_2(\boldsymbol{Q}) \right]^2 \sin^2 \vartheta_{\mathrm{pol}}. \qquad (21)$$

Although the nonresonant X-ray magnetic scattering is not strong, the polarization dependence irrelevant to any electron excitation process is useful to determine the contribution of an orbital angular momentum to each magnetic moment in the anti-ferromanget.

Fujiyama et al. [27] performed a measurement of the incident-polarization dependence of an off-resonant magnetic reflection in a Sr_2IrO_4 single crystal. The polarization direction of the incident X-ray beam was controlled by two diamond phase retarders (see Fig. 6). Figure 12 shows the dependence of the intensity on the polarization ϑ_{pol} of the incident X-ray beam for (1 0 22) magnetic scattering as well as for the background charge scattering. The spin–orbital ratio $L(\boldsymbol{Q})/S(\boldsymbol{Q})$ was evaluated to be 5.0 ± 0.7 from the fitting. This value clearly contradicts the quenched orbital angular momentum picture usually applied to $3d$ and $4d$ transition-metal oxide compounds, where the hole would occupy the $|xy \uparrow\rangle$ or $|xy \downarrow\rangle$ state.

The large enhancement of magnetic scattering at the L_3 edge enables one to detect the resonant diffuse scattering just above the magnetic transition temperature. The conventional way to estimate the exchange interaction between neighboring magnetic ions is to observe the magnon dispersion by inelastic neutron scattering. In the case of Sr_2IrO_4, unfortunately, large crystals suitable for such a measurement were not grown. Furthermore, strong absorption of neutron by Ir atoms makes inelastic neutron measurements difficult. In a three-dimensional magnetic network, the exchange interaction can also be estimated from the magnetic ordering temperature or the Weiss temperature in the paramagnetic phase. However, this analysis may not be applied to layered compounds, because of the possible larger inter-layer fluctuations. If the anisotropic magnetic fluctuation is detected, it would be useful for estimating the intralayer exchange interaction.

Figure 13 shows the temperature evolution of the profiles of a magnetic reflection. Magnetic diffuse scattering is clearly discernible above T_N. The out-of-plane profiles in the middle panel show that the inter-plane magnetic correlation rapidly decays with heating. In contrast, the in-plane magnetic correlation survives at least up to 250 K. The anisotropic correlation should be attributed to the two-dimensional nature of this layered magnetic system. The in-plane and out-of-plane correlation lengths were estimated from the profiles of the diffuse scattering approximately (1 0 22) and are plotted in the right panel of Fig. 13. The temperature dependence of the in-plane correlation length was compared to the Heisenberg, Ising, and XY models. It has been concluded that Sr_2IrO_4 behaves as a square-lattice Heisenberg antiferromagnet. This

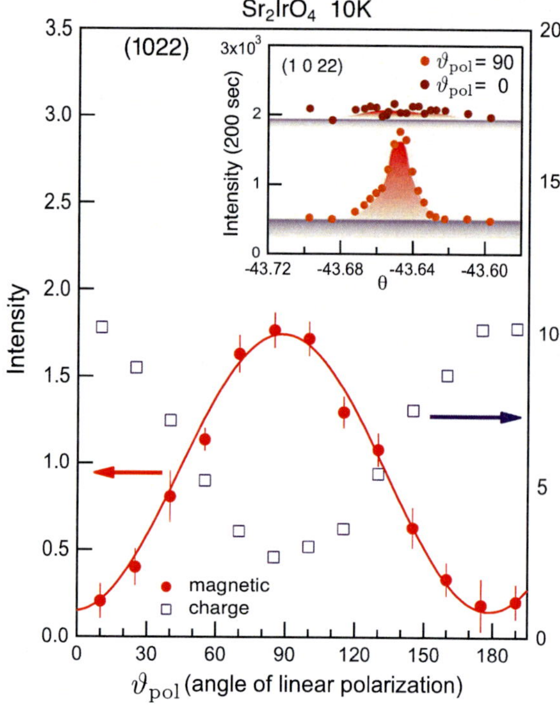

Fig. 12 The polarization dependence of the intensity of an X-ray magnetic reflection (1 0 24) and background charge scattering near (1 0 22) in the antiferromagnetic phase of Sr$_2$IrO$_4$. The experimental data are best fitted to the theoretical curve with $L/S = 5$. Excerpt from Ref. [27] by courtesy of American Physical Society

is rather surprising because the $J_{\mathrm{eff}} = 1/2$ states give an anisotropic spin distribution approximately the Ir atom. This behavior agrees with the prediction by Jackeli and Khaliullin that the Heisenberg-type antiferromagnetic exchange interaction should work between the corner-shared two IrO$_6$ octahedra [28].

5.2 Sr$_3$Ir$_2$O$_7$

Sr$_3$Ir$_2$O$_7$ is a bilayered compound in the Ruddlesden–Popper series of strontium iridates. Each IrO$_6$ octahedron is rotated about the c axis (stacking axis) as in the case of single-layered Sr$_2$IrO$_4$, as shown in Fig. 14a. As a result, the compound is distorted to form an orthorhombic unit cell with space group $Bbcb$. Cao et al. [30] found a magnetic transition at $T_C = 285$ K. Their resistivity measurement exhibited an insulating behavior at whole temperature range with an anomaly at T_C. Kim and coworkers performed a measurement of resonant X-ray scattering at the Ir L$_{2,3}$ edges on a single crystal [31]. They found a series of reflections (1 0 even) and (0 1 odd), which do not satisfy the reflection conditions in $Bbcb$, in the magnetically ordered phase below T_C. The ℓ dependence of the intensity suggests that the Ir moments are arranged to form the simplest antiferromagnetic order within each Ir bilayer. A large resonant enhancement was observed at the L$_3$ edge, while the resonance at the L$_2$

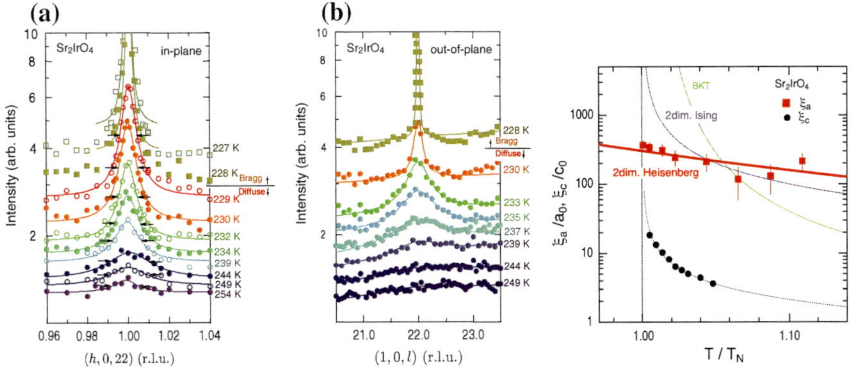

Fig. 13 Profiles of (1 0 22) magnetic reflection along the (*left*) $a*$ and (*middle*) $c*$ axes in Sr_2IrO_4 at various temperatures between 227 K ($= T_N - 1.5$ K) and 254 K ($= T_N + 25.5$ K). *Arrows in* **a** indicate the half maximum for each peak. The estimated correlation lengths along the a (*solid squares*) and c (*solid circles*) axes are plotted against temperature in the *right panel. Solid lines* are fitted curves to temperature dependence of the correlation length in a square-lattice antiferromagnet based on $s = 1/2$ Heisenberg, Ising, and XY models. Excerpt from Ref. [29] by courtesy of American Physical Society

edge was very weak. This contrast supports the $J_{eff} = 1/2$-like state of Ir^{4+} ion as in Sr_2IrO_4. They also investigated the polarization of forbidden X-ray reflections at (1 0 18) and (0 1 19) for the π-polarized incident. The signals were observed in the π–σ' channel but not in the π–π' channel. A detailed analysis revealed that the ordered magnetic moments point along the c axis in contrast to the case of Sr_2IrO_4. Figure 14b shows the magnetic structure. In principle, antisymmetric exchange interaction does not tilt the moments along the c axis, which might agree with much smaller induced magnetization in $Sr_3Ir_2O_7$ than in Sr_2IrO_4. It was proposed that the contrast should be ascribed to the slight enhancement of the elongation of IrO_6 octahedra in the bilayered compound [31].

5.3 CaIrO₃

$CaIrO_3$ crystallizes in the post-perovskite type structure shown in Fig. 15a under high pressures at high temperatures. An orthorhombic unit cell of space group *Cmcm* contains four iridium atoms. IrO_6 octahedron shares a corner (edge) with the neighbors along the c (a)axis to form a two-dimensional rectangular network of Ir ions. The sheets are stacked along the b axis with Ca ions intervening between neighboring sheets. The compound is an insulator, which undergoes a weak ferromagnetic transition at $T_c \sim 115$ K [32]. Ohgushi and coworkers [33] investigated the magnetic structure and electronic state by using resonant X-ray diffraction at the Ir $L_{2,3}$ edges. They found that (0 0 odd) reflections appear at the L_3 edge below T_c, as shown in Fig. 15b. When the scattering plane is perpendicular to the a axis, the (0 0 5) reflection

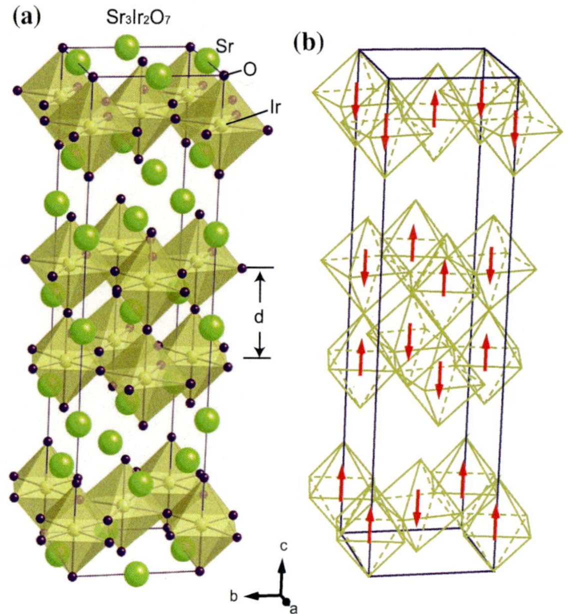

Fig. 14 Schematic drawing of the **a** crystallographic and **b** magnetic structures in Sr$_3$Ir$_2$O$_7$. Excerpt from Ref. [31] by courtesy of American Physical Society

was observed in the $\sigma-\pi'$ channel but not in the $\sigma-\sigma'$ channel (see insets of Fig. 16b, c). This indicates the magnetic origin of the reflections and the ordered moment has a component perpendicular to the a axis. Taking into account the weak spontaneous magnetization along the b axis, a stripe-type magnetic order shown in Fig. 15a was proposed. The Ir moments approximately orient along the c axis and canted toward the b axis.

The magnetic signal was not observed at the L$_2$ edge, as shown in Fig. 16b, suggesting the $J_{\text{eff}} = 1/2$ character of the 5d state of Ir^{4+}. The stripe-type magnetic order agrees with the previous theory based on the $J_{\text{eff}} = 1/2$ model, suggesting the exchange interaction of a quantum-compass type between two edge-shared Ir^{4+}O$_6$ [28, 33].

The resonant X-ray scattering study also provides detailed information on the electronic energy structure of Ir ions. When the scattering plane is perpendicular to the b axis, the ATS scattering due to the low local symmetry approximately Ir^{4+} also contributes to (0 0 odd) reflections at Ir absorption edges. Photon-energy dependence of (0 0 5) scattering of the two origins was clarified by changing the azimuth angle Ψ, as shown in Fig. 16. In the spectra approximately the L$_3$ edge, peaks are observed at 11.214, 11.221, 11.237, and 11.249 keV. The magnetic signal is peaked at 11.214 keV, which should be assigned to the transition between magnetic $j_{\text{eff}} = 1/2$ and $2p_{3/2}$. On the other hand, the signal related to the local anisotropy is largest at 11.221 keV, as shown in (c). his indicates that the levels located higher than t_{2g} should be split by the anisotropy. Although a possible candidate is the 5d e_g state, the observed energy separation of 7 eV would be larger than the widely accepted value of $10Dq$.

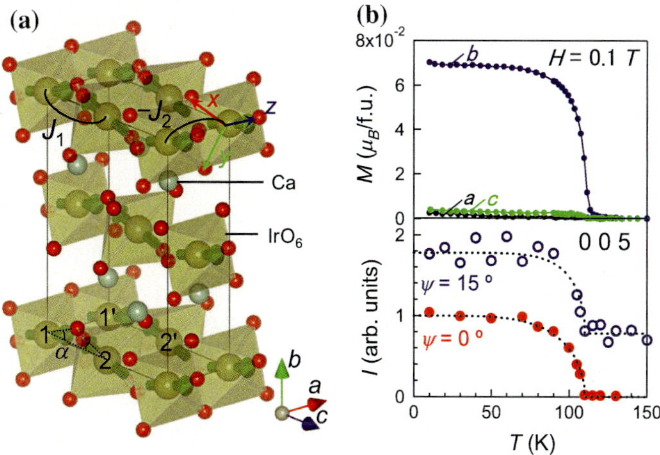

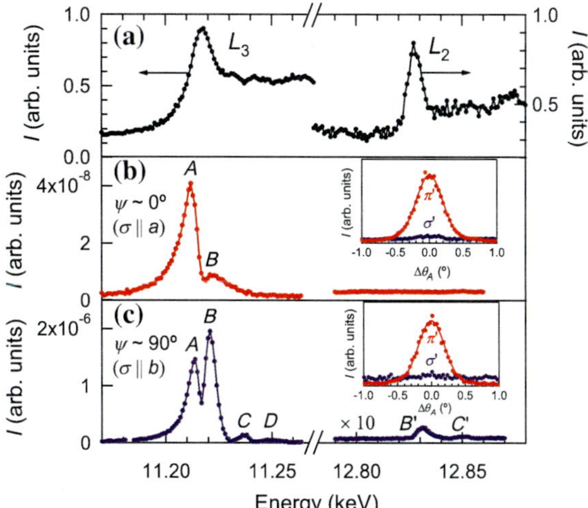

Fig. 15 a Crystallographic and magnetic structures of post-perovskite type $CaIrO_3$. *Solid lines* indicate the orthorhombic unit cell. **b** Temperature dependence of (*upper*) magnetization in the external magnetic field of 0.1 T, and (*lower*) intensity of the (005) reflection at the iridium L_3 absorption edge. Ψ is the azimuthal angle approximately the scattering vector (i.e., c axis). $\Psi = 0$ when the a axis is perpendicular to the scattering plane. Excerpt from Ref. [33] by courtesy of American Physical Society

Fig. 16 a X-ray absorption spectrum of $CaIrO_3$ near the Ir $L_{2,3}$ edges. **b** Spectrum of the intensity of (0 0 5) scattering with the scattering plane perpendicular to the a axis. **c** Spectrum of the intensity of (0 0 5) scattering with the scattering plane perpendicular to the b axis. The result of polarization analysis of the resonant peaks is shown in insets to **b** and **c** Excerpt from Ref. [33] by courtesy of American Physical Society

5.4 $Cd_2Os_2O_7$

Pyrochlore-type $Cd_2Os_2O_7$ was first synthesized by Sleight and coworkers [34]. They found a metal-insulator transition at $T_{MI} \sim 240$ K. Temperature dependence of magnetic susceptibility exhibits an anomaly at T_{MI}, implying the emergence of magnetic order. However, no successful neutron diffraction measurement has been reported so far due in part to the high neutron absorption coefficient of Cd. Yamaura and coworkers performed a study of resonant X-ray scattering and found (0 0 6) and (0 0 10) reflections at the Os L_3 absorption edge, as shown in Fig. 17 [35]. One should note here that the $\{4n + 2\ 0\ 0\}$ reflections do not appear in the regular pyrochlore structure with a space group of $Fd\bar{3}m$. The ATS scattering term of $(0\ 0\ 4n + 2)$ reflections at absorption edges should exhibit the following polarization dependence [14]:

- If the [110] or [1$\bar{1}$0] axis is in the scattering plane, the resonant reflection appears in σ–σ' and π–π' channels.
- If the [100] or [010] axis is in the scattering plane, the resonant reflection appears in σ–π' and π–σ' channels.

They measured the polarization of the (006) reflection as well as the temperature evolution if the scattering intensity, and found that another resonant component appears below T_{MI}. Although the azimuth angle was set that the ATS term of (006) should appear in the π–π' channel, the (006) reflection was observed mainly in the π–σ' channel, as shown in the inset of Fig. 17 [35]. The observed resonant signal should be mainly attributed to magnetic scattering. Their detailed polarization analysis indicated that the magnetic modulation is longitudinal. By considering the result of the NMR experiment [36], it has been concluded that Os magnetic moments are arranged as shown in Fig. 18 on the pyrochlore lattice. This magnetic structure is termed 'all-in/all-out'. The pyrochlore network is composed of regular tetrahedra. All four of the Os moments at the vertices of half of the Os tetrahedra point to the center (i.e., all-in). For the other half of the Os tetrahedra, all four of the Os moments at the vertices point away from the center (i.e., all-out).

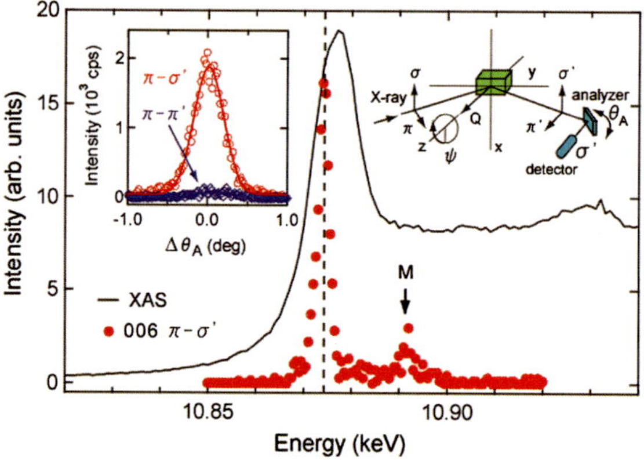

Fig. 17 Spectra of X-ray absorption at room temperature (*solid line*) and intensity of (006) reflection (*filled circles*) of $Cd_2Os_2O_7$ near the Os L_3 absorption edge in the π–σ' channel at 10 K. The crystal [100] axis was rotated from the x axis (normal to the scattering plane) by 47 degrees. The left inset shows the polarization dependence of the 006 reflection at 10.874 keV. The right inset shows the experimental setting. Excerpt from Ref. [35] by courtesy of Am. Phys. Soc

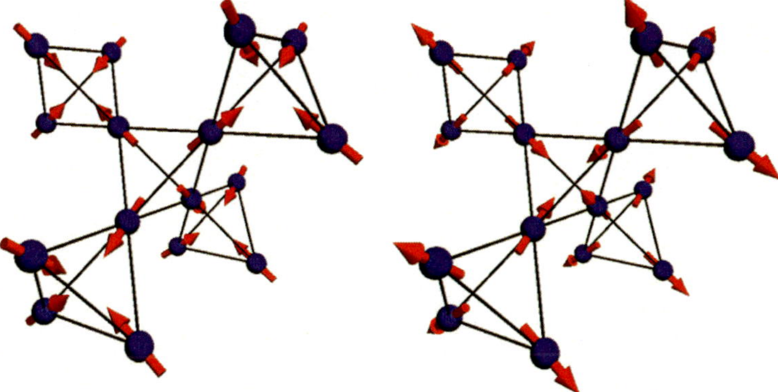

Fig. 18 Two different variants of the all-in/all-out magnetic order on the pyrochlore lattice

The all-in/all-out magnetic order breaks the time-reversal symmetry. One should note that there are two kinds of Os tetrahedra with different orientations in the pyrochlore network. In the all-in/all-out magnetic order, all the tetrahedra in the one orientation are of the all-in type, and all the other tetrahedra are of the all-out type. The time-reversal operation interchanges all-in tetrahedron and all-out tetrahedron without changing the orientation, and switches the spin arrangement between the two variants depicted in Fig. 18. The two variants cannot be superimposed by a translational operation.

The time-reversal breaking causes some unique physical responses, such as nonlinear magnetization, linear magnetostriction, linear magnetocapacitance, linear magnetoresistance, and so on [37]. In order to explore such phenomena, a single-domain sample should be measured. Nonetheless, it is not straightforward to observe and/or control the spatial distribution of domains of antiferromagnetic matter. Tardif et al. [38] succeeded in obtaining the domain image by a resonant X-ray diffraction technique. In the all-in/all-out magnetic order, magnetic scattering may interfere with the ATS scattering at (0 0 even), and results in an asymmetry between the right- and left-handed circularly polarized incidence. Because the magnetic scattering of all-in/all-out order is opposite in sign to each other between the two magnetic variants, the sign of asymmetry is also opposite between the two variants. The antiferromagnetic domain image can be obtained by scanning-type microscopy of focused X-ray beam with a photon energy resonant at the Os L3 edge. Figure 19 shows magnetic domain images of a sample after cooling in several different magnetic fields. The result implies that the application of a magnetic field would affect the domain distribution. It is noteworthy that metallic conduction is proposed on the domain wall between the two variants [39]. The domain imaging will provide important information for exploring such an exotic domain-wall state.

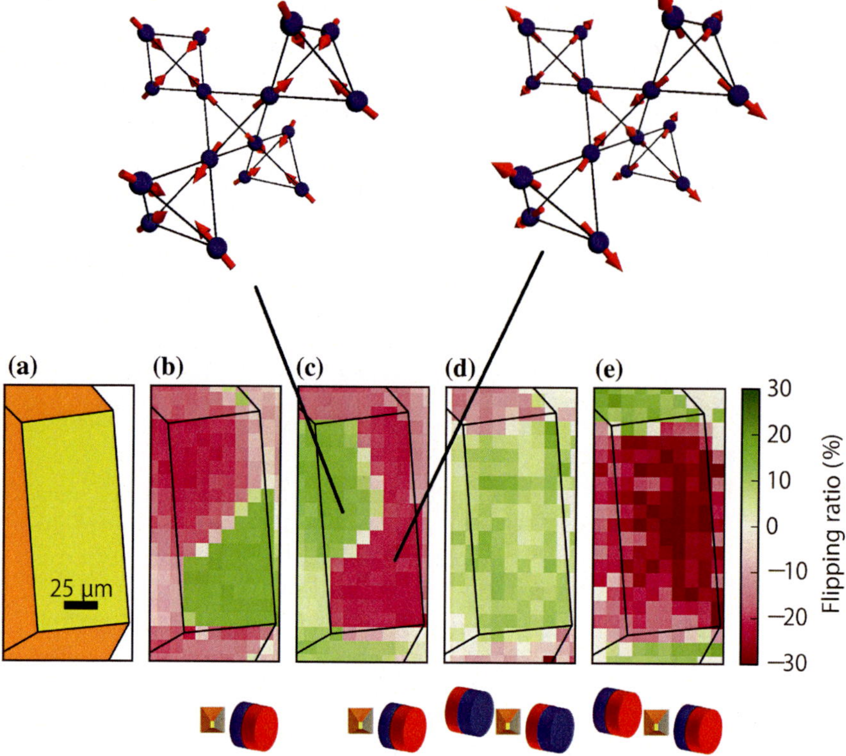

Fig. 19 Images of domain distributions on a rectangular (001) surface of a $Cd_2Os_2O_7$ crystal obtained by investigating the spatial variation of the asymmetry in circularly polarized X-ray scattering at 100 K. The incident beam was focused down to 500 nm by using a Kirkpatrick–Baez mirror system. Excerpt from Ref. [38] by courtesy of Am. Phys. Soc.

5.5 $Eu_2Ir_2O_7$

The phase diagram of the pyrochlore-type iridium oxide system $R_2Ir_2O_7$, where R is a rare earth element, was investigated by Matsuhira et al. [40, 41]. The electronic state of the iridate family is dominated by temperature and rare-earth ionic radius, as shown in Fig. 20. $Pr_2Ir_2O_7$ is metallic in the whole temperature range. $R_2Ir_2O_7$ pyrochlore compounds with other R elements exhibit an insulating behavior at low temperatures. The metal-insulator transition temperature T_{MI} shows a systematic dependence on the ionic radius of R. T_{MI} increases as the ionic size of R becomes smaller. As the magnetic susceptibility also exhibits an anomaly at the metal-insulator transition temperature, it is predicted that Ir magnetic moments are in an ordered state in the insulating phase. The first successful magnetic structure analysis was reported by Tomiyasu and coworkers [42]. They performed a powder neutron diffraction measurement on $Nd_2Ir_2O_7$ and found that Nd moments are arranged in the all-

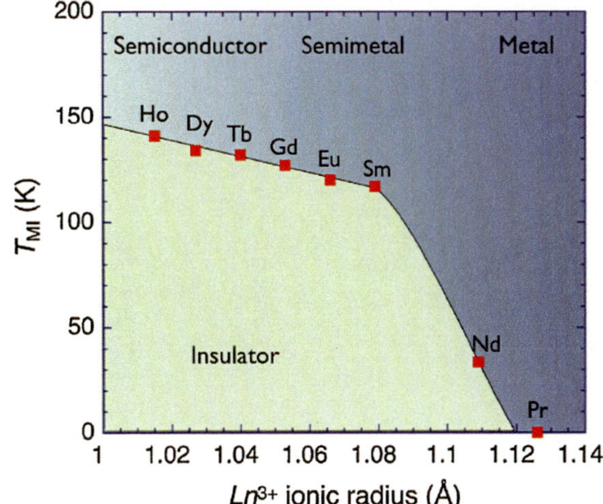

Fig. 20 Electronic phase diagram of the pyrochlore-type rare-earth (R) iridium oxide system $R_2Ir_2O_7$. Excerpt from Ref. [41] by courtesy of J. Phys. Soc. Jpn.

in/all-out type. Considering the coupling between Nd moments and Ir moments, they claimed that Ir moments should also be arranged in the same all-in/all-out pattern. More direct evidence was obtained by resonant X-ray magnetic scattering. Sagayama et al. [43] performed a synchrotron X-ray measurement on a single crystal of $Eu_2Ir_2O_7$. They found that the (006) and (0 0 10) reflections appear at iridium L_3 absorption edge in the insulating phase. Figure 21 shows that (10 0 0) reflection in the $\sigma-\pi'$ channel appear below T_{MI}. The azimuth angle was chosen to allow the ATS scattering to appear in $\sigma-\pi'$ and $\pi-\sigma'$ channels. Therefore, the (10 0 0) reflection can be assigned to magnetic scattering, as in the case of $Cd_2Os_2O_7$. Because the cubic symmetry is retained in the insulating phase, they concluded that the magnetic order is of the all-in/all-out type, as shown in Fig. 18.

The pyrochlore iridate with the all-in/all-out magnetic order has attracted considerable attention, since some theories predict novel electronic states such as Weyl semimetal or topological Mott insulator based on the $J_{\mathrm{eff}} = 1/2$ atomic states [44, 45]. However, in the real pyrochlore system, the atomic states could be deviated from the ideal $J_{\mathrm{eff}} = 1/2$ state because of the trigonal distortion of the IrO_6 octahedron. The effect of the trigonal field has also been examined by X-rays. Spectrum of resonant inelastic X-ray scattering clearly shows a splitting of the fully occupied $J_{\mathrm{eff}} = 3/2$ states [46, 47]. Comparison of the result with the IrO_6 cluster model implies that the effective trigonal field is as large as the spin–orbit coupling [47]. Such a combined study of elastic and inelastic X-ray scattering at the absorption edge may provide information on the detail of the electronic state in general, because resonant elastic scattering is sensitive to the unoccupied states, while resonant inelastic scattering is closely related to the occupied states.

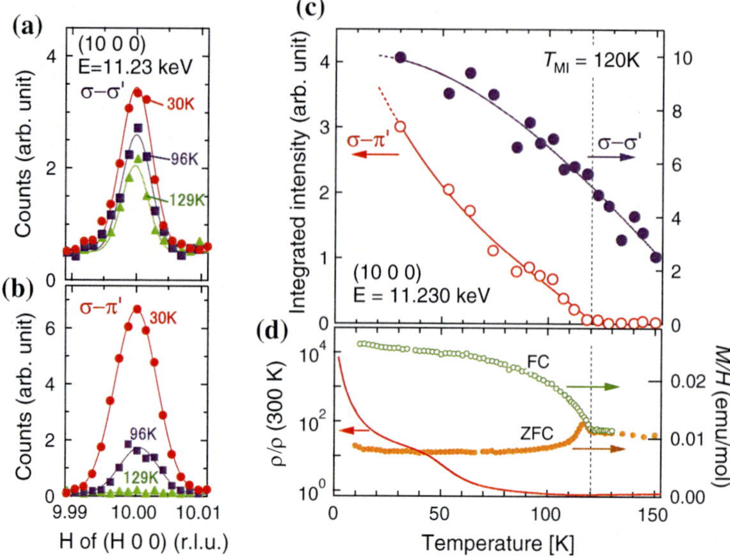

Fig. 21 Profiles of the (10 0 0) reflections of $Eu_2Ir_2O_7$ scanned along (h00) in reciprocal space at Ir L_3 edge (11.230 keV) through the **a** σ–σ' and **b** σ–π' channels at several temperatures. The scattering plane was perpendicular to [0 1 1]. **c** Temperature dependence of the integrated intensities of the (10 0 0) reflection for the two polarization channels. **d** Temperature dependence of the electrical resistivity (*solid line*) and magnetic susceptibility in a magnetic field of 0.1 T (*circles*). Excerpt from Ref. [43] by courtesy of Am. Phys. Soc.

6   Interference Between E1 and E2 Processes

E1 and E2 processes play major parts in resonant X-ray scattering, as discussed in Sect. 1. It has recently been proposed that some additional phenomena may occur through an interference between the E1 and E2 processes. In the isolated atom, the orbital part of the one-electron wave function is expressed as

$$\Psi(\mathbf{r}) = R(r)Y_\ell^m(\theta, \phi), \tag{22}$$

by using the spherical coordinate system (r, θ, ϕ). Here, $R(r)$ is a real function and $Y_\ell^m(\theta, \phi)$ is a spherical harmonic function. The E1 transition between two atomic states Y_ℓ^m and $Y_{\ell'}^{m'}$ is allowed only when $\ell - \ell' = \pm 1$, while the E2 transition is allowed only when $\ell - \ell' = \pm 2$. More generally, the E1 transition changes the parity of the orbital state, while the E2 transition takes place between two states with the same parity. The two types of transitions cannot interfere with each other in the centrosymmetric system.

Let us consider in more detail the condition for the E1–E2 interference. If the incident beam with $\mathbf{k} \parallel z$ is circularly polarized, the matrix elements between the g and n states for the E1 and E2 processes are expressed as

$$E \langle n|(x \pm iy)|g\rangle , \tag{23}$$

and

$$i\frac{Ek}{2} \langle n|(x \pm iy)z|g\rangle , \tag{24}$$

respectively. Here, E is the amplitude of electric field. The above two processes are different in phase by $\pm\pi/2$, and do not usually interfere with each other. An effective interference can take place in chiral matter [48]. It is dependent on the circular polarization of the incident beam, whether the interference is constructive or destructive. From this viewpoint, the interference is closely connected with the natural optical rotation and natural circular dichroism, while the major origin of these phenomena would be an interference between the electric-dipole (E1) and magnetic-dipole (M1) transitions.

If the incident beam is linearly polarized along x, the matrix elements for the E1 and E2 processes are represented as

$$E \langle n|x|g\rangle , \tag{25}$$

and

$$i\frac{Ek}{2} \langle n|xz|g\rangle , \tag{26}$$

respectively. While the polarization along the z axis may make both the matrix elements nonzero simultaneously, the phase factor i would prevent effective interference. Some time-reversal-odd polar moment along the z axis may give rise to E1–E2 interference in this configuration. The typical example is the so-called toroidal moment [48]. The classical picture of a toroidal moment is a loop of magnetic moments, as schematically shown in Fig. 22. In a solid, the vector product of atomic displacement δr_i and magnetic moment μ_i can be considered as a toroidal moment of $\delta r_i \times \mu_i/2$. If toroidal moments are oriented in the same direction in magnetic matter, the material is termed ferrotoroidic [49]. A simple example is a ferromagnetic ferroelectric. The ferrotoroidic moment in a ferromagnetic ferroelectric is considered to be parallel to the vector product of the spontaneous polarization P_s and magnetization M_s, $P_s \times M_s$.

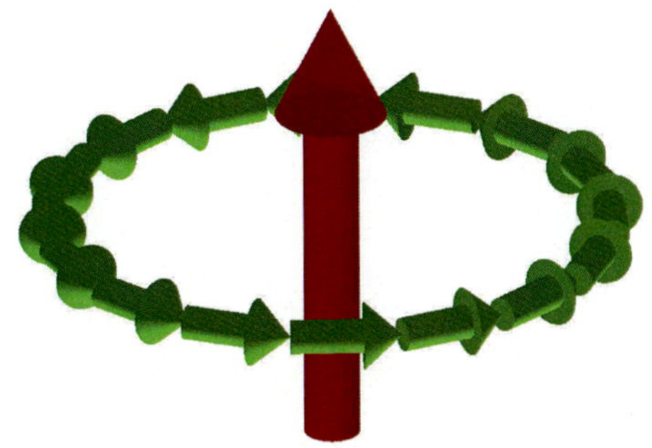

Fig. 22 Schematic drawing of a toroidal moment. *Small arrows* forming a loop are magnetic moments. A *central large arrow* shows a toroidal moment

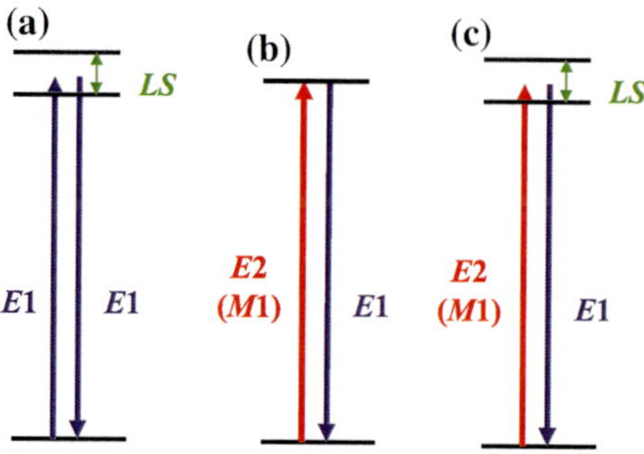

Fig. 23 Dichroic phenomena for X-ray and relevant electronic processes. **a** X-ray magnetic circular dichroism, **b** X-ray natural circular dichroism, and **c** X-ray nonreciprocal directional dichroism. E1, E2, M1, and LS denote electric-dipole transition, electric-quadrupole transition, magnetic-dipole transition, and spin–orbit interaction, respectively

It may be helpful to summarize several X-ray dichroic phenomena in the transmission geometry in terms of the microscopic processes of electron transitions. Ferromagnetic materials may exhibit XMCD at absorption edges of magnetic elements. In most cases, two E1 transitions with perpendicular polarizations interfere with each other through the spin–orbit coupling in the excited states. Chiral materials may exhibit X-ray natural circular dichroism (XNCD), which arises from interference between E1 and E2 transitions for perpendicular polarizations. The interference between E1 and E2 transitions for a linear polarization may take place in ferrotoroidic materials. This appears as X-ray nonreciprocal directional dichroism (XNDD). The comparison is schematically shown in Fig. 23.

In a ferrotoroidic, the E1–E2 interference term at each atom has the common phase in the average of the transmission geometry. The consequent macroscopic interference appears as the directional dichroism of the electromagnetic wave. Such directional dichroism of hard X-rays was first observed in a polar ferrimagnet $GaFeO_3$ and termed X-ray nonreciprocal directional dichroism (XNDD) by Kubota and coworkers [50]. $GaFeO_3$ is an orthorhombic system with a space group $Pc2_1n$, where a spontaneous polarization lies along the b axis. Below $T_C \approx 200$ K, Fe spin moments are arranged parallel to the c axis and a spontaneous ferrimagnetic moment appears. As a result, an effective toroidal moment appears in one direction along the a axis at each Fe site [51]. Compared with Eq. (26), the electromagnetic wave propagating along the toroidal moment (a axis) is predicted to exhibit directional dichroism.

Figure 24 shows the spectra of absorption μt and directional dichroism $\Delta\mu t$ of $GaFeO_3$ for two linear polarizations, $E^\omega \parallel b$ and $E^\omega \parallel c$ at the Fe K absorption edge. A strong absorption edge at approximately 7.125 keV in (a) is assigned to the E1-allowed 1s–4p transitions, while a small bump around 7.114 keV is ascribed to 1s–3d transitions. The bump is relatively strong compared with other ferrites due in part to the breaking of the inversion symmetry. Clear signals of dichroism are observed at the 1s–3d excitation in Fig. 24b. In addition, smaller peaks are also discernible around the 1s–4p transition. These $\Delta\mu$ signals disappear above T_C, indicating the magnetic origin.

Fig. 24 Spectra of the X-ray (**a**) absorption and (**b**) directional dichroism in a GaFeO$_3$ crystal at 50 K for the X-ray linearly polarized along b and c axes. The directional dichroism was defined as the difference in absorption coefficients when the direction of an applied magnetic field along the c axis is flipped. The spectrum of the second-harmonic component of the magnetic-field modulation for the c polarization is also shown by open circles in **b**. Excerpt from Ref. [50] by courtesy of Am. Phys. Soc.

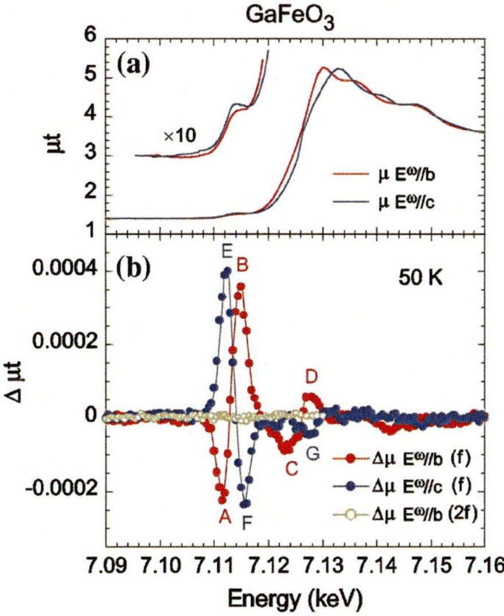

Although the Fe ions have no symmetry operation, there should only be two essential factors in symmetry breaking of an FeO$_6$ cluster to explain the spectra. An Fe site is coordinated by six oxygen atoms. The b axis can be approximated as the trigonal axis. Therefore, the 3d orbitals are roughly split into lower-lying $3y^2 - r^2$, zx, and $z^2 - x^2$ and higher-lying xy and yz in the first approximation. Here, the local x, y, and z axes are parallel to the a, b, and c axes, respectively. One must replace z in Eq. (26) with x in the measurement. As the d orbitals are represented by real functions, the information of magnetic order should be introduced through the spin–orbit coupling $\lambda L_z S_z$. As a result, the yz and zx as well as $3y^2 - r^2$ and xy orbitals are hybridized by the spin–orbit coupling. The spin–orbit coupling also acts between the $4p_x$ and $4p_y$. The inversion breaking is represented by the breaking of the b mirror, $E'y$, which would be essential in this polar ferrite. This asymmetric potential may mix the yz and $4p_z$ as well as xy and $4p_x$, irrelevant to the magnetism. As a result, the E1 and E2 processes can interfere with each other at the 3d and 4p levels. The correspondence between the observed spectra and schematic energy diagram is shown by indices from A to G in Figs. 24 and 25 [50]. For example, at the excitation from $1s$ to the lowest-lying zx labelled G, the absorption intensity $I_{1s \to zx}$ can be expressed as

$$I_{1s \to zx} \propto \left| \frac{\langle zx \,|\lambda S_z \ell_z|\, yz \rangle \, \langle yz \,|E'y|\, 4p_z \rangle \, \langle 4p_z \,|z|\, 1s \rangle}{(E_{zx} - E_{yz})(E_{zx} - E_{4p_z})} + \langle zx \,|izkx|\, 1s \rangle \right|^2$$

$$= \left| \frac{\langle zx \,|\lambda S_z \ell_z|\, yz \rangle \, \langle yz \,|E'y|\, 4p_z \rangle \, \langle 4p_z \,|z|\, 1s \rangle}{(E_{zx} - E_{yz})(E_{zx} - E_{4p_z})} \right|^2 + |\langle zx \,|zkx|\, 1s \rangle|^2$$

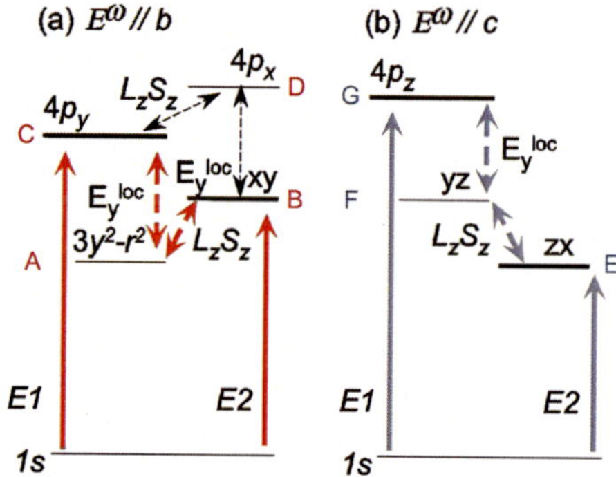

Fig. 25 Energy levels at Fe site and electron transitions between them relevant to the directional dichroism in GaFeO$_3$ shown in Fig. 24b. In the vicinity of the Fe K absorption edge, both the electric dipole (E1) process from Fe 1 s to Fe 4p and the electric quadrupole (E2) process from Fe 1 s to Fe 3d appear. For E1–E2 interference, spin–orbit coupling and breaking symmetry of the *b* mirror is crucial. The indices A–G for the levels correspond to peaks A–G of the spectra shown in Fig. 24b, respectively. Excerpt from Ref. [50] by courtesy of Am. Phys. Soc.

$$-2\lambda S_z E' k \frac{\langle zx\, |i\ell_z|\, yz \rangle \, \langle yz\, |y|\, 4p_z \rangle \, \langle 4p_z\, |z|\, 1s \rangle \, \langle zx\, |zx|\, 1s \rangle}{(E_{zx} - E_{yz})(E_{zx} - E_{4p_z})}. \tag{27}$$

The last term in Eq. (27) represents the interference between the E1 and E2 processes. The sign of the interference is dependent on the product of S_z, E', and k. This is associated with the scalar product of the toroidal moment on the atom and the propagation vector k of the X-ray. These processes relevant to the XNDD signals are schematically drawn in Fig. 25.

The E1–E2 interference should also induce unique X-ray scattering at the absorption edge of the atoms hosting toroidal moments. Such resonant X-ray scattering was first observed in the ferrotoroidic GaFeO$_3$ [52]. A magnetic-field modulation of the resonant X-ray scattering at the Fe K pre-edge (1s–3d absorption edge) revealed that the magnetic scattering amplitude is larger at (040) than at (020). Apparently, the result contradicts the layered ferrimagnetic structure with the fundamental magnetic modulation of (020). It was suggested that the toroidal moments on two Fe sites should contribute much to the resonant magnetic scattering.

The similar resonant X-ray diffraction was observed even in the ferrimagnetic phase of a centrosymmetric spinel oxide MnCr$_2$O$_4$. Matsubara et al. [53] examined the resonant X-ray magnetic scattering at Mn K edge, and found a signal at (222) reflection in a magnetic field applied along the [1 1 $\bar{2}$] axis, as shown in Fig. 26c. In the spinel oxide, Mn^{2+} ions occupy the tetrahedral sites which are arranged in the diamond structure. The Mn spin moments are ferromagnetically arranged below approximately $T_C = 51$ K [54]. In addition, a cycloidal incommensurate modulation is superimposed below 14 K. In both phases, the magnetic form factor of Mn moments at (222) is calculated to be exactly zero. The (222) reflection can appear only if the scattering matrix at two Mn ions in a primitive cell is different from each other. The

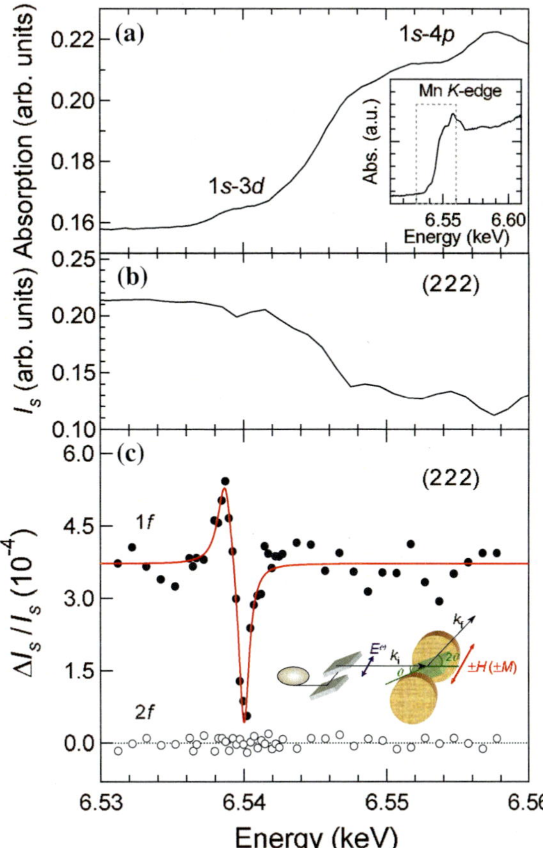

Fig. 26 Several X-ray spectra of $MnCr_2O_4$ near the Mn K edge. **a** X-ray absorption spectrum at room temperature. **b** Intensity of (222) reflection with σ-polarized incidence at 20 K. **c** Modulation of the intensity of (222) reflection at 20 K in an ac external magnetic field of 80 mT along the $[1\,1\,\bar{2}]$ axis. The frequency f of ac magnetic field is 5 Hz. *Solid (open) symbols* show the $1f$ ($2f$) component of modulation. The incident X-ray is polarized parallel to the magnetic field. Excerpt from Ref. [53] by courtesy of Am. Phys. Soc.

local inversion symmetry is lacking at Mn sites because of the tetrahedral ligand field. The tetrahedral coordinations at the two Mn sites are opposite each other in sign, while the magnetic moments are parallel to each other. The simultaneous breaking of local spatial inversion and time reversal can work as an effective toroidal moment for X-rays. As a result, magnetic scattering through the E1–E2 interference at the two Mn sites in a primitive cell should be opposite in sign and may hence give rise to a resonant magnetic signal at (222). A similar signal was also observed in magnetite (Fe_3O_4) at the Fe K edge [55].

7 Conclusion

X-ray magnetic scattering may exhibit a large resonant enhancement at the absorption edges of the magnetic elements. The resonant enhancement does not only make the signal detection easier but also provides unique information on the electronic states.

The element-selective probe of orbital and spin states can be complementary to quantitative information on the spatial distribution of magnetic moments provided by neutron diffraction studies. Progress in the development of X-ray sources in the near future may enable us to perform a time- and position-resolved measurements of magnetic and electronic structures in matter.

Acknowledgements The author expresses his gratitude to Dr. H. Ohsumi and Mr. Sakai for fruitful discussions and providing Fig. 5. He would also like to thank Editage (www.editage.jp) for English language editing.

References

1. C.G. Shull, J.S. Smart, Phys. Rev. **76**, 1256 (1949)
2. F. de Bergevin, M. Brunel, Phys. Lett. **39A**, 141 (1972)
3. W.C. Röntgen, Sitzungs-Berichte der Physikalisch-Medicinischen Gesellschaft zu Würzburg **9**, 132 (1895)
4. J. Chadwick, Nature **129**, 312 (1932)
5. M. Blume, J. Appl. Phys. **57**, 3615 (1985)
6. K. Namikawa, M. Ando, T. Nakajima, H. Kawata, J. Phys. Soc. Jpn. **54**, 4099 (1985)
7. G. Schütz, W. Wagner, W. Wilhelm, P. Kienle, R. Zeller, R. Frahm, G. Materik, Phys. Rev. Lett. **58**, 737 (1987)
8. G. Schütz, R. Wienke, W. Wilhelm, W. Wagner, R. Frahm, P. Kienle, Phys. B **158**, 284 (1989)
9. J.P. Hannon, G.T. Trammell, M. Blume, D. Gibbs, Phys. Rev. Lett. **61**, 1245 (1988)
10. S.W. Lovesey, S.P. Collins, *X-ray Scattering and Absorption by Magnetic Materials* (Clarendon Press, Oxford, 1996)
11. J.A. Bearden, Rev. Mod. Phys. **39**, 125 (1967)
12. D.H. Templeton, L.K. Templeton, Acta Crystallogr. A **36**, 237 (1980)
13. D.H. Templeton, L.K. Templeton, Acta Crystallogr. A **38**, 62 (1982)
14. V.E. Dmitrienko, Acta Crystallogr. A **39**, 29 (1983)
15. G. Moliere, Ann. Phys. **35**, 297 (1939)
16. K. Hirano, K. Izumi, T. Ishikawa, S. Annaka, S. Kikuta, Jpn. J. Appl. Phys. **30**, L407 (1991)
17. H. Ohsumi, M. Mizumaki, S. Kimura, M. Takata, H. Suematsu, Physica **345B**, 258 (2004)
18. W. Chao, J. Kim, S. Rekawa, P. Fischer, E.H. Anderson, Opt. Express **17**, 17669 (2008)
19. P. Kirkpatrick, A.V. Baez, J. Opt. Soc. Am. **38**, 766 (1948)
20. H. Mimura, S. Handa, T. Kimura, H. Yumoto, D. Yamakawa, H. Yokoyama, S. Matsuyama, K. Inagaki, K. Yamamura, Y. Sano, K. Tamasaku, Y. Nishino, M. Yabashi, T. Ishikawa, K. Yamauchi, Nat. Phys. **6**, 122 (2010)
21. D. Gibbs, D.R. Harshman, E.D. Isaacs, D.B. McWhan, D. Mills, C. Vettier, Phys. Rev. Lett. **61**, 1241 (1988)
22. D. Gibbs, G. Grübel, D.R. Harshman, E.D. Isaacs, D.B. McWhan, D. Mills, C. Vettier, Phys. Rev. B **43**, 5663 (1990)
23. L. Sève, N. Jaouen, J.M. Tonnerre, D. Raoux, F. Bartolomé, M. Arend, W. Felsch, A. Rogalev, J. Goulon, C. Gautier, J.F. Bérar, Phys. Rev. B **60**, 9662 (1999)
24. G. Cao, J. Bolivar, S. McCall, J.E. Crow, R.P. Guertin, Phys. Rev. B **57**, R11039 (1998)
25. B.J. Kim, H. Ohsumi, T. Komesu, S. Sakai, T. Morita, H. Takagi, T. Arima, Science **323**, 1329 (2009)
26. M. Blume, D. Gibbs, Phys. Rev. B **37**, 1779 (1988)
27. S. Fujiyama, H. Ohsumi, K. Ohashi, D. Hirai, B.J. Kim, T. Arima, M. Takata, H. Takagi, Phys. Rev. Lett. **112**, 016405 (2014)
28. G. Jackeli, G. Khaliullin, Phys. Rev. Lett. **102**, 017205 (2009)

29. S. Fujiyama, H. Ohsumi, T. Komesu, J. Matsuno, B.J. Kim, M. Takata, T. Arima, H. Takagi, Phys. Rev. Lett. **108**, 247212 (2012)
30. G. Cao, Y. Xin, C.S. Alexander, J.E. Crow, P. Schlottmann, M.K. Crawford, R.L. Harlow, W. Marshall, Phys. Rev. B **66**, 214412 (2002)
31. J.W. Kim, Y. Choi, J. Kim, J.F. Mitchell, G. Jackeli, M. Daghofer, J. van den Brink, G. Khaliullin, B.J. Kim, Phys. Rev. Lett. **109**, 037204 (2012)
32. K. Ohgushi, H. Gotou, T. Yagi, Y. Kiuchi, F. Sakai, Y. Ueda, Phys. Rev. B **74**, 241104(R) (2006)
33. K. Ohgushi, J. Yamaura, H. Ohsumi, K. Sugimoto, S. Takeshita, A. Tokuda, H. Takagi, M. Takata, T. Arima, Phys. Rev. Lett. **110**, 217212 (2013)
34. A. Sleight, J. Gillson, J. Weiher, W. Bindloss, Solid State Commun. **14**, 357 (1974)
35. J. Yamaura, K. Ohgushi, H. Ohsumi, T. Hasegawa, I. Yamauchi, K. Sugimoto, S. Takeshita, A. Tokuda, M. Takata, M. Udagawa, M. Takigawa, H. Harima, T. Arima, Z. Hiroi, Phys. Rev. Lett. **108**, 247205 (2012)
36. K. Yamauchi, M. Takigawa et al., Private communication
37. T. Arima, J. Phys. Soc. Jpn. **82**, 013705 (2013)
38. S. Tardif, S. Takeshita, H. Ohsumi, J. Yamaura, D. Okuyama, Z. Hiroi, M. Takata, T. Arima, Phys. Rev. Lett. **114**, 147205 (2015)
39. Z. Hiroi, J. Yamaura, T. Hirose, I. Nagashima, Y. Okamoto, APL Mater. **3**, 041501 (2015)
40. K. Matsuhira, M. Wakeshima, R. Nakanishi, T. Yamada, A. Nakamura, W. Kawano, S. Takagi, Y. Hinatsu, J. Phys. Soc. Jpn. **76**, 043706 (2007)
41. K. Matsuhira, M. Wakeshima, Y. Hinatsu, S. Takagi, J. Phys. Soc. Jpn. **80**, 094701 (2011)
42. K. Tomiyasu, K. Matsuhira, K. Iwasa, M. Watahiki, S. Takagi, M. Wakeshima, Y. Hinatsu, M. Yokoyama, K. Ohoyama, K. Yamada, J. Phys. Soc. Jpn. **81**, 034709 (2012)
43. H. Sagayama, D. Uematsu, T. Arima, K. Sugimoto, J.J. Ishikawa, E. O'Farrell, S. Nakatsuji, Phys. Rev. B **87**, 100403(R) (2013)
44. D. Pesin, L. Balents, Nat. Phys. **6**, 376 (2010)
45. X. Wan, A.M. Turner, A. Vishwanath, S.Y. Savrasov, Phys. Rev. B **83**, 205101 (2011)
46. L. Hozoi, H. Gretarsson, J.P. Clancy, B.-G. Jeon, B. Lee, K.H. Kim, V. Yushankhai, P. Fulde, D. Casa, T. Gog, J. Kim, A.H. Said, M.H. Upton, Y.-J. Kim, J. van den Brink, Phys. Rev. B **89**, 115111 (2014)
47. D. Uematsu, H. Sagayama, T. Arima, J.J. Ishikawa, S. Nakatsuji, H. Takagi, M. Yoshida, J. Mizuki, K. Ishii, Phys. Rev. B (to appear) **92**, 094405 (2015)
48. S. Di Matteo, Y. Joly, C.R. Natoli, Phys. Rev. Lett. **72**, 144406 (2005)
49. H. Schmid, Ferroelectrics **252**, 41 (2001)
50. M. Kubota, T. Arima, Y. Kaneko, J.P. He, X.Z. Yu, Y. Tokura, Phys. Rev. Lett. **92**, 137401 (2004)
51. T. Arima, D. Higashiyama, Y. Kaneko, J.P. He, T. Goto, S. Miyasaka, T. Kimura, K. Oikawa, T. Kamiyama, R. Kumai, Y. Tokura, Phys. Rev. B **70**, 064426 (2004)
52. T. Arima, J.H. Jung, M. Matsubara, M. Kubota, J.P. He, Y. Kaneko, Y. Tokura, J. Phys. Soc. Jpn. **74**, 1419 (2005)
53. M. Matsubara, Y. Shimada, K. Ohgushi, T. Arima, Y. Tokura, Phys. Rev. B **79**, 180407R (2009)
54. K. Tomiyasu, J. Fukunaga, H. Suzuki, Phys. Rev. B **70**, 214434 (2004)
55. M. Matsubara, Y. Shimada, T. Arima, Y. Taguchi, Y. Tokura, Phys. Rev. B **72**, 220404R (2005)

Resonant Soft X-Ray Scattering Studies of Transition-Metal Oxides

Hiroki Wadati

1 Introduction

Transition-metal (TM) oxides have been attracting great interest in these decades because of their intriguing physical properties such as metal-insulator transition, colossal magnetoresistance, and ordering of spin, charge, and orbitals [1]. In such systems where d or f orbitals are partially filled, electrons have three degrees of freedom, charge, orbital, and spins, and these also interact with lattice. Various anomalous behaviors are usually accompanied by the ordering of charge, orbital, and spins, and the observation of the ordered state has been a big issue in modern condensed-matter physics. In conventional X-ray diffraction, the incident photons interact with all the electrons in a material. The atomic scattering cross section for X-rays is proportional to Z^2, where Z is the atomic number. This means that signal is always dominated by heavy elements.

Resonant X-ray scattering has emerged as a new experimental tool for investigating the ordering of charge, orbital, and spins [2]. The sensitivity is achieved by combining X-ray diffraction and X-ray absorption spectroscopy in one experiment.

This chapter is structured as follows. In Sect. 2, I describe the principles of resonant soft X-ray scattering (RSXS). Sections 3–5 contain interfaces, charge and orbital ordering, and multiferroics, respectively. In Sects. 6 and 7, I show current developments about measurements under magnetic fields and time-resolved measurements. Section 8 is devoted to summary and future prospects.

H. Wadati (✉)
Institute for Solid State Physics, University of Tokyo, Kashiwa, Japan
e-mail: wadati@issp.u-tokyo.ac.jp

© Springer-Verlag Berlin Heidelberg 2017
Y. Murakami and S. Ishihara (eds.), *Resonant X-Ray Scattering
in Correlated Systems*, Springer Tracts in Modern Physics 269,
DOI 10.1007/978-3-662-53227-0_5

2 Principles of Resonant Soft X-Ray Scattering

Below I will treat the phenomenon of resonant scattering both semiclassically and quantum mechanically [3–5].

2.1 Semiclassical Treatment

Resonant X-ray scattering process can be semiclassically treated by considering a multielectron atom as a collection of harmonic oscillator. The equation of motion is

$$m\ddot{x} + m\gamma\dot{x} + m\omega_s^2 x = -eE. \tag{1}$$

The first term is the acceleration, the second term is a dissipative force term with $\gamma \ll \omega_s$, and the third term is the restoring force for an oscillator of resonant frequency ω_s. $-e(E + v \times B)$ is the Lorentz force and the term $v \times B$ can be neglected in the non-relativistic case of $v/c \ll 1$.

With an incident electric field

$$E = E_i e^{-i\omega t}, \tag{2}$$

the displacement x, velocity, and acceleration all have the same $e^{-i\omega t}$ time dependence, and the time derivative can be replaced by $-i\omega t$. By substituting $x(t) = x_0 e^{-i\omega t}$, the equation of motion becomes

$$m(-i\omega)^2 x_0 + m\gamma(-i\omega)x_0 + m\omega_s^2 x_0 = -eE_i. \tag{3}$$

The harmonic displacement is given by

$$x_0 = \frac{1}{\omega^2 - \omega_s^2 + i\gamma\omega} \frac{eE_i}{m}, \tag{4}$$

The radiated field E_{rad} at distance r and time t is given by the acceleration $\ddot{x}(t - r/c)$ at the retarded time $t - r/c$

$$E_{rad}(r, t) = \frac{e}{4\pi\varepsilon_0 c^2 r}\ddot{x}(t - r/c) \tag{5}$$

By inserting $\ddot{x}(t - r/c) = -\omega^2 x_0 e^{-i\omega t} e^{i\omega r/c}$,

$$E_{rad}(r, t) = \frac{-\omega^2}{\omega^2 - \omega_s^2 + i\gamma\omega} \frac{e^2}{4\pi\varepsilon_0 mc^2} \frac{e^{ikr}}{r} E \tag{6}$$

The atomic scattering factor f_s is given in units of the electron classical radius $r_0 = e^2/4\pi\varepsilon_0 mc^2 = 2.82 \times 10^{-5}$ Å,

$$f_s = \frac{\omega^2}{\omega^2 - \omega_s^2 + i\gamma\omega}. \tag{7}$$

The cross section σ is given by

$$\sigma = \frac{8\pi}{3}r_e^2 \frac{\omega^4}{(\omega^2 - \omega_s^2)^2 + (\gamma\omega)^2}. \tag{8}$$

This value shows a strong resonance at resonance $\omega = \omega_s$.

In the limit $\omega^2 \ll \omega_s^2$ and $\gamma \ll \omega_s$,

$$\sigma = \frac{8\pi}{3}r_e^2 \left(\frac{\omega}{\omega_s}\right)^4 = \frac{8\pi}{3}r_e^2 \left(\frac{\lambda_s}{\lambda}\right)^4, \tag{9}$$

which has a λ^{-4} wavelength dependence, known as Rayleigh's law.

2.2 Optical Parameters

In general for X-rays, the refractive index $n(\omega)$ deviates only a small amount from 1, and can be expressed as

$$n(\omega) = 1 - \delta + i\beta. \tag{10}$$

δ and β are expressed in terms of real and imaginary parts of the complex atomic scattering factor.

$$\delta = \frac{n_a r_e \lambda^2}{2\pi} f_1(\omega) \tag{11}$$

$$\beta = \frac{n_a r_e \lambda^2}{2\pi} f_2(\omega) \tag{12}$$

At off-resonance, the index of refraction is tabulated [6]. Optical parameters near resonance energies can be determined from measurements of the absorption cross-section μ, which is proportional to the imaginary part of the scattering factor. δ is then determined via Kramers–Kronig transformation of β.

$$f_1(\omega) - Z = -\frac{2}{\pi}P\int_0^\infty \frac{uf_2(u)}{u^2 - \omega^2}du \tag{13}$$

$$f_2(\omega) = \frac{2\omega}{\pi}P\int_0^\infty \frac{f_1(u) - Z}{u^2 - \omega^2}du, \tag{14}$$

where $P\int_0^\infty$ is Cauchy principal value.

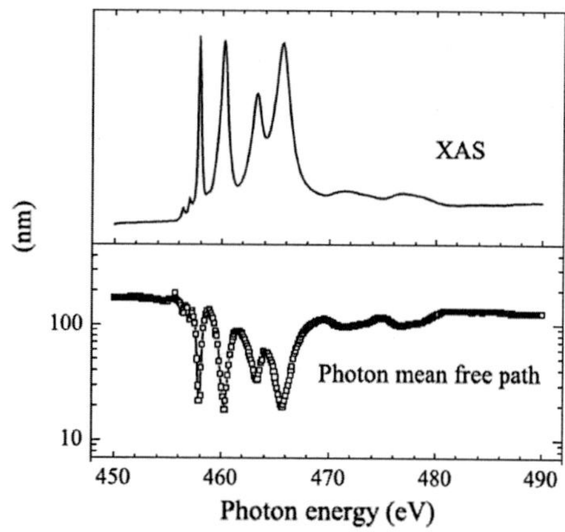

Fig. 1 XAS spectrum and photon mean free path of SrTiO$_3$ at the Ti $2p \rightarrow 3d$ edge [7]

2.3 Photon Mean Free Path

Figure 1 shows the mean free path l

$$l = \frac{\lambda}{4\pi\beta} \tag{15}$$

in comparison with the X-ray absorption spectroscopy (XAS) spectrum at the Ti $2p \rightarrow 3d$ edge in SrTiO$_3$ (STO) [7]. Below and above the resonance, mean free path has values well above 100 nm, meaning that the signal is quite bulk sensitive. At on-resonance, the photon mean free path becomes as small as 20 nm, which is moderately surface sensitive. Even this minimum value is one order of magnitude larger than the probing depth in typical surface-sensitive techniques such as photoelectron spectroscopy [8].

2.4 Dynamical Theory of X-Ray Scattering in Parratt's Recursive Method

There are two theories which account for X-ray diffraction: kinematical and dynamical theories. The kinematical theory neglects absorption and multiple scattering, whereas the dynamical theory takes into account all the interactions. Here we introduce the dynamical theory of X-ray scattering from multilayers, which is Parratt's recursive method to calculate reflectivity.

Let us consider a multilayer composed of materials A and B, arranged alternately in thickness Δt_A and Δt_B, respectively (Fig. 2). The multilayers period is thus $d = \Delta t_A + \Delta t_B$. It is assumed that N is the number of bilayer pairs, so that the total number of media involved, including vacuum and substrate, is $n = 2N + 2$.

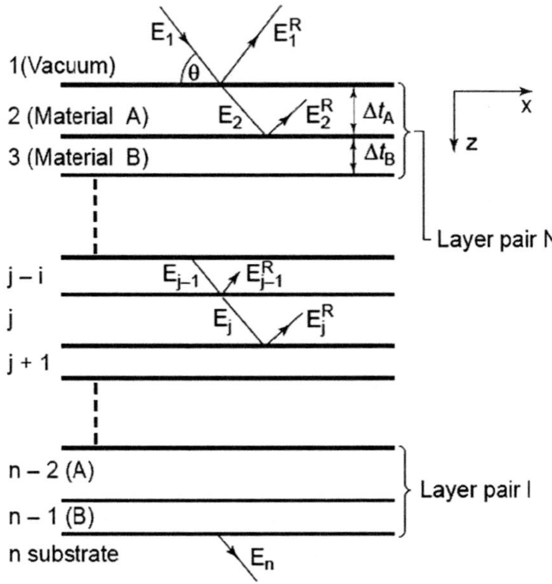

Fig. 2 Diffraction of X-rays from a multilayer structure, N bilayer pairs of two materials A and B [9, 10]

Maxwell's equation requires that the tangical components of the electric and magnetic fields should be continuous at each interface. Let E_j be the amplitude of the incident electric field in the j-th layer with the refractive index $\hat{n}_j$, and E_j^R be the amplitude of the reflected electric field. The coefficients F_j^σ and F_j^π are given as

$$F_j^\sigma = \frac{g_j - g_{j+1}}{g_j + g_{j+1}} \tag{16}$$

$$F_j^\pi = \frac{\hat{n}_{j+1}^2 g_j - \hat{n}_j^2 g_{j+1}}{\hat{n}_{j+1}^2 g_j + \hat{n}_j^2 g_{j+1}}, \tag{17}$$

where

$$g_j = \hat{n}_j \sin\theta_j = (\hat{n}_j^2 - \cos^2\theta)^{1/2} \tag{18}$$

θ_j is the grazing incidence angle at the interface between j-th and $(j+1)$-th layer and θ is the one from the vacuum. R_j is defined as

$$R_j = a_j^2 \frac{E_j^R}{E_j} = a_j^4 \frac{F_j + R_{j+1}}{1 + F_j R_{j+1}}, \tag{19}$$

where

$$a_j = \exp\left(-i\frac{\pi \hat{n}_j \sin\theta_j d_j}{\lambda}\right) = \exp\left(-ig_j\pi \frac{d_j}{\lambda}\right). \tag{20}$$

From Eq. (19), on can obtain the value of R_{j-1} from that of R_j. Since the substrate is considered to be infinitely thick, the amplitude of the reflected wave in the substrate is zero: $R_{2N+1} = 0$. The successive application of Eq. (19) ($2N$ times) will eventually lead to R_1, and the reflectivity is then given by

$$\frac{I(\theta)}{I_0} = |R_1|^2, \tag{21}$$

where I_0 and $I(\theta)$ is the incident and reflected intensities, respectively.

Near the absorption edges, one has to use the following modified Bragg's law, which takes into account the effects of refraction

$$m\lambda = 2d \sin\theta \left(1 - \frac{4\bar{\delta}d^2}{m^2\lambda^2}\right) \tag{22}$$

$\bar{\delta}$ is defined as the average. The values of δ at the Fe $2p$ edges were obtained (without using the Kramers–Kronig relation) from the photon-energy dependence of Bragg peak positions in the case of Fe/V superlattice [11].

2.5 Quantum Mechanical Treatment

RSXS is a combination of X-ray scattering and soft X-ray absorption spectroscopy. By performing X-ray diffraction measurements at the absorption edge of the constituent atom, one can obtain information about the ordering with the sensitivity to the specific atom. Such sensitivity can be understood from the schematic illustration of RSXS process in the case of the Mn $2p \rightarrow 3d$ edge, shown in Fig. 3. In the initial ground state $|i\rangle$ (a), an incoming photon is absorbed and a Mn $2p$ core-level electron is excited into an unoccupied state close to the Fermi level. This results in the intermediate state of the scattering process $|n\rangle$ (b). Then the excited electron recombines with the core hole and a photon is emitted, making the resulting final state $|f\rangle$ identical to the initial state. Here, a large spin-orbit splitting of $\sim 10\,\mathrm{eV}$ between Mn $2p_{3/2}$ and Mn $2p_{1/2}$ states is the origin of the sensitivity to the spin ordering, which also causes big magnetic circular dichroism in absorption spectroscopy [12].

The interaction Hamiltonian H' is obtained by replacing the momentum operator $\boldsymbol{p}$ by $\boldsymbol{p} - e\boldsymbol{A}$ in the free-electron Hamiltonian $H_0 = p^2/2m_e$, which gives

$$H' = -\frac{e}{m_e}\boldsymbol{p}\cdot\boldsymbol{A} + \frac{e^2\boldsymbol{A}^2}{2m_e}. \tag{23}$$

$$\boldsymbol{A} = \sqrt{\frac{2\pi\hbar c}{V}} \sum_{i,k} \frac{1}{\sqrt{k}} \boldsymbol{e}_i \left(a_{k,i}e^{i(\boldsymbol{k}\cdot\boldsymbol{r}-\omega t)} + a^{\dagger}_{k,i}e^{-i(\boldsymbol{k}\cdot\boldsymbol{r}-\omega t)}\right) \tag{24}$$

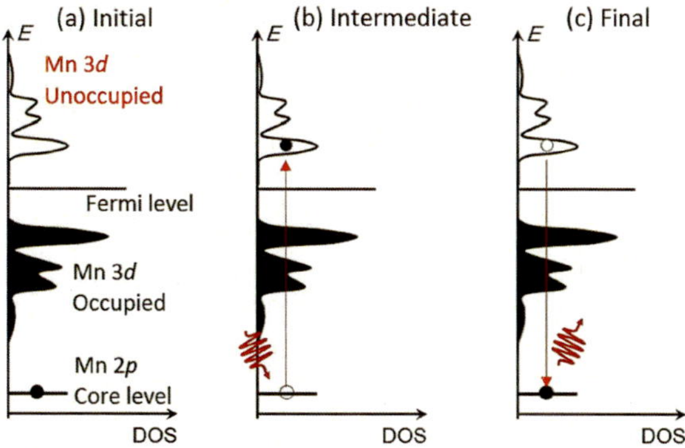

Fig. 3 Schematic illustration of resonant soft X-ray scattering process in the case of the Mn $2p \rightarrow 3d$ edge. The initial state (*left*), the intermediate state (*middle*), and the final states (*right*) are shown

The operators $a_{k,i}$ and $a_{k,i}^{\dagger}$ are the annihilation and creation operators of a photon with polarization e_i and wavevector k. The term $p \cdot A$ is linear in $a_{k,i}$ and $a_{k,i}^{\dagger}$, thus changes the number of photons by ± 1. The term A^2 is quadratic in $a_{k,i}$ and $a_{k,i}^{\dagger}$, and changes the number of photons by 0 or ± 2.

The transition probability per unit time, T_{fi}, from the initial state $|i\rangle$ to the final state $|f\rangle$ via intermediate states $|n\rangle$ is given up to second order by the Kramers–Heisenberg relation

$$T_{fi} = \frac{2\pi}{\hbar} \left| \langle f|H'|i\rangle + \sum_n \frac{\langle f|H'|n\rangle\langle n|H'|i\rangle}{E_i - E_n} \right|^2 \delta(E_i - E_f)\rho(E_f). \tag{25}$$

$\rho(E_f)$ is the density of states in the final state.

The first term describes the non-resonant Thomson scattering of photons through A^2 interaction (Fig. 4a), and the second term is from two successive $p \cdot A$ interactions (Fig. 4b, c).

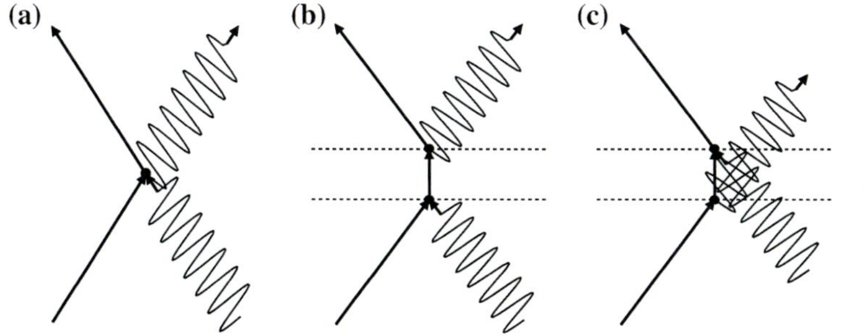

Fig. 4 Space-time diagrams for the scattering of a photon by an atom. The time runs from the *bottom* to the *top*. **a** The first-order perturbation of A^2. Absorption and emission of one photon, **b**, **c** the second-order perturbation of $p \cdot A$. Absorption and emission of one photon via a virtual intermediate state

The cross section in the dipole approximation is given by

$$\frac{d\sigma}{d\Omega} = \frac{\hbar^2 \omega^4}{c^2} \alpha_f{}^2 \left| \sum_n \frac{\langle f | \mathbf{r} \cdot \boldsymbol{\varepsilon}_f{}^* | n \rangle \langle n | \mathbf{r} \cdot \boldsymbol{\varepsilon}_i | i \rangle}{(\hbar\omega - E_{0n}) + i(\Delta_n/2)} \right|^2, \tag{26}$$

where $\boldsymbol{\varepsilon}_i$ and $\boldsymbol{\varepsilon}_i$ are the polarizations of incident and scattered X-rays. The resonant scattering factor is given by

$$f_{\mathrm{res}} = \left(\boldsymbol{\varepsilon}_f{}^* \cdot \boldsymbol{\varepsilon}_i \right) F^{(0)} - i \left(\boldsymbol{\varepsilon}_f{}^* \times \boldsymbol{\varepsilon}_i \right) \cdot \mathbf{m} F^{(1)} + \left(\boldsymbol{\varepsilon}_f{}^* \cdot \mathbf{m} \right) (\boldsymbol{\varepsilon}_i \cdot \mathbf{m}) F^{(2)}. \tag{27}$$

There are three different terms in Eq. (27). The first term does not depend on the spin direction, and corresponds to nonmagnetic resonant scattering. The polarization dependence $(\boldsymbol{\varepsilon}_f{}^* \cdot \boldsymbol{\varepsilon}_i)$ is the same as for Thomson scattering. The second term is linear in $\mathbf{m}$ and therefore describes magnetic resonant scattering. This term is related to the X-ray magnetic circular dichroism (XMCD) in absorption. The polarization dependence is $(\boldsymbol{\varepsilon}_f{}^* \times \boldsymbol{\varepsilon}_i)$, so the polarization is rotated by scattering [13, 14]. The third term is quadratic in $\mathbf{m}$ and also corresponds to resonant magnetic scattering. This is related to the X-ray magnetic linear dichroism (XMLD) in absorption.

Non-resonant elastic X-ray diffraction can also give information on magnetic structures, but magnetic X-ray scattering is a relativistic effect. The magnetic scattering cross section compared to the charge scattering is reduced by a factor of $(\hbar\omega/m_0 c^2)^2$ where $m_0 c^2 = 511\,\mathrm{keV}$ (electron rest mass). In the conventional X-ray diffraction, photon energies $\hbar\omega \sim 10\,\mathrm{keV}$, giving the factor as small as $\sim 10^{-4}$.

2.6 Experimental Setup

Figure 5 shows the schematic picture of RSXS. An incoming photon with a wave vector $\mathbf{k}$ hits on the sample and is scattered into the direction defined by $\mathbf{k}'$. This gives a scattering vector $\mathbf{Q} = \mathbf{k}' - \mathbf{k}$. In the specular geometry, both the incoming and the outgoing photons are at an angle θ with respect to the sample surface, and the scattering angle between $\mathbf{k}$ and $\mathbf{k}'$ is 2θ. The scattering plane is defined by the

Fig. 5 Schematic picture of resonant soft X-ray scattering with the definition of $\mathbf{k}$, $\mathbf{k}'$, $\mathbf{Q}$, θ, π, σ, π', σ', and φ

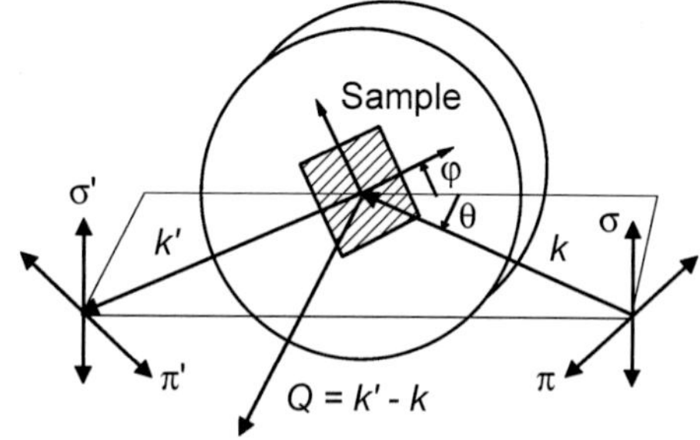

vectors k and k'. Polarization directions parallel or perpendicular to this scattering plane are referred to as π and σ polarizations, respectively.

Soft X-rays are subject to strong absorption by ambient atmosphere. Therefore, RSXS systems are designed as ultra-high vacuum (UHV) instruments. Let us show a 4-circle diffractometer with a horizontal scattering which was installed at beamline 10ID-2 (REIXS) of the Canadian Light Source, as shown in Fig. 6 [15].

The UHV diffractometer has a full range of sample (θ) and detector (2θ) rotations, spanning from $-25°$ to $+265°$, and limited tilted (χ) and azimuthal (φ) rotations, each spanning $\pm 4.5°$. All of these motions are achieved using in-vacuum stepper motors.

The detector arm has three detectors: photodiode, channeltron, and micro-channelplate (MCP). A photodiode is a most standard detector. Here the used photodiode is an AXUV100 Si photodiode from International Radiation Detectors, Inc. and has a $10\,\text{mm} \times 10\,\text{mm}$ square detection area. With the exception of a dark current

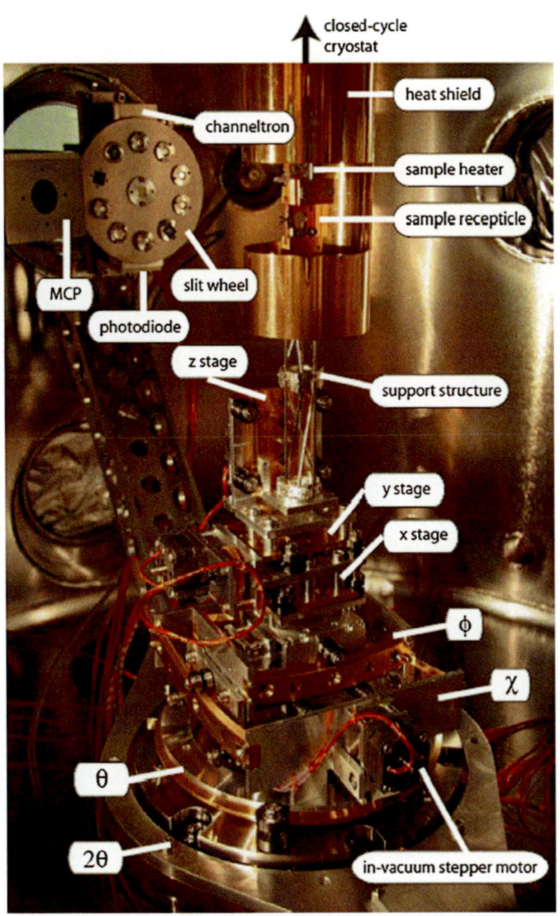

Fig. 6 UHV diffractometer installed at the Canadian Light Source [15]

of order 0.1–4 pA, the response of the photodiode is linear in photon intensities over a large dynamic range, extending from $\sim 10\,\mu$A for the direct beam to $\sim$ pA range for X-ray fluorescence, which is particularly suitable for reflectivity measurements. A channeltron is used for the detection of weak signals. MCP has a property for the two dimensional detection of scattering signals.

The azimuthal scan is a rotation around the Bragg scattering wavevector, and gives one the information of the direction of the ordered spins. The azimuthal rotation in the Canadian Light Source is limited to $\pm 4.5°$ [15], whereas the one in the Swiss Light Source is manual but covers the whole $360°$. Figure 7 shows the sample holder with the pins attached in a threefold symmetry. One can rotate the sample holder manually by using the sample transfer line with an accuracy of $\pm 2°$.

In order to obtain the complete information from RSXS experiments, polarization analysis of the scattered X-rays is necessary. This method uses multilayers with suitable lattice spacings as analyzers. With a Bragg diffraction angle corresponding to

Fig. 7 Sample holder with the pins attached in a threefold symmetry

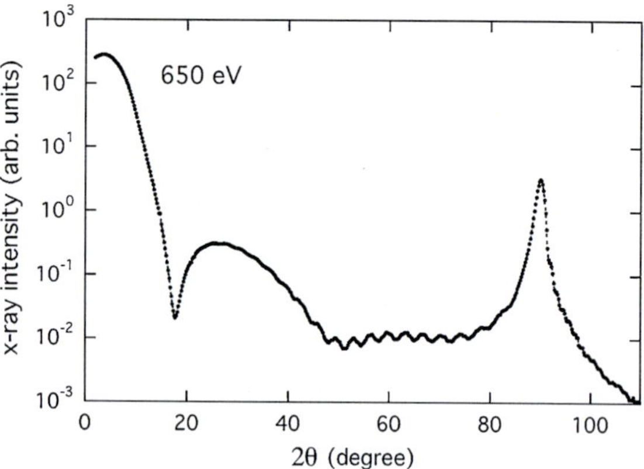

Fig. 8 Reflectivity of a W/C multilayer at a photon energy of 650 eV (in the vicinity of the Mn 2p edges) with σ incident polarization [16]

the Brewster angle ($2\theta = 90°$) at the given photon energy, X-rays with π polarization are ideally completely suppressed. By rotating the analyzer crystal around the X-ray beam scattered from the sample, one can characterize the polarization of the scattered X-rays. Staub et al. [16] reported an RSXS instrument with a polarization analyzer. Polarization analysis is performed by an artificial W/C multilayer. Figure 8 shows the reflectivity of the multilayer, demonstrating that the efficiency of such a multilayer at the first-order Bragg peak is of the order of a few percent.

3 Interfaces

Interfaces between two different transition-metal oxides often exhibit novel electronic properties that are not present in the component materials. For example, a major reconstruction of the orbital occupation and orbital symmetry in the interfacial CuO_2 layers at the $(Y,Ca)Ba_2Cu_3O_7/La_{0.67}Ca_{0.33}MnO_3$ interface has been observed [17]. Another particularly prominent topic is the interfaces between two band insulators STO and $LaAlO_3$ (LAO). This system is especially interesting due to the metallic conductivity [18] and even superconductivity [19] found at the interface. RSXS is quite suitable for extracting quantitative information on the electronic structures at these interfaces as will be shown below.

3.1 Cuprate Thin Films and Multilayers

Abbamonte et al. studied the carrier distribution in $La_2CuO_{4+\delta}$ thin films on STO substrates by RSXS [20].

In the Born approximation, the scattering amplitude can be expressed by

$$\chi(q) = \int_{-\infty}^{\infty} dx\, \chi(x) e^{iqx}, \tag{28}$$

where $\chi(x)$ is the spatially varying susceptibility and $q = 4\pi \sin\theta/\lambda$ is the momentum transfer. When the thin film is anisotropic, the susceptibility is a tensor

$$\chi(q) = \begin{pmatrix} \chi_{ab}(q) & 0 \\ 0 & \chi_c(q) \end{pmatrix}. \tag{29}$$

In this case the scattering matrix element, incorporating polarization effects, has the form

$$M = \boldsymbol{\varepsilon} \cdot \chi(q) \cdot \boldsymbol{\varepsilon}', \tag{30}$$

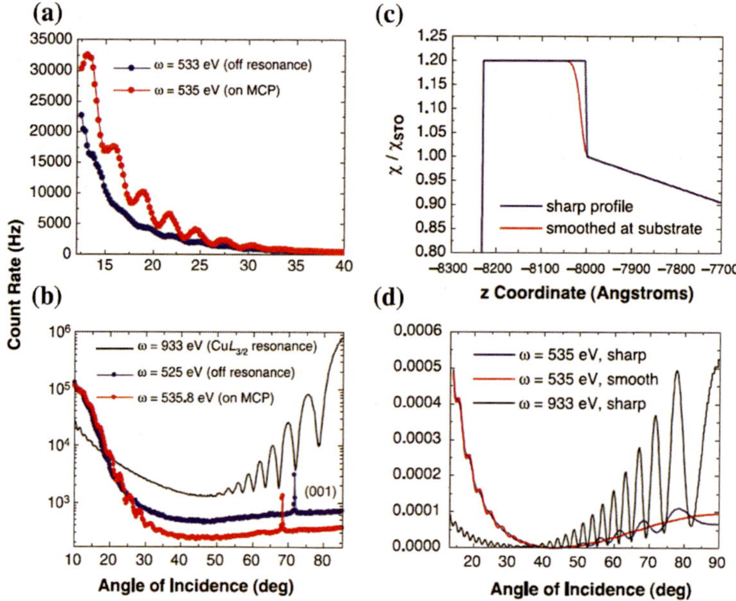

Fig. 9 **a**, **b** Reflectivity of $La_2CuO_{4+\delta}$ thin films on $SrTiO_3$ substrates [20]. **c** Susceptibility profiles for a perfect film on a substrate (*blue line*) and one that has been smoothed at the interface (*red line*). **d** Interference fringes resulting from the two profiles showing damping at high scattering angles for the smoothed layer. This is for comparison to (**b**)

where ε and ε' are the polarizations of incident and scattered X-rays. They plotted $|M|^2$ in Fig. 9. The thin-film reflectivities of these samples are characterized by well-developed thickness oscillations when measured at the Cu $2p$ edges, proving high sample homogeneity and surface/interface quality, as shown in Fig. 9. In contrast, with a photon energy corresponding to the mobile carrier peak, the oscillations vanish at higher angles of incidence. This peak in the O $1s$ pre-edge region represents a characteristic energy for scattering from the doped carriers. Hence, the reflectivity measured at this photon energy is almost entirely determined by the distribution of the doped carriers. The strong damping of the thickness oscillations at this energy is explained by assuming a smoothing of the carrier density at the interface between the thin film and the substrate.

Smadici et al. studied the distribution of doped holes in superlattices consisting of double layers of insulating La_2CuO_4 and overdoped $La_{1.64}Sr_{0.36}CuO_4$ [21], They determined the amount of charge redistribution at the interface by RSXS, and successfully obtained the origin of superconductivity in the multilayer of two non-superconducting materials.

3.2 *LaAlO₃/SrTiO₃ Interfaces*

There has been an intense debate about the origin of this metallicity; that is, whether it is due to oxygen vacancies "extrinsic" [22, 23] or due to the polar nature of the LAO structure [24], which could result in an "electronic reconstruction".

Wadati et al. investigated the electronic structure of the STO/LAO superlattice (SL) by RSXS [25, 26]. The SL sample consisted of seven periods of 12 ucs of STO and six ucs of LAO. Figure 10 shows the X-ray reflectivity spectra measured at 455 eV (Ti $2p$ off-resonance) (a) and 458.4 eV (Ti $2p$ on resonance) (b).

The reflectivity spectra in Fig. 10a, b show finite-size Fresnel oscillations corresponding to the total thickness of seven periods, and (002) and (003) Bragg peaks. The oscillations are clear at off resonance (b), but not evident at on-resonance (a). This is because the attenuation length of photons at 455 eV is about 100 nm, which is comparable to the total thickness when including the incident angle of about 20°–30°, but at 458.4 eV the attenuation length of only about 20 nm is much shorter than the total thickness and therefore the oscillations are not observed. Since the ratio of STO and LAO thicknesses is 2:1, the (003) peak would be forbidden in the infinitely thick and the zero absorption limit for samples of the ideal structure. The structure factors for (003) peaks are given as

$$S(003) = f_{TiO_2,int} - f_{TiO_2,bulk} + f_{AlO_2,int} - f_{AlO_2,bulk}, \tag{31}$$

where $f_{TiO2,int}$ and $f_{AlO2,int}$ are the scattering factors at the interface. $f_{TiO2,bulk}$ and $f_{AlO2,bulk}$ are those for bulk structures (SrO/TiO$_2$/SrO and LaO/AlO$_2$/LaO), and f_{SrO} and f_{LaO} are those for SrO and LaO layers, respectively. From Eq. (31) one can see that (003) is an interface-sensitive peak, which will be forbidden if the interface has the same scattering factor as bulk. The (003) forbidden peak was also used in the case of LaMnO$_3$/SrMnO$_3$ interface [28].

To analyze the reflectivity spectra in Fig. 10a, b, they used the recursive Parratt's method [9, 10] and simulated the reflectivity. Here four models were considered as shown in Fig. 10c. Model A is the case where all interfaces are sharp. Model C is the case without any sharp interfaces with or here the refractive index n is equal to $1 + \delta + i\beta$ at the interface taken to be the average of that of STO and LAO. These two models are considered as "symmetric models" because they do not consider the difference of LaO|TiO$_2$/SrO and SrO|AlO$_2$/LaO interfaces. Asymmetric models are models B1 and B2. These models include different interfaces as far as the refractive index is concerned. In model B1, only the SrO|AlO$_2$/LaO interfaces are sharp and in model B2, only the LaO|TiO$_2$/SrO interfaces are sharp. It was previously reported that metallic behavior is only observed at LaO|TiO$_2$/SrO and also the LaO|TiO$_2$/SrO interfaces are atomically less sharp than the SrO|AlO$_2$/LaO seeming to support model B1. Motivated by these studies, they investigated which model can best describe the reflectivity spectra. The comparison of the reflectivity spectra between experiment and calculation are shown in Fig. 10a, b. One should

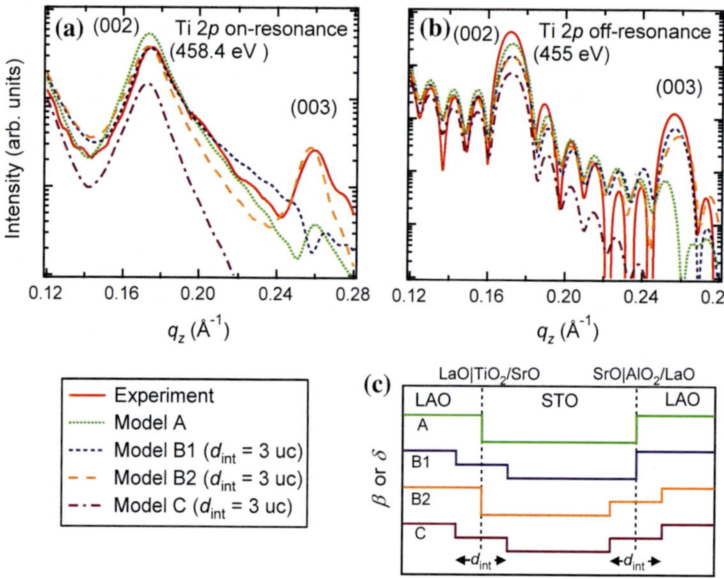

Fig. 10 Comparison of the reflectivity spectra between experiment and calculation. **a** Ti $2p$ on-resonance, **b** Ti $2p$ off-resonance, **c** models of the SL samples [25]

note here that the energy dependence of the attenuation length of photons (i.e., on-resonance and off-resonance) has been taken into account by the energy dependence of the imaginary part of the refractive index. From these figures one can see that the strong (003) peaks are only present in asymmetric models. In the two asymmetric models, model B1 reproduces the experimental results fairly well for Ti $2p$ off-resonance (b), but for Ti $2p$ on resonance (a), model B2 gives a better description of the experiment. Also they obtained the best fitting when the thickness of the interface (d_{int}) is taken to be about 3 uc. It should be noted here that in model A, with no reconstruction, the (003) peak is still present at on-resonance in Fig. 10a. This is due to the short penetration depth of the x rays at resonance, providing imperfect extinction as described before. From these results, they concluded that the SL is a highly asymmetric system with two different types of interfaces, and the thickness of the interface is about 3 uc, but they cannot conclude which model (B1 or B2) is better to describe the experiment.

There was another recent study of the LAO/STO SL by RSXS [27]. They obtained similar results but concluded that oxygen vacancies play a major role. The origin of the metallic interface is still controversial.

4 Charge and Orbital Ordering

4.1 $La_{0.5}Sr_{1.5}MnO_4$

The prototypical material is a layered manganite $La_{0.5}Sr_{1.5}MnO_4$. The average valence of Mn sites was 3.5+ at room temperature, and below $T_{CO} = 240\,$K, a charge ordering into Mn^{3+} and Mn^{4+} sites occurs. Below $T_N = 120$ K, so called CE-type antiferromagnetic ordering is realized. The orbital-ordered states in $La_{0.5}Sr_{1.5}MnO_4$ were studies by using resonant hard X-ray diffraction at the Mn $1s \to 4p$ edge [29]. The resonant enhancement of the orbital ordering peak (1/4 1/4 0) was observed at the Mn $1s \to 4p$ edge, and there was an effort to obtain the information of the Mn $3d$ states through the Coulomb interaction between $4p$ and $3d$ states [30].

Figure 11 shows a model for charge and orbital ordering of $Mn^{3+}:Mn^{4+} =1:1$ manganites. In this model, atoms 1 and 3 are Mn^{3+}, and atoms 2 and 4 are Mn^{4+} ions. Atomic scattering tensors for atoms 1 and 3 are given by

$$f_1 = \begin{pmatrix} \frac{f_\alpha+f_\beta}{2} & \frac{f_\alpha-f_\beta}{2} & 0 \\ \frac{f_\alpha-f_\beta}{2} & \frac{f_\alpha+f_\beta}{2} & 0 \\ 0 & 0 & f_\gamma \end{pmatrix} \quad f_3 = \begin{pmatrix} \frac{f_\alpha+f_\beta}{2} & \frac{f_\beta-f_\alpha}{2} & 0 \\ \frac{f_\beta-f_\alpha}{2} & \frac{f_\alpha+f_\beta}{2} & 0 \\ 0 & 0 & f_\gamma \end{pmatrix}, \tag{32}$$

and for atoms 2 and 4 by

$$f_2 = f_4 = \begin{pmatrix} f_{xx} & 0 & 0 \\ 0 & f_{yy} & 0 \\ 0 & 0 & f_{zz} \end{pmatrix}. \tag{33}$$

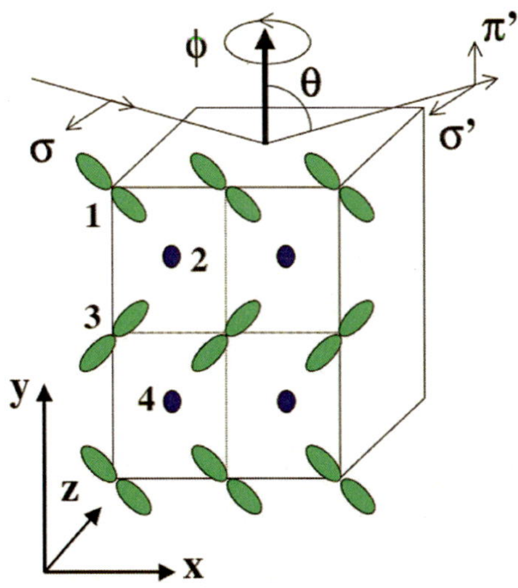

Fig. 11 Model for charge and orbital ordering of $Mn^{3+}:Mn^{4+} =1:1$ manganites. Atoms 1 and 3 are Mn^{3+}, and atoms 2 and 4 are Mn^{4+}. Only Mn atoms are shown [31]

The structure factor for the (1/4, 1/4, 0) orbital ordering is given by

$$F(1/4,\ 1/4,\ 0) = f_1 + if_2 - f_3 - if_4 = \begin{pmatrix} 0 & f_\alpha - f_\beta & 0 \\ f_\alpha - f_\beta & 0 & 0 \\ 0 & 0 & 0 \end{pmatrix}. \qquad (34)$$

The intensity of the orbital-ordering peak is

$$I_{\sigma\sigma}(1/4,\ 1/4,\ 0) = I_{\sigma\sigma}(1/4,\ 1/4,\ 0) = 0 \qquad (35)$$

$$I_{\sigma\pi}(1/4,\ 1/4,\ 0) = I_{\pi\sigma}(1/4,\ 1/4,\ 0) = |\left(f_\alpha - f_\beta\right)\cos\theta\sin\varphi|^2. \qquad (36)$$

However, it was theoretically pointed out that $4p$ states are delocalized and are rather sensitive to lattice distortions than orbital ordering [32]. It was then theoretically proposed that RSXS at the Mn $2p \rightarrow 3d$ edge, can directly give the information of the ordering of Mn $3d$ ortibals [33]. RSXS studies of $La_{0.5}Sr_{1.5}MnO_4$ were subsequently reported in Refs. [34–37].

Wilkins et al. studied the orbital and magnetic reflections at wave vectors of $\boldsymbol{q}_{OO} = (1/4, 1/4, 0)$ and $\boldsymbol{q}_{AF} = (1/4, 1/4, 1/2)$, respectively [34, 37]. Figure 12 (left) shows the energy dependence of the scattered intensity at the orbital order wave vector $\boldsymbol{q}_{OO}$ at the Mn $2p$ edges. Figure 12 (right) shows analogous data for the antiferromagnetic order peak $\boldsymbol{q}_{AF}$. It can be clearly observed in these figures that the orbital and magnetic reflection exhibit strong resonances at the Mn $2p$ edges. The insets show the temperature dependences of the orbital and magnetic order parameters, respectively. The orbital order parameter decreases continuously with increasing temperature and disappears at $T_{OO} = 230\,$K. The intensity of the antiferromagnetic peak shows a similar behaviour but disappears at a lower temperature of $T_N = 120\,$K. In both figures the experimental data are compared with theoretical calculations of the RSXS spectra based on atomic multiplet calculations in a crystal field. In the calculations the crystal fields were modified in such a way as to fit the experimental data. In the panels (a) and (b) the cubic and the tetragonal crystal field parameters were adjusted for a $d_{x^2-z^2}$ and $d_{3x^2-r^2}$ type of orbital ordering, respectively. In the panel (c) for the $d_{x^2-z^2}$ type ordering a small orthorhombic crystal field component was added. From the comparison of the experimental data with the theoretical calculations, they concluded that the orbital ordering below T_{OO} is predominantly of $d_{x^2-z^2}$ type and that the inclusion of a small orthorhombic crystal field component slightly improved the fits. However, details of the complicated experimental lineshape could not be reproduced by the calculation, especially in the case of the antiferromagnetic peak. Recent two linear dichroism studies in X-ray absorption also gave contradictory results on the type of orbital ordering of this material [38, 39]. Determining $d_{3x^2-r^2}$ or $d_{x^2-z^2}$ type still remains an open issue.

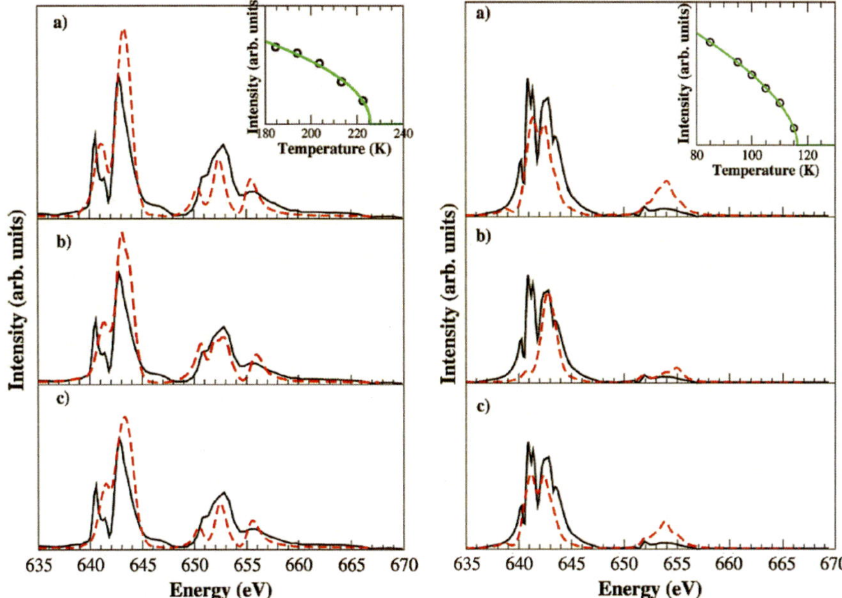

Fig. 12 *Left* Photon-energy dependence of the orbital-ordering (1/4, 1/4, 0) peak at 63 K (*full black line*) with the theoretical fits (*dashed red lines*) for **a** $d_{x^2-z^2}$-type, **b** $d_{3x^2-r^2}$-type, and **c** $d_{x^2-z^2}$-type with an orthorhombic crystal field. The *inset* shows the temperature evolution of the orbital order parameter. *Right* Photon-energy dependence of the antiferromagnetic ordering (1/4, 1/4, 1/2) peak at 63 K (*full black line*) with the theoretical fits (*dashed red lines*) for **a** $d_{x^2-z^2}$-type, **b** $d_{3x^2-r^2}$-type, and **c** $d_{x^2-z^2}$-type with an orthorhombic crystal field. The *inset* shows the temperature evolution of the antiferromagnetic order parameter [37]

4.2 $Pr_{1-x}Ca_xMnO_3$

The hole-doped perovskite manganites $R_{1-x}A_xMnO_3$, where R is a rare-earth ($R =$ La, Nd, Pr) and A is an alkaline-earth atom ($A =$ Sr, Ba, Ca), exhibit charge, and orbital ordering around $x = 0.5$ [40]. $Nd_{1-x}SrMnO_3$ shows ordering only in the vicinity of $x = 0.5$, whereas $Pr_{1-x}CaMnO_3$, which has a smaller bandwidth, has a particularly stable ordered state in a wide hole concentration range $0.3 \leq x \leq 0.75$. RSXS studies of $Nd_{0.5}Sr_{0.5}MnO_3$ and $Pr_{1-x}CaMnO_3$ were reported in Refs. [41–43].

Figure 13 shows resonance profiles of the ordering peak (1/4, 1/4, 0) measured at 65 K on (a) $Pr_{0.7}Ca_{0.3}MnO_3$ and (b) $Pr_{0.5}Ca_{0.5}MnO_3$ [43]. The XAS spectra are overall very similar, but there are striking differences in the resonance profiles. This can be explained by the existence (absence) of spin canting in $Pr_{0.7}Ca_{0.3}MnO_3$ ($Pr_{0.5}Ca_{0.5}MnO_3$).

The magnetic and electronic phases of Mn oxides can be controlled in thin films grown on substrates with various lattice parameters. Wadati et al. performed an RSXS measurements of $Pr_{0.5}Ca_{0.5}MnO_3$ thin films grown epitaxially on $(LaAlO_3)_{0.3}$-

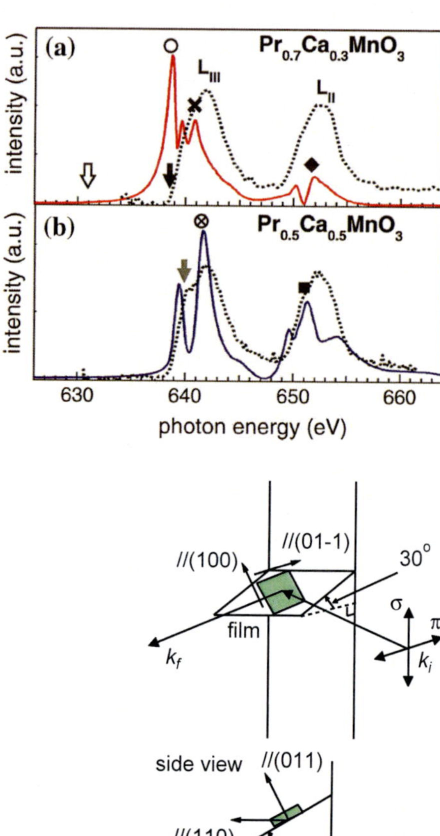

Fig. 13 Resonance profiles of the ordering peak (1/4, 1/4, 0) measured at 65 K on **a** $Pr_{0.7}Ca_{0.3}MnO_3$ and **b** $Pr_{0.5}Ca_{0.5}MnO_3$. The *dotted lines* are the XAS spectra measured in the total fluorescence yield (TFY) mode (rescaled by 100 times) [43]

Fig. 14 The experimental geometry with the [110] axis in the scattering plane and using a simple cubic perovskite cell notation [45, 46]

$(SrAl_{0.5}Ta_{0.5}O_3)_{0.7}$ (LSAT) (011) [44] to study how the microscopic magnetic ordering in thin films can differ significantly from that of the corresponding bulk materials [45]. Figure 14 shows the experimental geometry to align the [110] axis in the scattering plane.

Figure 15 shows the photon-energy dependence of the intensity observed across the Mn $2p$ edges at various temperatures using π and σ polarizations. To facilitate a comparison of the lineshapes, panels (c) and (d) show the same data as given in (a) and (b), but this time normalized to the area. As can be observed in Fig. 15, I_π and I_σ are of very similar magnitude and exhibit the same lineshapes for 150 K < T < 200 K. These lineshapes are reminiscent to that of corresponding bulk materials [34–37], supporting interpretation in terms of orbital ordering. However, the present data for the film display differences from those for the bulk, indicative of a modified orbital state. The condition $I_\pi = I_\sigma$ clearly breaks down at $T_N = 150$ K upon cooling: while I_π shows a strong increase by a factor of 10 accompanied by a clear lineshape change, I_σ remains almost unaltered. They conclude that it is resonant magnetic scattering of the Mn-sublattice. They observed another dramatic change of the RSXS linshapes

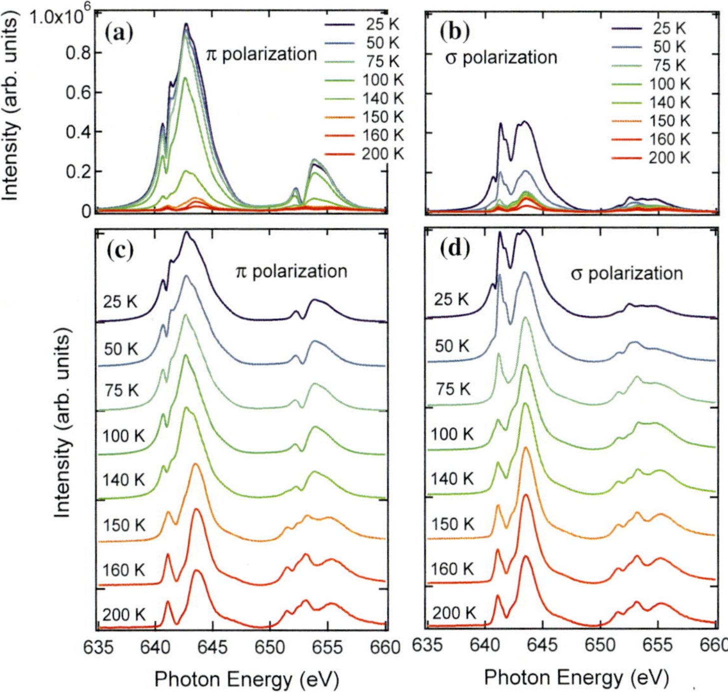

Fig. 15 Photon-energy dependence of the (1/4, 1/4, 0) peak intensity at Mn $2p$ edges at various temperatures using polarization (**a**) and polarization (**b**). To compare lineshapes, all data in the **a** and **b** have been normalized to each spectrum area, shown at the respective **c** and **d** as a function of photon energies [45]

at $T_2 = 75\,\text{K}$ well below T_N. This time, I_σ displays clear changes of lineshape and intensity, as shown in Fig. 15b, d. The variation of I_σ across T_2 signals a third phase transition, which has not been reported earlier and is absent in the bulk material. From further analyses, they concluded that T_2 are due to a spin reorientation.

Nakamura et al. fabricated superlattices of $La_{0.5}Sr_{0.5}MnO_3$ (LSMO) and $Pr_{0.5}Ca_{0.5}MnO_3$ (PCMO) on LSAT (011) substrates [47]. Pure LSMO and PCMO thin films on LSAT (011) substrates show FM ($T_C \sim 310$ K) and CO ($T_{CO} \sim 220\,\text{K}$), respectively. They found that they could control the phase boundary at the interface between the ferromagnetic metallic LSMO and the antiferromagnetic and CO/OO PCMO by applying magnetic fields or changing the individual layer thicknesses. In addition, they observed superlattice reflections by hard X-rays, indicative of the structural distortions induced by the charge and orbital order in these systems. To gain more insight into these problems of phase transitions and phase competitions, Wadati et al. investigated the charge-orbital ordering in [PCMO (5 layers)/LSMO (5 layers)]$_{15}$ superlattices by RSXS.

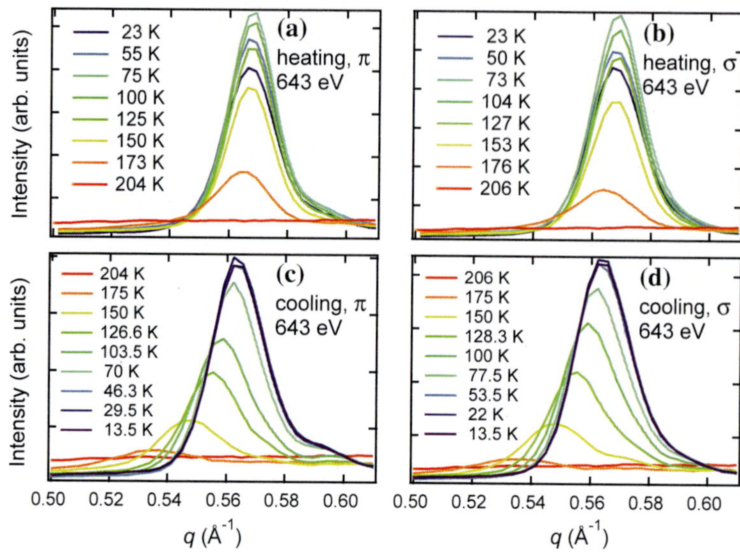

Fig. 16 Temperature dependence of the (*hh*0) peak. **a** and **b** (**c, d**) were measured in the heating (cooling) cycle. Incident X-ray polarizations were π in **a** and **c** and σ **b** and **d**. All the data were taken at $h\nu = 643\,\text{eV}$ (Mn $2p_{3/2} \rightarrow 3d$ absorption edge) [46]

Figure 16a, b show the q dependence of the orbital order (*hh*0) reflection in a simple cubic perovskite notation for various temperatures at the Mn $2p_{3/2}$ edge (643 eV) for both π and σ incoming X-ray polarizations. The temperature dependence of this reflection in a heating run [panel (a)] significantly deviates from that in a cooling run [panel (b)]. The reflection appears below $\sim$200 K, which is almost the same as the pure PCMO thin film [45].

The temperature variation of the corresponding peak intensity, the correlation length, and positions are summarized in Fig. 17. The correlation length ξ is obtained from the full width at half-maximum (FWHM) Δq through $\xi = 2\pi/\Delta q$. In all of these quantities, a large hysteresis is observed between cooling and heating cycles. The peak intensity monotonically increases for decreasing temperatures in contrast to the heating cycle, where, interestingly, it first increases and then decreases before it disappears around 200 K. The peak position is incommensurate and temperature dependent in the full range of the cooling cycle. In contrast, the position remains frozen until temperatures of approximately 170 K for the heating cycle. The correlation length also shows an interesting behavior. In the heating cycle between 50 and 150 K, the peak intensity reaches a maximum around 70 K, and the correlation length shows the largest values at a slightly higher temperature. This result shows that there is a clear connection among the improved order (maximal intensity), the correlation (maximal correlation length), and the largest h. To test if this relation is quantitatively the same for cooling and heating cycles, they show in panel (d) the

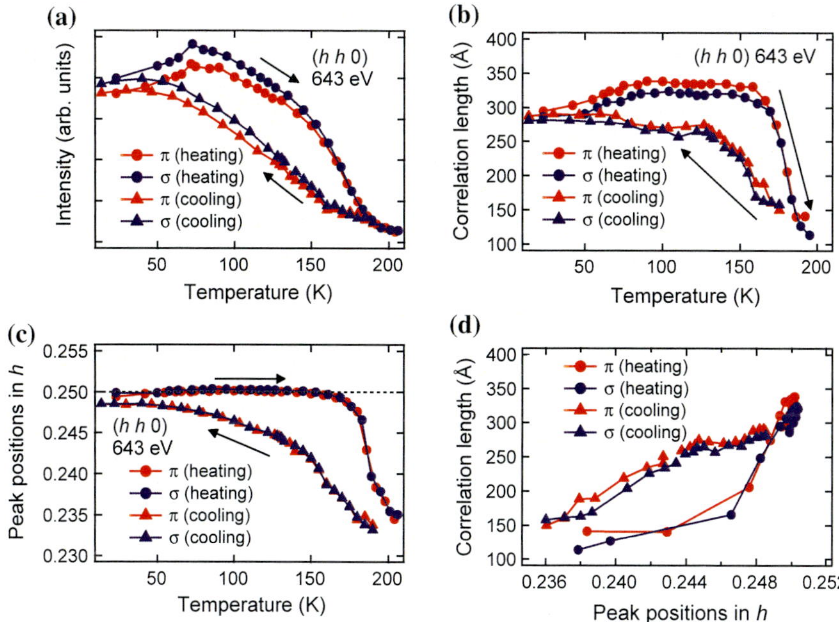

Fig. 17 Temperature dependence of the (hh0) peak intensity (**a**), correlation length (**b**), and positions (**c**). In (**d**), the correlation length is plotted as a function of peak positions [46]

correlation length as a function of peak position. The length is linearly dependent on the peak position for both the cooling and the heating cycles, with different slopes.

These results now lead to the following picture in Fig. 18. At 25 K, a correlation length is about $\sim$300 Å from Fig. 17b, giving the effective correlation length perpendicular to the film surface of $\sim$300cos(60°) Å = 150 Å, where cos(60°) comes from the geometry in Fig. 14. The correlation length of $\sim$150 Å indicates that the correlation between the CO/OO states perfectly bridges the FM layers built in between, making the CO/OO states three dimensional (one unit of [PCMO (5 layers)/LSMO (5 layers)] is 30 Å). This is visualized in the left panel of Fig. 18. This orbital coupling bridging over the FM LSMO layers becomes weak when the temperature increases, because the interface CO/OO states disappear first, leading to an effective increase of the separation of the CO/OO layers as shown in the middle panel of Fig. 18. Near 200 K, the effective correlation length becomes as short as $\sim$100 cos(60°) Å = 50 Å as seen in Fig. 17b. This can cover fewer than two units of [PCMO (5 layers)/LSMO (5 layers)], making the CO/OO states 2D-like.

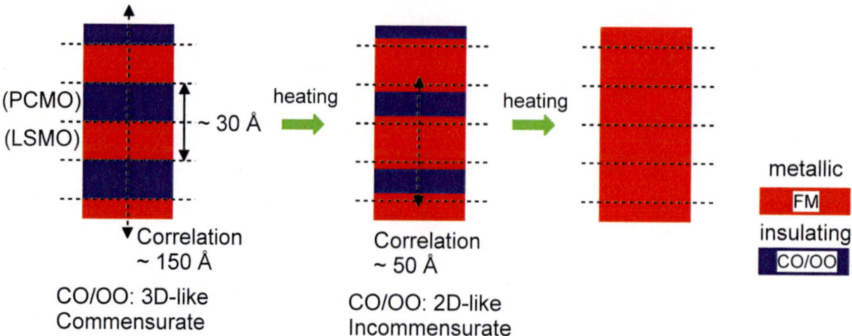

Fig. 18 Proposed real-space sketches for the multilayer at different temperatures (or temperature cycle history). Ferromagnetic interfaces lead to charge transfer and to a reduction of the $3d$ coupling of the orbital order, represented by incommensurability and dynamics reducing the correlation length [46]

5 Multiferroics

5.1 *Bulk Studies*

Recently, there has been a lot of interest in multiferroic materials displaying both ferroelectric and magnetic orders. It is of particular importance to control magnetization (electric polarization) by an electric (magnetic) field as this has large potential for novel device applications. This can be most easily achieved by materials with giant magnetoelectric couplings. Orthorhombic (o) $RMnO_3$ (R denotes rare earth metal) with perovskite structure belongs to this category and can be viewed as prototypical multiferroic materials. For example, in $TbMnO_3$, ferroelectricity occurs below 28 K [48], concomitant with the onset of cycloidal spin ordering. The ferroelectricity in the cycloidal states is realized by the shifts of the oxygen ions through the inverse Dzyaloshinskii–Moriya interaction [49, 50].

$$P \propto \sum_{i,j} e_{i,j} \times \left(S_i \times S_j \right) \tag{37}$$

This is in contrast to E-type antiferromagnetic structures, where ferroelectricity is caused by symmetric exchange striction. E-type magnetic structures occur in o-$RMnO_3$ with smaller R ions. It is predicted that the E-type structure leads to a larger polarization, which has been experimentally confirmed in the o-$RMnO_3$ systems.

TbMnO$_3$ is the prototypical material. Cycloidal magnetic ordering was observed by neutron diffraction, which is the origin of the ferroelectric polarization [51]. Figure 19 shows magnetic diffraction of $TbMnO_3$ taken at 20 K along the (0, 1, 0) direction. Three superlattice reflections are observed at $(0, \tau, 0)$, $(0, 2\tau, 0)$, and $(0, 1 - 2\tau, 0)$ with $\tau \sim 0.28$. The $(0, \tau, 0)$ was assigned to incommensurate ordering of

Fig. 19 Magnetic diffraction from TbMnO$_3$. *Top* Scan along the [010] direction. *Bottom* Azimuthal dependence of the three magnetic peaks, providing information on the magnetic-moment direction [52]

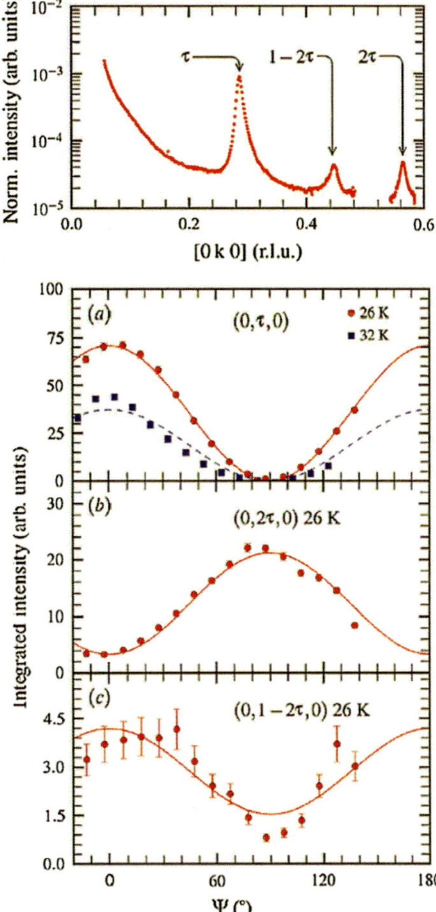

the magnetic moments on the Mn sites and the $(0, 2\tau, 0)$ to be the second harmonic from the resonant X-ray cross section. The $(0, 1 - 2\tau, 0)$ reflection corresponds to the second harmonic from the $(0, 1 - \tau, 0)$ reflection.

Figure 19 shows the azimuthal dependence of the three magnetic peaks. In the top panel, the $(0, \tau, 0)$ reflection is shown at 26 K (red circles) and 32 K (blue squares). While the maximum integrated intensity changes between these two temperatures, the functional form is identical, indicating that the direction of this Fourier component does not change across $T_{N2} = 28$ K, where a spontaneous electric polarization appears. This azimuthal dependence can be fitted to the form

$$I_\tau(\varphi) = A \cos^2 \varphi, \tag{38}$$

where A is a constant and φ is the azimuthal angle. Thus this Fourier component is along the c axis. This result is confirmed by the azimuthal dependence of the $(0, 2\tau, 0)$ signal at 26 K. This can be fitted to the form

$$I_{2\tau}(\varphi) = A' + B' \cos^2 \varphi. \tag{39}$$

With the assumption that this reflection is the second harmonic from the resonant X-ray cross section, this is also consistent with a Fourier component along the c axis.

5.2 Thin Films

The fabrication of the o-RMnO$_3$ thin films has been especially important for device application of the multiferroic materials. Moreover, bulk o-RMnO$_3$ samples with smaller R ions ($R =$ Y, Ho, Lu) can only be synthesized under high oxygen pressure [53], which strongly limits studies on the most interesting materials due to the absence of significantly large high-quality single crystals. Recently, Nakamura et al. reported the fabrication of o-YMnO$_3$ thin films onto the YAlO$_3$ (010) substrate [54]. Their thin film showed a ferroelectric transition at 40 K with a large saturation polarization of 0.8 μC/cm^2. The ferroelectric polarization could be controlled by magnetic fields, demonstrating magnetoelectric behaviors. There are three steps in phase transitions: the antiferromagnetic transition at $T_N = 45$ K, the ferroelectric transition at $T_C = 40$ K, and the inflection of electric polarization at 35 K. Wadati et al. investigated the magnetic structure by RSXS and hard X-ray diffraction to clarify the exact magnetic structure.

Figure 20 shows the temperature dependence of the $(0\, q_b\, 0)$ ($q_b \sim 0.5$) peak with π (a) and σ (b) incident X-ray polarizations. Here the diffraction data were taken at $h\nu = 643.1$ eV (Mn $2p_{3/2} \to 3d$ absorption edge). This peak, which is indicated by vertical bars, appears at 45 K, which coincides with the antiferromagnetic transition temperature T_N determined from magnetization measurements [54]. Weaker peaks are observed on both sides of the reflection. These are antiferromagnetic Kiessig fringes, and describe the limited thickness of the magnetic contrast of the film. There is almost no difference between π (a) and σ (b) polarizations. The intensity of the peaks increases monotonically with cooling. The peak position deviates from the commensurate $q_b = 1/2$ position for all temperatures.

The azimuthal-angle dependence allows one to gain information on the directions of the Mn spins. Figure 21 shows the φ (azimuthal angle) dependence of the intensity of the magnetic $(0\, q_b\, 0)$ reflection. The intensities for π and σ incident polarizations are fitted by

$$I(\pi) = A + B \cos^2 \varphi \tag{40}$$

$$I(\sigma) = A' \cos^2 \varphi, \tag{41}$$

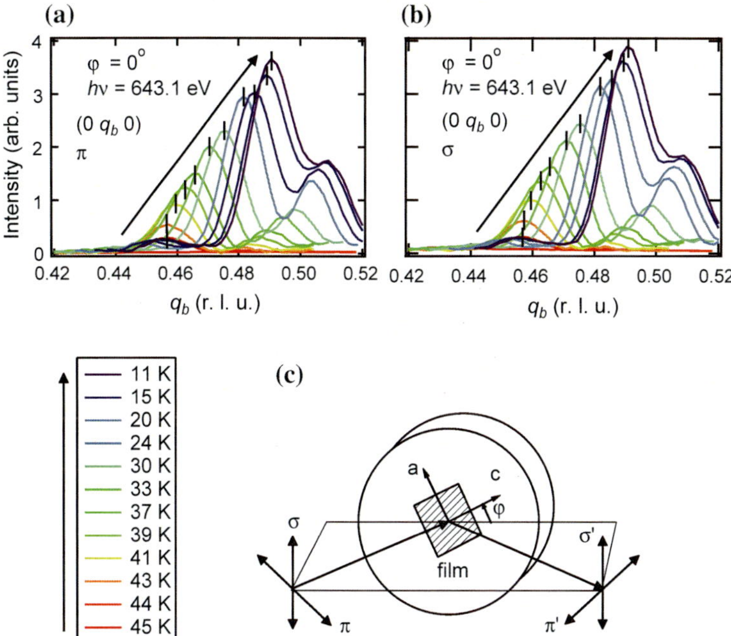

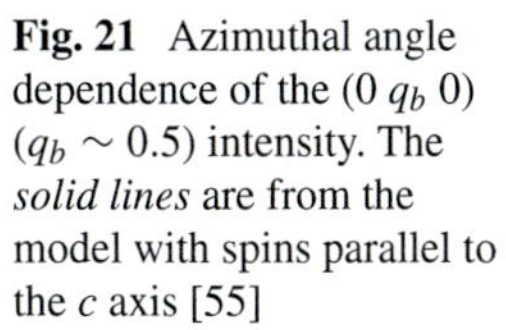

Fig. 20 Temperature dependence of the $(0\,q_b\,0)$ $(q_b \sim 0.5)$ peak in π (**a**) and σ (**b**) incident X-ray polarizations. **c** The experimental geometry with the definition of the azimuthal angle φ [55]

Fig. 21 Azimuthal angle dependence of the $(0\,q_b\,0)$ $(q_b \sim 0.5)$ intensity. The *solid lines* are from the model with spins parallel to the c axis [55]

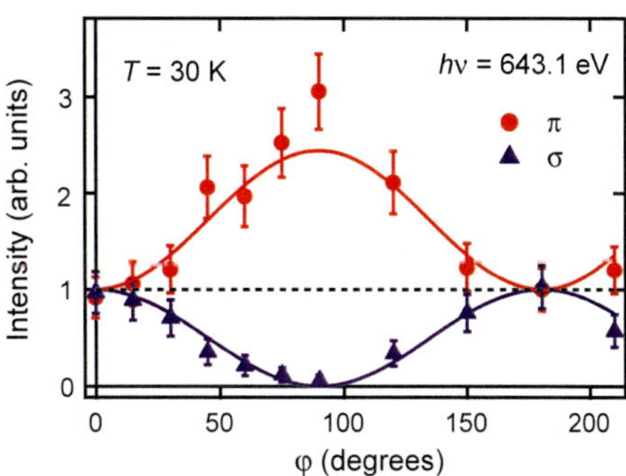

indicating that the magnetic component is along c axis. This reflects an *ab* cycloid with a spin canting along the c axis, and indicates that the experiment is only sensitive to its magnetic sinusoidal c axis component.

In order to investigate the lattice distortions associated with magnetic order and electric polarization, hard X-ray diffraction measurements were also performed. The commensurate (0 1 0) reflection appears below 35 K as shown in Fig. 22a. This reflection is structurally forbidden in the chemical high-temperature structure (Pbnm) and caused by the lattice distortion accompanying ferroelectricity. Interestingly, no incommensurability of this reflection is observed by hard X-ray diffraction, in clear contrast to the observed magnetic reflection. Moreover, this reflection does appear below 35 K, at lower temperatures than the onset of the magnetic reflection, in accord with the step onset of the spontaneous electric polarization [54], as can be seen from the temperature-dependent integrated intensity shown in Fig. 22b.

They obtained a full picture of the magnetic states of the epitaxial $YMnO_3$ thin film by combining the above results with the macroscopic measurements of magnetization and electric polarization [54]. The ground state of the $YMnO_3$ thin film can be explained by the coexistence of the cycloidal and the E-type states as shown in Fig. 23 and as theoretically predicted [56]. In this coexistence region, magnetic reflection is incommensurate and lattice peaks are commensurate because the E-type phase has a much larger lattice distortion than the cycloidal phase. The existence of the E-

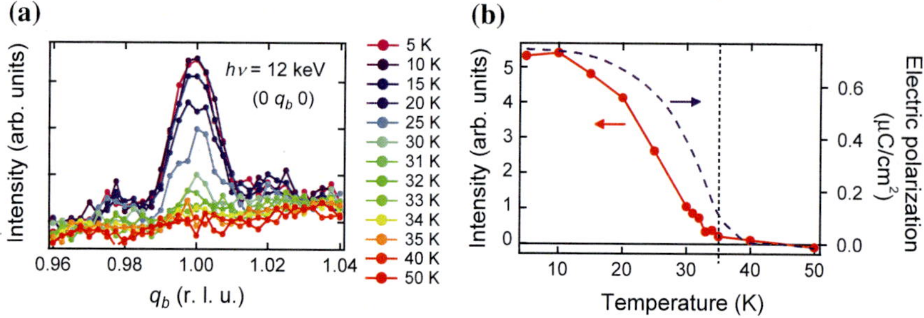

Fig. 22 **a** Temperature dependence of the (0 1 0) peak taken at $h\nu = 12\,\text{keV}$. In **b**, peak intensities are plotted as a function of temperature together with the electric polarization (*broken lines*) taken from Ref. [54]. The temperature of 35 K is also indicated as the onset of the (0 1 0) peak and the step onset of the spontaneous electric polarization [55]

Fig. 23 Spin structures in the E-type (**a**) and the *ab*-cycloidal (**b**) states. Spin canting along the c axis makes the magnetic (0 q_b 0) peak have some intensity [55]

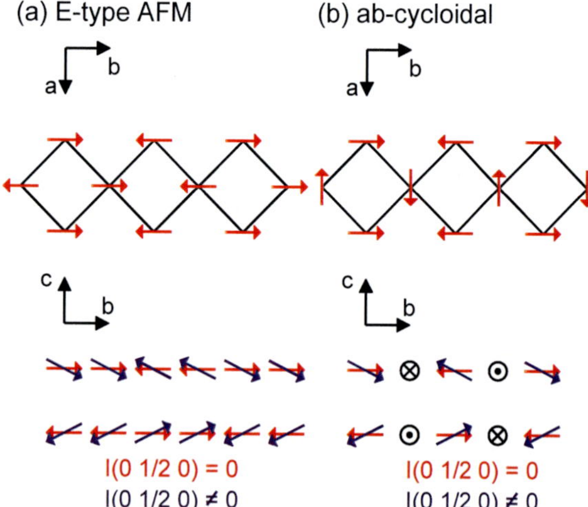

type phase causes the large electric polarization of $0.8\,\mu C/cm^2$ due to the symmetric exchange striction.

Similar study of o-LuMnO$_3$ thin films on YAlO$_3$ substrates were reported in Ref. [57]. The results also showed incommensurate magnetic peaks and commensurate lattice-distortion peaks.

6 Measurements Under Magnetic Fields

In order to study magnetic phase transitions, it is necessary to apply an external magnetic field. RSXS is a photon-in-photon-out type experiment, so one can apply a magnetic field during measurements. This is a great advantage over a photon-in-photon-out type photoelectron spectroscopy [8] because there outgoing electrons are affected by magnetic fields. RSXS systems with a superconducting magnet are now operated in several synchrotron facilities such as BESSY and SPring-8.

6.1 Charge Density Wave in Cuprates

In present RSXS measurements, it is not technically possible to apply magnetic fields high enough to completely suppress superconducting states, but the behavior of the observed features in the lower fields that are already available is instructive. Figure 24 shows the effects of magnetic fields on the charge-density wave (CDW) peak (0, 0.325) in ortho-II YBa$_2$Cu$_3$O$_{6.55}$ [58]. Below $T_c = 61$ K, the integrated intensity of the peak increases linearly with field for both H and K directions. At 6 T, they observed an increase of the integrated intensity by a factor ~ 2 in both directions (either due to an enhanced CDW modulation amplitude, or due to an enhanced volume coverage of CDW domains), and they do not observe any sign of saturation. Above T_c, the zero-field and high-field intensities match, in agreement with the notion that the field enhancement of the CDW fluctuations arises from a competition between CDW and superconducting instabilities. This is also supported by the weak decrease of the peaks FWHM with field for $T < T_c$ (Fig. 24d).

6.2 Devil's Staircase

The recently discovered Co oxide SrCo$_6$O$_{11}$ is a particularly interesting material with intrinsic magnetoresistance [59]. One of the most striking magnetic features is plateaus in the magnetization as a function of the applied magnetic field along the c-axis. These plateaus correspond to 1/3 and 3/3 of the saturated moment and were found to reflect different stackings along c, namely an $\uparrow\uparrow\uparrow$ structure for the

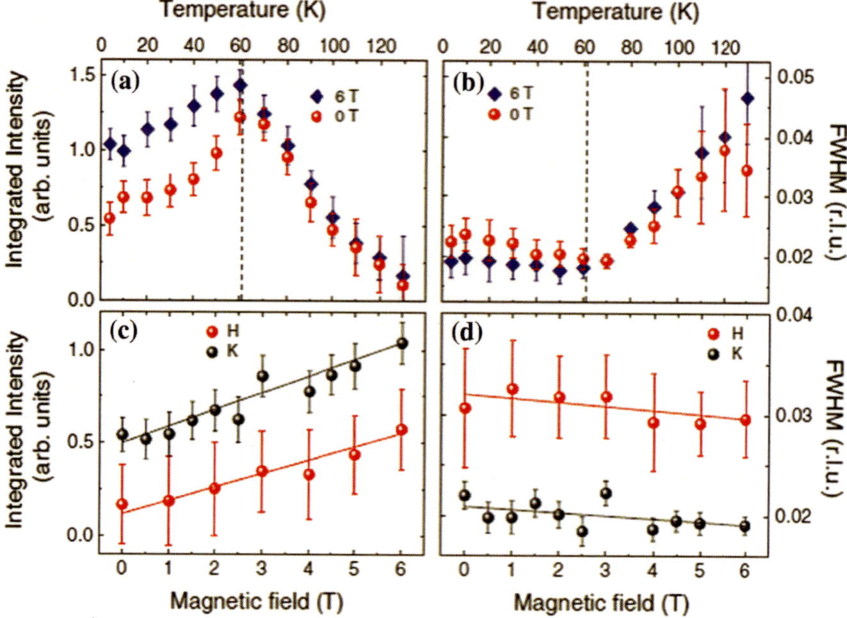

Fig. 24 **a** Temperature dependence of the integrated intensity of the CDW peak in ortho-II YBa$_2$Cu$_3$O$_{6.55}$ along (0, K, 0) for magnetic fields 0 and 6 T. **b** FWHM of the peak at two fields. Magnetic field dependence of **c** the intensity and **d** the FWHM at 10 K [58]

3/3 phase and an ↑↑↓ configuration for the 1/3 phase. Matsuda et al. performed an RSXS study as a function of temperature and applied magnetic field [60].

Figure 25b shows the diffraction peaks of SrCo$_6$O$_{11}$ at zero magnetic field for various temperatures. Quite surprisingly and very uncommon for RSXS experiments, a large number of superlattice reflections at $L = 2/3$, 5/6, 1, 7/6, 4/3 and 3/2 is observed. The small and temperature independent peak at $L = 1.37$ is assigned to some impurity in the sample because it does not show temperature dependence. $L = 1$ commensurate (CM) peak and two incommensurate (ICM) peaks around $L = 0.8$, 1.2 appear at 20 K (T_{c1}). These ICM peaks move to $L = 5/6$ and 7/6, respectively, as the temperature is decreased, and finally are locked at these values at 12 K (T_{c2}), respectively. At T_{c2}, there appear $L = 5/6$ and 8/7 shoulders of the $L = 7/6$, and simultaneously $L = 2/3$, 4/3 and 3/2 peaks. Intensities of all the magnetic peaks were independent of the polarizations σ and π, as shown in Fig. 25d for the case of $L = 6/5$. For a trigonal local symmetry and spins along the c direction, the magnetic scattering factor can be expressed as

$$f_{mag} = \begin{matrix} \sigma' \\ \pi' \end{matrix} \begin{pmatrix} & \overset{\sigma}{0} & \overset{\pi}{m_c \sin\theta} \\ & m_c \sin\theta & 0 \end{pmatrix} \tag{42}$$

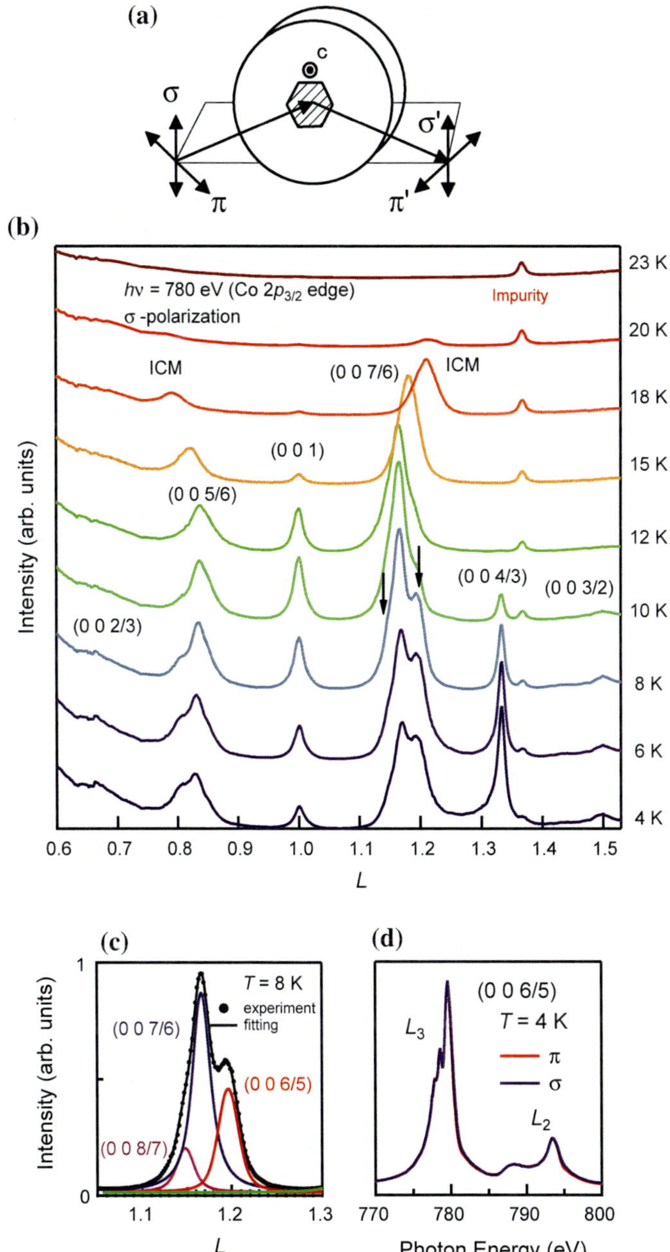

Fig. 25 **a** Experimental geometry for RSXS measurements. The *arrows* indicate the directions of polarizations of X-rays. **b** Magnetic peak profile of $SrCo_6O_{11}$ at various temperatures at zero magnetic field. **c** Magnetic peak fitting of $SrCo_6O_{11}$ around $L = 7/6$. **d** Photon-energy dependence of intensity of the $L = 6/5$ reflection for σ and π polarizations [60]

Fig. 26 Magnetic phase diagram of $SrCo_6O_{11}$ determined by RSXS measurements. The phase boundary between ↑↑↑ and ↑↑↓ states was determined by magnetization measurements [60]

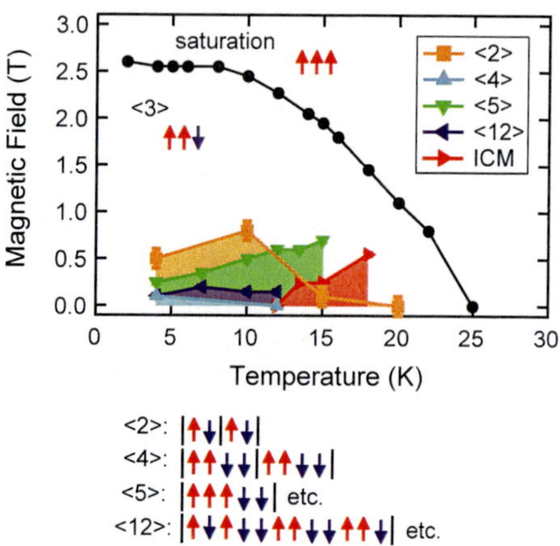

where m_c is the components of the spins along the c-axis and θ is the scattering angle. In this case, pure σ-π' and π-σ' channels have the same intensity, which agrees very well with the experimental results and verifies the interpretation in terms of magnetic scattering. The emergence of the magnetic $L = 2/3$ and $4/3$ peaks agrees well with the powder neutron diffraction measurement at 2 T. In order to assign the shoulder peaks around $L = 7/6$, the data were fitted by three components of $L = 7/6$, $8/7$ and $6/5$ as shown in Fig. 25c. These results therefore directly reveal that a large number of magnetic phases coexist in zero magnetic field and, in particular that the ↑↑↓ configuration is realized even at zero magnetic field. They observed the magnetic peaks of $L = n/6$ with $n = 4, 5, 6, 7, 8$, and 9. However, the temperature dependence varies for different n and the peaks with $n = 5$ and 7 show the shift from ICM to CM peak position and the others do not. This indicates that all the different peaks cannot be due to one magnetic modulation with $L = 1/6$ but belong to different magnetic stacking sequences. This is also reflected in the observed field dependent behaviour.

They have been able to explore the entire H-T diagram of this complex magnetic system as shown in Fig. 26. The phase boundary between ↑↑↑ and ↑↑↓ states was determined by magnetization measurements, and the other boundaries were determined by the present RSXS results. Here, $\langle n \rangle$ represents the magnetic periodicities. Since the $SrCo_6O_{11}$ unit cell contains two equivalent Co(3) Bragg planes along c, (002) is the first allowed structural reflections and (001) corresponds to a simple ↑↓↑↓ antiferromagnetic order. Therefore $\langle 2 \rangle$ corresponds to $L = 1$, $\langle 4 \rangle$ to $L = 3/2$, $\langle 5 \rangle$ to $L = 4/5$, and $\langle 12 \rangle$ to $L = 5/6$. The phase diagram demonstrates that various magnetic orderings with different periodicities are formed in the low temperature and low field region. Obviously, the energies of these magnetic structures are quite close, and the corresponding energy differences sensitively depend on temperature and magnetic fields. This phenomenon is called "the devil's staircase".

7 Time-Resolved Measurements

So far all the RSXS results were about static properties of materials, and did not utilize the time structures of synchrotron radiation (SR), which has pulse width of several 10 ps. Figure 27a summarizes the time scales in electron storage rings at various SR facilities. The time structure is composed of three time scales: the revolution time ($T_{rev} \sim 1\,\mu s = 10^{-6}\,s$), the bunch interval ($T_{RF} \sim 1\,ns = 10^{-9}\,s$), and the bunch length ($T_d \sim 10\,ps$) ($1\,ps = 10^{-12}\,s$). The bunch length is directly related to the SR pulse width of several 10 ps.

Using the SR pulses as a probe, a pump-probe method has been performed with a femtosecond laser pulse as a pump, as shown in Fig. 27b. The time resolution is limited by the SR pulse-width. In order to reach the time scales of elementary excitations such as phonon and magnons in materials, it is necessary to achieve the time resolution better than picoseconds. One method of obtaining fs X-ray pulse is the use of laser slicing [63]. Electron bunch is shot by a high intensity ultrashort laser pulse to extract a very short electron bunch from a longer electron bunch. The intense fs light pulses are resonantly matched to electron periodic motion in an undulator. The excited electrons are spatially separated from their neighbors (laser slicing). In the subsequent passage through a (bending) magnet field, the spiked fraction radiates ultrashort SR pulse.

Although the laser slicing technique stably generates fs light pulses, the actual measurement using the beam is quite difficult due to very low photon flux. The situation is easily understood by a simple estimation. Repetition frequency of the high intense laser is typically kHz, which is 1/100,000 of the SR frequency (typically on the order of 100 MHz) and a ratio of the spiked region to the total electron bunch is 100 fs/50 ps = 1/500. Thus, photon flux of the laser-slicing fs light pulses is about 10^{-8} of that of the usual SR pulses.

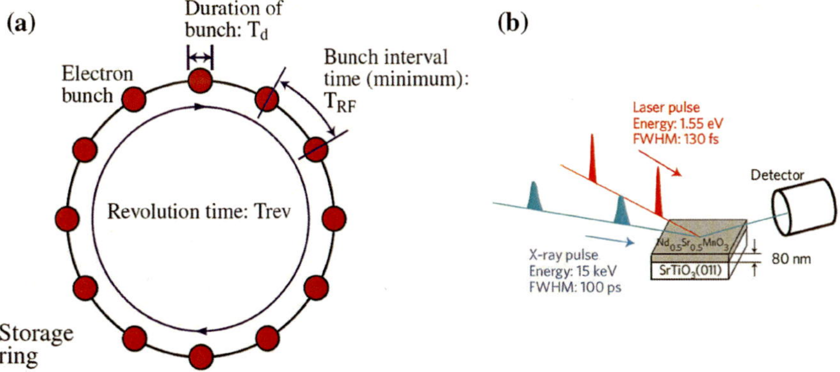

Fig. 27 **a** Time-structure of SR electron storage ring [61]. **b** Schematic setup of the optical-pump/X-ray-probe experiment [62]

For this reason, there has recently been an increased demand for the intense fs pulses of VUV, soft X-ray, and X-ray. A scheme to generate X-ray laser without mirrors is a single-pass free-electron laser (FEL). Traveling through a sufficiently long undulator, electrons in a bunch are subjected to the combined force of undulator magnets and their own synchrotron radiation. Then, they begin to form microbunches, spaced by exactly one wavelength of the dominant synchrotron radiation. Constructive interference of the radiation from these microbunches produces an X-ray beam with full transverse coherence. The instantaneous brightness is ten billion-fold greater and the pulse width (<10–100 fs) is thousand-fold shorter than SR, emitted from a typical undulator in the storage ring. Such FEL amplification is traditionally called "self-amplified spontaneous emission" (SASE) and overview mechanism. The SASE-type FELs have now been operated at LCLS (USA) [64] and SACLA (Japan) [65] for X-ray region.

Figure 28 shows the time dependence of the SL Bragg reflection from $La_{0.42}Ca_{0.58}MnO_3$ thin films for several absorbed pump fluences by using the laser slicing in the Swiss Light Source. Here the thickness of the thin film was about 40 nm, comparable

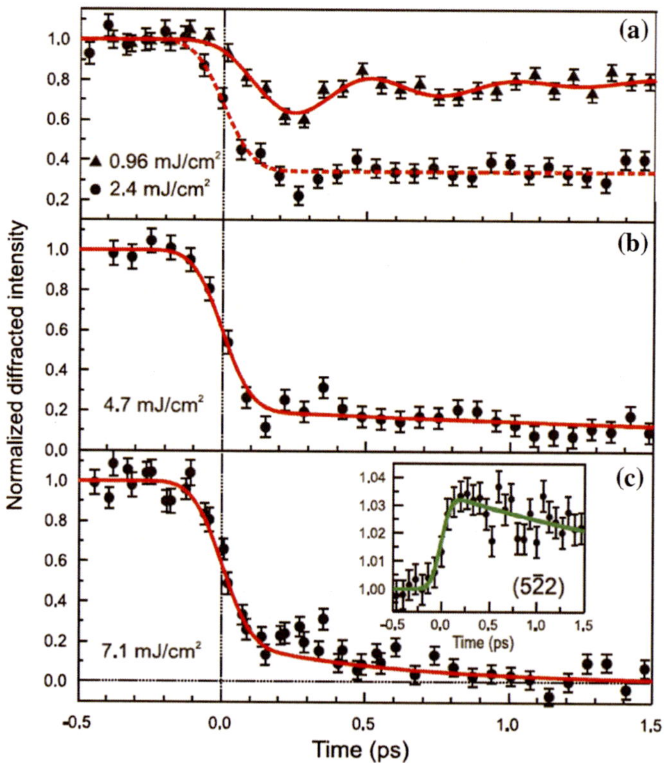

Fig. 28 Time dependence of the normalized diffracted intensity of the SL reflection from $La_{0.42}Ca_{0.58}MnO_3$ thin films are shown for several pump fluences [63]

to the penetration depth of the exciting laser light. Such thin films enable to excite and probe the same sample volume in spite of the difference of the penetration depths of the exciting laser light and the X-rays (70 nm and 4 μm, respectively). The exciting laser is a Ti:sapphire laser with a 800 nm wavelength (1.55 eV), which mainly excite the Mn^{3+} intra-site transition, and drives the relaxation of the Jahn–Teller distortion.

The data at the lowest fluence of 0.96 mJ/cm^2 show oscillations at approximately 2 THz corresponding to the displacive excitation of a coherent optical phonon. Figure 28a shows a fit of the normalized X-ray intensity $I(t)/I(0)$ to a simple model of displacive excitation

$$I(t)/I(0) = 1 - Ae^{-t/\tau_1}(1 - e^{-t/\tau_2}\cos 2\pi \nu t). \tag{43}$$

The fit yields a phonon frequency $\nu = 1.98$ THz, a relaxation time $\tau_1 = 4.5$ ps, and a damping time $\tau_2 = 560$ fs. They estimated the displacement of the La/Ca atoms, and the observed 20 % drop in diffraction intensity corresponds to a laser induced displacement of the La/Ca atomic positions by 0.0082(5) Å.

Figure 29 shows the evolution of the normalized diffracted X-ray intensity for three superlattice reflections obtained by using X-ray FEL (XFEL) in LCLS [66]. The ordering peaks can be categorized into three (in Pbnm notation with k odd): one type of reflection (0 $k/2$ 0) is sensitive directly to the orbital order and Jahn-Teller distortion, a second type (0 k 0) is more sensitive to the charge order, and a general

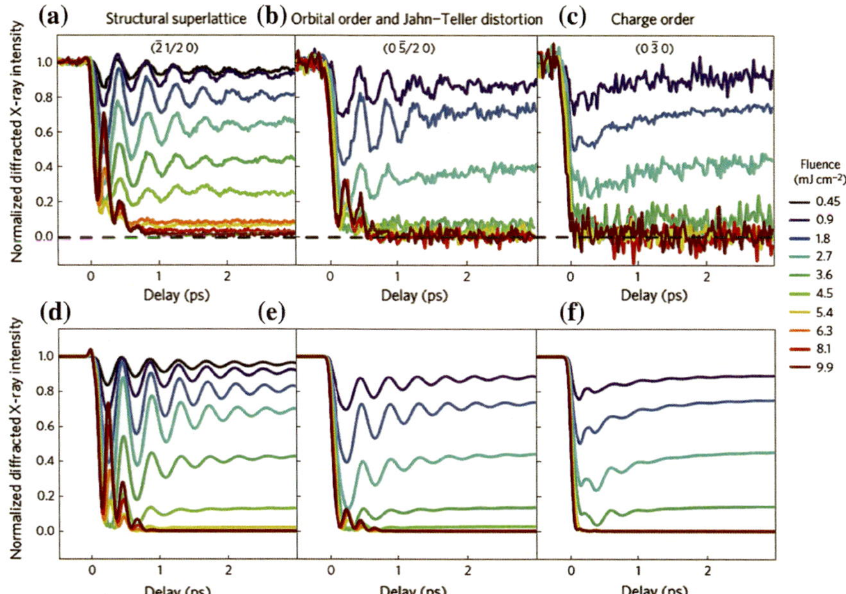

Fig. 29 Evolution of the normalized diffracted X-ray intensity for three superlattice reflections [66]

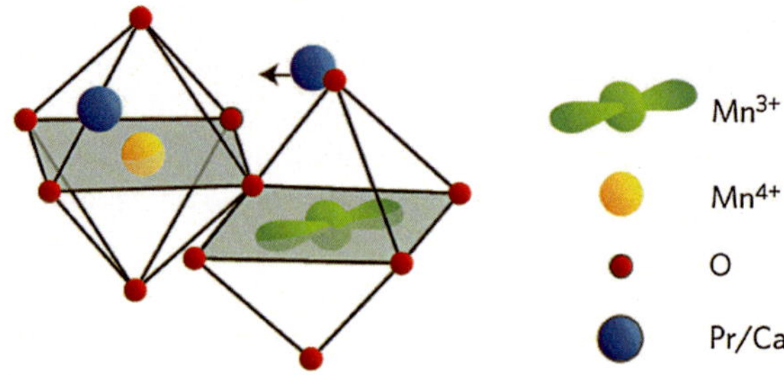

Fig. 30 Optical phonon modes related to the charge and orbital order phase: the motion of Pr/Ca [66]

type $(h\,k/2\,l)$ measures predominantly the overall structural distortion. The observed dynamics for lower pump fluences are dominated by a strong 2.45 THz oscillation, which is the slowest of a series of coherent optical phonon modes observed when these materials are excited with very short optical pulses [67]. The slowest mode with frequency 2.5 THz has been assigned solely to the motion of Pr/Ca as shown in Fig. 30.

Time-resolved RSXS study of orbital ordering was already reported from LCLS in Ref. [68], where $La_{0.5}Sr_{1.5}MnO_4$ was excited by 130-fs, 1.2 mJ/cm² midinfrared pulses, whose energy was tuned to a center wavelength of 13.5 μm (92 meV) with a 4.5-μm full width at half maximum (FWHM) bandwidth. The melting of the orbital (1/4, 1/4, 0) and magnetic (1/4, 1/4, 1/2) ordering peaks for midinfrared excitation was observed with a single exponential decay of 6.3 and 12.2 ps, respectively.

8 Summary and Future Prospects

In this chapter, I gave an overview of what a novel type of experimental technique RSXS has revealed in the condensed-matter physics: interfaces, charge and orbital ordering, and multiferroics, measurements under magnetic fields, and time-resolved measurements. I consider the future of this technique in the following three directions.

1. Looking for novel type of ordering: skyrmions
   One of the most representative such spin textures is the skyrmion lattice which can be described by multiple helical q-vectors. The formation of skyrmionic crystals has been observed in chiral cubic (but noncentrosymmetric) B20-type compounds such as $Fe_{0.5}Co_{0.5}Si$ [69] and Cu_2OSeO_3 [70]. The skyrmions in Cu_2OSeO_3 were already observed by RSXS [71], and the next step will be to look for skyrmions in a cubic (centrosymmetric) perovskites $SrFeO_3$ [72–74].
2. Space-resolved measurements
   Recently, the spin-chiral domains in helimagnetic $Ba_{0.5}Sr_{1.5}Zn_2Fe_{12}O_{22}$ were observed by scanning resonant X-ray microdiffraction using circularly polarized X-rays [75]. They focused the incident beam using Kirkpatrick-Baez configuration mirrors equipped just before the diffractometer [76]. The spot size was 30 μm in horizontal and 15 μm in vertical directions, and they observed the domains with

the size of the order of 0.1 mm. The future step will be to focus the X-rays below 100 nm by using a Fresnel zone plate, as was already used in nanoscale three-dimensional spatial-resolved photoelectron spectroscopy [77].

3. Time-resolved measurements

Time-resolved RSXS using XFEL will be a very important future directions. Ti:sapphire has a high energy of 1.5 eV and thermal effects are also expected, so midinfrared or terahertz pulses as a pump will be more suitable for exciting phonons and magnons. Recent time-resolved RSXS study of $La_{1.875}Ba_{0.125}CuO_4$ revealed a decay of CDW states by exciting with midinfrared pulses [78]. The decay time constant was found to be ~300 fs. From the study of multiferroic $TbMnO_3$, the oscillations due to electromagnons were observed by the excitations with terahertz pulses [79]. It is highly necessary to extend this time-resolved technique to novel types of ordering such as devil's staircase and skyrmions.

Acknowledgements This work was supported by the Japan Society for the Promotion of Science (JSPS) through the "Funding Program for World-Leading Innovative R&D on Science and Technology (FIRST Program)" initiated by the Council for Science and Technology Policy (CSTP), and by the Ministry of Education, Culture, Sports, Science and Technology of Japan (X-ray Free Electron Laser Priority Strategy Program).

References

1. M. Imada, A. Fujimori, Y. Tokura, Rev. Mod. Phys. **70**, 1039 (1998)
2. J. Fink, E. Schierle, E. Weschke, J. Geck, Rep. Prog. Phys. **76**, 056502 (2013)
3. J. Als-Nielsen, D. McMorrow, *Elements of Modern X-Ray Physics* (Wiley, New York, 2001)
4. J. Stohr, H.C. Siegmann, *Magnetism* (Springer, Berlin, 2006)
5. D. Attwood, *Soft X-Rays and Extreme Ultraviolet Radiation* (Cambridge University Press, Cambridge, 2007)
6. B.L. Henke, E.M. Gullikson, J.C. Davis, At. Data Nucl. Data Tables **54**, 181 (1993)
7. J. Schlappa, C.F. Chang, Z. Hu, E. Schierle, H. Ott, E. Weschke, G. Kaindl, M. Huijben, G. Rijnders, D.H.A. Blank, L.H. Tjeng, C. Schussler-Langeheine, J. Phys. Condens. Matter **24**, 035501 (2012)
8. S. Hufner, *Photoelectron Spectroscopy* (Springer, Berlin, 2003)
9. L.G. Parratt, Phys. Rev. **95**, 359 (1954)
10. J.H. Underwood, J.T.W. Barbee, Appl. Opt. **20**, 3027 (1981)
11. M. Sacchi, C.F. Hague, L. Pasquali, A. Mirone, J.-M. Mariot, P. Isberg, E.M. Gullikson, J.H. Underwood, Phys. Rev. Lett. **81**, 1521 (1998)
12. F.M.F. de Groot, A. Kotani, *Core Level Spectroscopy of Solids, Advances in Condensed Matter Science* (CRC, Boca Raton, FL, 2008)
13. J.P. Hannon, G.T. Trammell, M. Blume, D. Gibbs, Phys. Rev. Lett. **61**, 1245 (1988)
14. J.P. Hill, D.F. McMorrow, Acta Crystallogr. Sect. A **52**, 236 (1996)
15. D.G. Hawthorn, F. He, L. Venema, H. Davis, A.J. Achkar, J. Zhang, R. Sutarto, H. Wadati, A. Radi, T. Wilson, G. Wright, K.M. Shen, J. Geck, H. Zhang, V. Novak, G.A. Sawatzky, Rev. Sci. Instrum. **82**, 073104 (2011)
16. U. Staub, V. Scagnoli, Y. Bodenthin, M. Garcia-Fernandez, R. Wetter, A.M. Mulders, H. Grimmer, M. Horisberger, J. Synchrotron. Radiat. **15**, 469 (2008)
17. J. Chakhalian, J.W. Freeland, H.-U. Habermeier, G. Cristiani, G. Khaliullin, M. van Veenendaal, B. Keimer, Science **318**, 1114 (2007)

18. A. Ohtomo, H.Y. Hwang, Nature **427**, 423 (2004)
19. N. Reyren, S. Thiel, A.D. Caviglia, L.F. Kourkoutis, G. Hammer, C. Richter, C.W. Schneider, T. Kopp, A.-S. Ruetschi, D. Jaccard, M. Gabay, D.A. Muller, J.-M. Triscone, J. Mannhart, Science **317**, 1196 (2007)
20. P. Abbamonte, L. Venema, A. Rusydi, G.A. Sawatzky, G. Logvenov, I. Bozovic, Science **297**, 581 (2002)
21. S. Smadici, J.C.T. Lee, S. Wang, P. Abbamonte, G. Logvenov, A. Gozar, C. Deville Cavellin, I. Bozovic, Phys. Rev. Lett. **102**, 107004 (2009)
22. W. Siemons, G. Koster, H. Yamamoto, W.A. Harrison, G. Lucovsky, T.H. Geballe, D.H.A. Blank, M.R. Beasley, Phys. Rev. Lett. **98**, 196802 (2007)
23. A. Kalabukhov, R. Gunnarsson, J. Borjesson, E. Olsson, T. Claeson, D. Winkler, Phys. Rev. B **75**, 121404(R) (2007)
24. N. Nakagawa, H.Y. Hwang, D.A. Muller, Nat. Mater. **5**, 204 (2006)
25. H. Wadati, D.G. Hawthorn, J. Geck, T. Higuchi, Y. Hikita, H.Y. Hwang, L. Fitting Kourkoutis, A. Muller, S.-W. Huang, D.J. Huang, H.-J. Lin, C. Schussler-Langeheine, H.-H. Wu, E. Schierle, E. Weschke, N.J.C. Ingle, G.A. Sawatzky, J. Appl. Phys. **106**, 083705 (2009)
26. H. Wadati, J. Geck, D.G. Hawthorn, T. Higuchi, M. Hosoda, C. Bell, Y. Hikita, H.Y. Hwang, C. Schussler-Langeheine, E. Schierle, E. Weschke, G.A. Sawatzky, *IOP Conference Series Material Science Engineering*, vol. 24, p. 012012 (2011)
27. J. Park, B.-G. Cho, K.D. Kim, J. Koo, H. Jang, K.-T. Ko, J.-H. Park, K.-B. Lee, J.-Y. Kim, D.R. Lee, C.A. Burns, S.S.A. Seo, H.N. Lee, Phys. Rev. Lett. **110**, 017401 (2013)
28. S. Smadici, P. Abbamonte, A. Bhattacharya, X. Zhai, A. Rusydi, J.N. Eckstein, S.D. Bader, J.-M. Zuo, Phys. Rev. Lett. **99**, 196404 (2007)
29. Y. Murakami, H. Kawada, H. Kawata, M. Tanaka, T. Arima, Y. Moritomo, Y. Tokura, Phys. Rev. Lett. **80**, 1932 (1998)
30. S. Ishihara, S. Maekawa, Phys. Rev. Lett. **80**, 3799 (1998)
31. J. Garcia, M.C. Sanchez, G. Subias, J. Blasco, J. Phys. Condens. Matter **13**, 3243 (2001)
32. I.S. Elfimov, V.I. Anisimov, G.A. Sawatzky, Phys. Rev. Lett. **82**, 4264 (1999)
33. C.W.M. Castleton, M. Altarelli, Phys. Rev. B **62**, 1033 (2000)
34. S.B. Wilkins, P.D. Spencer, P.D. Hatton, S.P. Collins, M.D. Roper, D. Prabhakaran, A.T. Boothroyd, Phys. Rev. Lett. **91**, 167205 (2003)
35. S.S. Dhesi, A. Mirone, C. De Nadai, P. Ohresser, P. Bencok, N.B. Brookes, P. Reutler, A. Revcolevschi, A. Tagliaferri, O. Toulemonde, G. van der Laan, Phys. Rev. Lett. **92**, 056403 (2004)
36. U. Staub, V. Scagnoli, A.M. Mulders, K. Katsumata, Z. Honda, H. Grimmer, M. Horisberger, J.M. Tonnerre, Phys. Rev. B **71**, 214421 (2005)
37. S.B. Wilkins, N. Stojia, T.A.W. Beale, N. Binggeli, C.W.M. Castleton, P. Bencok, D. Prabhakaran, A.T. Boothroyd, P.D. Hatton, M. Altarelli, Phys. Rev. B **71**, 245102 (2005)
38. D.J. Huang, W.B. Wu, G.Y. Guo, H.-J. Lin, T.Y. Hou, C.F. Chang, C.T. Chen, A. Fujimori, T. Kimura, H.B. Huang, A. Tanaka, T. Jo, Phys. Rev. Lett. **92**, 087202 (2004)
39. H. Wu, C.F. Chang, O. Schumann, Z. Hu, J.C. Cezar, T. Burnus, N. Hollmann, N.B. Brookes, A. Tanaka, M. Braden, L.H. Tjeng, D.I. Khomskii, Phys. Rev. B **84**, 155126 (2011)
40. Y. Tokura, Rep. Prog. Phys. **69**, 797 (2006)
41. K.J. Thomas, J.P. Hill, S. Grenier, Y.-J. Kim, P. Abbamonte, L. Venema, A. Rusydi, Y. Tomioka, Y. Tokura, D.F. McMorrow, G. Sawatzky, M. van Veenendaal, Phys. Rev. Lett. **92**, 237204 (2004)
42. U. Staub, M. Garcia-Fernandez, Y. Bodenthin, V. Scagnoli, R.A. De Souza, M. Garganourakis, E. Pomjakushina, K. Conder, Phys. Rev. B **79**, 224419 (2009)
43. S.Y. Zhou, Y. Zhu, M.C. Langner, Y.-D. Chuang, P. Yu, W.L. Yang, A.G. Cruz Gonzalez, N. Tahir, M. Rini, Y.-H. Chu, R. Ramesh, D.-H. Lee, Y. Tomioka, Y. Tokura, Z. Hussain, R.W. Schoenlein, Phys. Rev. Lett. **106**, 186404 (2011)
44. D. Okuyama, M. Nakamura, Y. Wakabayashi, H. Itoh, R. Kumai, H. Yamada, Y. Taguchi, T. Arima, M. Kawasaki, Y. Tokura, Appl. Phys. Lett. **95**, 152502 (2009)

45. H. Wadati, J. Geck, E. Schierle, R. Sutarto, F. He, D.G. Hawthorn, M. Nakamura, M. Kawasaki, Y. Tokura, G.A. Sawatzky, New J. Phys. **16**, 033006 (2014)
46. H. Wadati, J. Okamoto, M. Garganourakis, V. Scagnoli, U. Staub, E. Sakai, H. Kumigashira, T. Sugiyama, E. Ikenaga, M. Nakamura, M. Kawasaki, Y. Tokura, New J. Phys. **16**, 073044 (2014)
47. M. Nakamura, D. Okuyama, J.S. Lee, T. Arima, Y. Wakabayashi, R. Kumai, M. Kawasaki, Y. Tokura, Adv. Mater. **22**, 500 (2010)
48. T. Kimura, T. Goto, H. Shintani, K. Ishizaka, T. Arima, Y. Tokura, Nature (London) **426**, 55 (2003)
49. H. Katsura, N. Nagaosa, A.V. Balatsky, Phys. Rev. Lett. **95**, 057205 (2005)
50. M. Mostovoy, Phys. Rev. Lett. **96**, 067601 (2006)
51. M. Kenzelmann, A.B. Harris, S. Jonas, C. Broholm, J. Schefer, S.B. Kim, C.L. Zhang, S.-W. Cheong, O.P. Vajk, J.W. Lynn, Phys. Rev. Lett. **95**, 087206 (2005)
52. S.B. Wilkins, T.R. Forrest, T.A.W. Beale, S.R. Bland, H.C. Walker, D. Mannix, F. Yakhou, D. Prabhakaran, A.T. Boothroyd, J.P. Hill, P.D. Hatton, D.F. McMorrow, Phys. Rev. Lett. **103**, 207602 (2009)
53. S. Ishiwata, Y. Kaneko, Y. Tokunaga, Y. Taguchi, T. Arima, Y. Tokura, Phys. Rev. B **81**, 100411(R) (2010)
54. M. Nakamura, Y. Tokunaga, M. Kawasaki, Y. Tokura, Appl. Phys. Lett. **98**, 082902 (2011)
55. H. Wadati, J. Okamoto, M. Garganourakis, V. Scagnoli, U. Staub, Y. Yamasaki, H. Nakao, Y. Murakami, M. Mochizuki, M. Nakamura, M. Kawasaki, Y. Tokura, Phys. Rev. Lett. **108**, 047203 (2012)
56. M. Mochizuki, N. Furukawa, N. Nagaosa, Phys. Rev. Lett. **105**, 037205 (2010)
57. Y.W. Windsor, S.W. Huang, Y. Hu, L. Rettig, A. Alberca, K. Shimamoto, V. Scagnoli, T. Lippert, C.W. Schneider, U. Staub, Phys. Rev. Lett. **113**, 167202 (2014)
58. S. Blanco-Canosa, A. Frano, T. Loew, Y. Lu, J. Porras, G. Ghiringhelli, M. Minola, C. Mazzoli, L. Braicovich, E. Schierle, E. Weschke, M. Le Tacon, B. Keimer, Phys. Rev. Lett. **110**, 187001 (2013)
59. S. Ishiwata, I. Terasaki, F. Ishii, N. Nagaosa, H. Mukuda, Y. Kitaoka, T. Saito, M. Takano, Phys. Rev. Lett. **98**, 217201 (2007)
60. T. Matsuda, S. Partzsch, T. Tsuyama, E. Schierle, E. Weschke, J. Geck, T. Saito, S. Ishiwata, Y. Tokura, H. Wadati, Phys. Rev. Lett. **114**, 236403 (2015)
61. S. Yamamoto, I. Matsuda, J. Phys. Soc. Jpn. **82**, 021003 (2013)
62. H. Ichikawa, S. Nozawa, T. Sato, A. Tomita, K. Ichiyanagi, M. Chollet, L. Guerin, N. Dean, A. Cavalleri, S. Adachi, T. Arima, H. Sawa, Y. Ogimoto, M. Nakamura, R. Tamaki, K. Miyano, S. Koshihara, Nat. Mater. **10**, 101 (2011)
63. P. Beaud, S.L. Johnson, E. Vorobeva, U. Staub, R.A. De Souza, C.J. Milne, Q.X. Jia, G. Ingold, Phys. Rev. Lett. **103**, 155702 (2009)
64. P. Emma, R. Akre, J. Arthur, R. Bionta, C. Bostedt, J. Bozek, A. Brachmann, P. Bucksbaum, R. Coffee, F.-J. Decker, Y. Ding, D. Dowell, S. Edstrom, A. Fisher, J. Frisch, S. Gilevich, J. Hastings, G. Hays, P. Hering, Z. Huang, R. Iverson, H. Loos, M. Messerschmidt, A. Miahnahri, S. Moeller, H.-D. Nuhn, G. Pile, D. Ratner, J. Rzepiela, D. Schultz, T. Smith, P. Stefan, H. Tompkins, J. Turner, J. Welch, W. White, J. Wu, G. Yocky, J. Galayda, Nat. Photonics **4**, 641 (2010)
65. T. Ishikawa, H. Aoyagi, T. Asaka, Y. Asano, N. Azumi, T. Bizen, H. Ego, K. Fukami, T. Fukui, Y. Furukawa, S. Goto, H. Hanaki, T. Hara, T. Hasegawa, T. Hatsui, A. Higashiya, T. Hirono, N. Hosoda, M. Ishii, T. Inagaki, Y. Inubushi, T. Itoga, Y. Joti, M. Kago, T. Kameshima, H. Kimura, Y. Kirihara, A. Kiyomichi, T. Kobayashi, C. Kondo, T. Kudo, H. Maesaka, X.M. Marechal, T. Masuda, S. Matsubara, T. Matsumoto, T. Matsushita, S. Matsui, M. Nagasono, N. Nariyama, H. Ohashi, T. Ohata, T. Ohshima, S. Ono, Y. Otake, C. Saji, T. Sakurai, T. Sato, K. Sawada, T. Seike, K. Shirasawa, T. Sugimoto, S. Suzuki, S. Takahashi, H. Takebe, K. Takeshita, K. Tamasaku, H. Tanaka, R. Tanaka, T. Tanaka, T. Togashi, K. Togawa, A. Tokuhisa, H. Tomizawa, K. Tono, S. Wu, M. Yabashi, M. Yamaga, A. Yamashita, K. Yanagida, C. Zhang, T. Shintake, H. Kitamura, N. Kumagai, Nat. Photonics **6**, 540 (2012)

66. P. Beaud, A. Caviezel, S.O. Mariager, L. Rettig, G. Ingold, C. Dornes, S.-W. Huang, J.A. Johnson, M. Radovic, T. Huber, T. Kubacka, A. Ferrer, H.T. Lemke, M. Chollet, D. Zhu, J.M. Glownia, M. Sikorski, A. Robert, H. Wadati, M. Nakamura, M. Kawasaki, Y. Tokura, S.L. Johnson, U. Staub, Nat. Mater. **13**, 923 (2014)
67. H. Matsuzaki, H. Uemura, M. Matsubara, T. Kimura, Y. Tokura, H. Okamoto, Phys. Rev. B **79**, 235131 (2009)
68. M. Forst, R.I. Tobey, S. Wall, H. Bromberger, V. Khanna, A.L. Cavalieri, Y.D. Chuang, W.S. Lee, R. Moore, W.F. Schlotter, J.J. Turner, O. Krupin, M. Trigo, H. Zheng, J.F. Mitchell, S.S. Dhesi, J.P. Hill, A. Cavalleri, Phys. Rev. B **84**, 241104(R) (2011)
69. X.Z. Yu, Y. Onose, N. Kanazawa, J.H. Park, J.H. Han, Y. Matsui, N. Nagaosa, Y. Tokura, Nature **465**, 901 (2010)
70. S. Seki, X.Z. Yu, S. Ishiwata, Y. Tokura, Science **336**, 198 (2012)
71. M.C. Langner, S. Roy, S.K. Mishra, J.C.T. Lee, X.W. Shi, M.A. Hossain, Y.-D. Chuang, S. Seki, Y. Tokura, S.D. Kevan, R.W. Schoenlein, Phys. Rev. Lett. **112**, 167202 (2014)
72. S. Ishiwata, M. Tokunaga, Y. Kaneko, D. Okuyama, Y. Tokunaga, S. Wakimoto, K. Kakurai, T. Arima, Y. Taguchi, Y. Tokura, Phys. Rev. B **84**, 054427 (2011)
73. Y.W. Long, Y. Kaneko, S. Ishiwata, Y. Tokunaga, T. Matsuda, H. Wadati, Y. Tanaka, S. Shin, Y. Tokura, Y. Taguchi, Phys. Rev. B **86**, 064436 (2012)
74. S. Chakraverty, T. Matsuda, H. Wadati, J. Okamoto, Y. Yamasaki, H. Nakao, Y. Murakami, S. Ishiwata, M. Kawasaki, Y. Taguchi, Y. Tokura, H.Y. Hwang, Phys. Rev. B **88**, 220405(R) (2013)
75. Y. Hiraoka, Y. Tanaka, T. Kojima, Y. Takata, M. Oura, Y. Senba, H. Ohashi, Y. Wakabayashi, S. Shin, T. Kimura, Phys. Rev. B **84**, 064418 (2011)
76. H. Ohashi, Y. Senba, H. Kishimoto, T. Miura, E. Ishiguro, T. Takeuchi, M. Oura, K. Shirasawa, T. Tanaka, M. Takeuchi, K. Takeshita, S. Goto, S. Takahashi, H. Aoyagi, M. Sano, Y. Furukawa, T. Ohata, T. Matsushita, Y. Ishizawa, S. Taniguchi, Y. Asano, Y. Harada, T. Tokushima, K. Horiba, H. Kitamura, T. Ishikawa, S. Shin, *AIP Conference Proceedings*, vol. 879, p. 523 (2007)
77. K. Horiba, Y. Nakamura, N. Nagamura, S. Toyoda, H. Kumigashira, M. Oshima, K. Amemiya, Y. Senba, H. Ohashi, Rev. Sci. Instrum. **82**, 113701 (2011)
78. M. Forst, R.I. Tobey, H. Bromberger, S.B. Wilkins, V. Khanna, A.D. Caviglia, Y.-D. Chuang, W.S. Lee, W.F. Schlotter, J.J. Turner, M.P. Minitti, O. Krupin, Z.J. Xu, J.S. Wen, G.D. Gu, S.S. Dhesi, A. Cavalleri, J.P. Hill, Phys. Rev. Let. **112**, 157002 (2014)
79. T. Kubacka, J.A. Johnson, M.C. Hoffmann, C. Vicario, S. de Jong, P. Beaud, S. Grubel, S.-W. Huang, L. Huber, L. Patthey, Y.-D. Chuang, J.J. Turner, G.L. Dakovski, W.-S. Lee, M.P. Minitti, W. Schlotter, R.G. Moore, C.P. Hauri, S.M. Koohpayeh, V. Scagnoli, G. Ingold, S.L. Johnson, U. Staub, Science **343**, 1333 (2014)

Resonant Inelastic X-Ray Scattering in Strongly Correlated Copper Oxides

Kenji Ishii

1 Introduction

Resonant X-ray scattering is a technique combining spectroscopy and scattering. It is classified into two categories according to the energy transfer at the scattering process. The first is resonant elastic X-ray scattering (REXS), where the energy transfer is zero, i.e., incident and scattered X-rays have the same energy. It is mostly used for diffraction experiments, and a variety of applications of RXES are described in other chapters of this book. In the second technique, called resonant inelastic X-ray scattering (RIXS), the energy transfer is finite. While REXS is a tool for measuring static properties, one can investigate dynamics of electrons in materials through the observation of electronic excitations using RIXS. In the last decades, RIXS has achieved significant progress by the availability of very intense X-rays produced by third-generation synchrotron sources [1, 2]. The energy resolution of RIXS has gradually improved in accordance with the development of spectrometers and related techniques, and it has achieved several tens of meV recently. Nowadays, various types of excitation are studied using RIXS.

In principle, electronic excitations can be observed by nonresonant inelastic X-ray scattering (NIXS), where the incident photon energy is far from the absorption edge of the constituent elements of the sample. Even though the energy resolution of NIXS is below 1 meV, which is much better than that of RIXS, NIXS usually suffers from low intensity of electronic excitations near the Fermi energy. This shortcoming can be overcome by using the resonant process of RIXS, where the incident photon energy is tuned near the absorption edge of a constituent element, resulting in significant

K. Ishii (✉)
National Institutes for Quantum and Radiological Science and Technology,
1-1-1 Kouto, Sayo, Hyogo 679-5148, Japan
e-mail: kenji@spring8.or.jp

© Springer-Verlag Berlin Heidelberg 2017
Y. Murakami and S. Ishihara (eds.), *Resonant X-Ray Scattering
in Correlated Systems*, Springer Tracts in Modern Physics 269,
DOI 10.1007/978-3-662-53227-0_6

resonant enhancement of the electronic excitations. In addition, selection of the absorption edge and polarization dependence can provide fruitful information on the symmetry of electronic states. In this respect, the comparison of RIXS with NIXS is similar to the case of elastic scattering.

RIXS has several important characteristics as a spectroscopic method for measuring electronic excitations. The momentum resolution is a great advantage over conventional optical techniques. One can observe electronic excitations with finite momentum transfer ($\mathbf{Q}$) using RIXS, whereas the conventional optical spectroscopies are limited to measurement at $\mathbf{Q} = 0$. Angle-resolved photoemission spectroscopy (ARPES) is another momentum-resolved technique. While ARPES essentially gives one-particle excitation spectra for occupied states below the Fermi energy, RIXS gives two particle spectra, that is, it probes both occupied and unoccupied states through the presence of an excited electron into the unoccupied states accompanied by a hole in the occupied states in the final state of RIXS. From a technical point of view, RIXS is a photon-in photon-out technique that ensures bulk sensitivity and can be applied to measurement under extreme conditions [3]. We also point out some aspects of RIXS that are complementary to inelastic neutron scattering (INS). INS is a well-established tool for measuring spin excitations and the intensity of INS is given by the dynamical spin correlation function. In contrast, RIXS can probe all electronic degrees of freedom, i.e., charge, spin, and orbital. Notably, it has been proved recently that spin-flip excitations can be measured by RIXS at the L-edge of transition metals [4–6]. A high flux of incident photons and a large scattering cross section enable one to measure magnetic excitations and fluctuations in the entire accessible Brillouin zone, even in the paramagnetic state. In addition, reasonable intensities can be obtained from tiny samples and even from thin films. In this sense, RIXS at the L-edge compensates for the shortcomings of INS, providing a new tool for measuring magnetic excitations.

A main purpose of RIXS is to investigate the electron dynamics near the Fermi energy, that plays a principal role in determining the electric and magnetic properties of materials. The electron dynamics is crucial for understanding the mechanism of the properties. One can elucidate the interactions between electrons from the observation of electronic excitations. For example, an interband transition across the Mott gap is directly connected to the Coulomb interaction between electrons and the momentum dependence of the transition is related to the band structure of the lower and upper Hubbard bands. As for the spin excitation, the exchange interaction between spins can be determined from the dispersion relation of a spin-wave excitation.

In recent years, RIXS has been mostly used for observation of electronic excitations in strongly correlated electron systems. Especially, copper oxides including high-T_c superconductors, have been studied most intensively and systematically. This chapter will mainly address the copper oxides that exemplify the RIXS technique for the studies of electronic excitations in strongly correlated electron systems.

2 Basics of Resonant Inelastic X-Ray Scattering

2.1 *Theoretical Outline*

In this section, the theoretical framework of RIXS is outlined. If necessary, you can find further detailed theoretical treatment of RIXS in Sects. 3 and 4 of Ref. [2] and Sect. 2 of Ref. [7].

In the process of X-ray scattering, an incident photon with energy $\hbar\omega_i$, momentum $\hbar\mathbf{k}_i$, and polarization $\boldsymbol{\varepsilon}_i$ is injected into a material and then a photon with energy $\hbar\omega_f$, momentum $\hbar\mathbf{k}_f$, and polarization $\boldsymbol{\varepsilon}_f$ is emitted from the material. Energy $\hbar\omega = \hbar\omega_i - \hbar\omega_f$ and momentum $\hbar\mathbf{Q} = \hbar\mathbf{k}_i - \hbar\mathbf{k}_f$ are transferred to the material. The kinetic energy term of the Hamiltonian of electrons interacting with an electromagnetic field is given by

$$H = \sum_j \frac{1}{2m}\left(\mathbf{p}_j - \frac{e}{c}\mathbf{A}(\mathbf{r}_j)\right)^2, \tag{1}$$

where $\mathbf{p}_j$ and $\mathbf{r}_j$ represent the momentum and the position operators for the jth electron, $\mathbf{A}(\mathbf{r}_j)$ is the vector potential at $\mathbf{r}_j$, and e, m and c are respectively the electron charge, mass and the velocity of light. The Hamiltonian can be expanded by taking the Coulomb gauge ($\nabla \cdot \mathbf{A} = 0$)

$$\begin{aligned}
H &= H_{\text{el}} + H_{\text{int}}^{(1)} + H_{\text{int}}^{(2)} \\
&= \sum_j \frac{1}{2m}\mathbf{p}_j^2 - \sum_j \frac{e}{mc}\mathbf{A}(\mathbf{r}_j) \cdot \mathbf{p}_j + \sum_j \frac{e}{2mc}\mathbf{A}^2(\mathbf{r}_j).
\end{aligned} \tag{2}$$

The transition probability per unit time (W) is given by the first order of $H_{\text{int}}^{(2)}$ and the second order of $H_{\text{int}}^{(1)}$ according to the Fermi's golden rule, and the double differential cross section is obtained by multiplying W by the density of final states of photon $\rho(\hbar\omega_f) = \frac{V\omega_f^2}{8\pi^3\hbar c^3}$ and dividing by the incident flux $I_0 = \frac{c}{V}$:

$$\begin{aligned}
\frac{d^2\sigma}{d\Omega_f d\hbar\omega_f} &= W\rho(\hbar\omega_f)/I_0 \\
&= \frac{V^2\omega_f^2}{8\pi^3\hbar c^4}W \\
&= \left(\frac{e^2}{mc^2}\right)^2\left(\frac{\omega_f}{\omega_i}\right)\sum_f\left|\langle f|\sum_j \exp\left(i\mathbf{Q}\cdot\mathbf{r}_j\right)|i\rangle\left(\boldsymbol{\varepsilon}_i\cdot\boldsymbol{\varepsilon}_f^*\right)\right. \\
&\quad + \left(\frac{\hbar}{m}\right)\sum_n\sum_{j,j'}\left[\frac{\langle f|\boldsymbol{\varepsilon}_f^*\cdot\mathbf{p}_j\exp\left(-i\mathbf{k}_f\cdot\mathbf{r}_j\right)|n\rangle\langle n|\boldsymbol{\varepsilon}_i\cdot\mathbf{p}_{j'}\exp\left(i\mathbf{k}_i\cdot\mathbf{r}_{j'}\right)|i\rangle}{E_i - E_n + \hbar\omega_i + i\Gamma_n}\right. \\
&\quad\quad \left.\left.+ \frac{\langle f|\boldsymbol{\varepsilon}_i\cdot\mathbf{p}_j\exp\left(i\mathbf{k}_i\cdot\mathbf{r}_j\right)|n\rangle\langle n|\boldsymbol{\varepsilon}_f^*\cdot\mathbf{p}_{j'}\exp\left(-i\mathbf{k}_f\cdot\mathbf{r}_{j'}\right)|i\rangle}{E_i - E_n - \hbar\omega_f}\right]\right|^2 \\
&\quad \times \delta\left(E_i - E_f + \hbar\omega_i - \hbar\omega_f\right)
\end{aligned} \tag{3}$$

where V is the volume of the system and $|i\rangle$, $|n\rangle$, $|f\rangle$ represent the initial, intermediate, state and final states with energies of E_i, E_n, and E_f, respectively [8, 9]. An imaginary component $i\Gamma_n$ is added to account for the lifetime of the intermediate state. The cross section of NIXS is given by the first term of Eq. (3). When the incident photon energy is tuned close to the energy difference between $|n\rangle$ and $|i\rangle$ ($\hbar\omega_i \simeq E_n - E_i$), we can obtain a resonantly enhanced large cross section of the second term of Eq. (3) that yields RIXS.

In most cases, we can take the dipole limit for the absorption and emission of the photon in the RIXS process. For example, at the Cu K-edge, the matrix element $\langle n| \, \boldsymbol{\varepsilon}_i \cdot \mathbf{p}_{j'} \exp\left(i\mathbf{k}_i \cdot \mathbf{r}_{j'}\right) |i\rangle$ in Eq. (3) is finite within the size of $1s$ wave function given by a_B/Z with Bohr radius a_B ($=0.529\,\text{Å}$) and atomic number Z ($=29$), and is much smaller than the wavelength of the photon $2\pi/k_i$ ($\simeq 1.4\,\text{Å}$). The dipole approximation is also valid at the Cu L_3-edge, where the wavelength of the photon is $\simeq 13.3\,\text{Å}$. Then, we consider the $1s \rightarrow 4p$ and $2p_{3/2} \rightarrow 3d$ transitions for the Cu K- and L_3-edges, respectively.

Schematic figures of RIXS process for copper oxides are shown in Fig. 1. Copper $3d$ orbitals are relevant for the electronic states near the Fermi energy. At the Cu L_3-edge (or L_2-edge), we can directly access the $3d$ orbitals through the dipole transition, as shown in Fig. 1a. An incident photon with ω_i, $\mathbf{k}_i$, and $\boldsymbol{\varepsilon}_i$ induces the dipole transition from the $2p$ orbital to an unoccupied $3d$ orbital. Subsequently, a $3d$ electron in an occupied state fills the $2p$ core hole and a photon with ω_f, $\mathbf{k}_f$, and $\boldsymbol{\varepsilon}_f$ is emitted. As a final state, one electron is excited from the occupied state to the unoccupied state. Cu K-edge RIXS is more complicated because the Cu $4p$ orbital is empty in the ground state. In this case, an incident photon brings about a dipole transition from the $1s$ orbital to the $4p$ orbital. If the $4p$ electron decays back to the $1s$ orbital without changing the valance electrons, it is solely elastic scattering. In inelastic scattering,

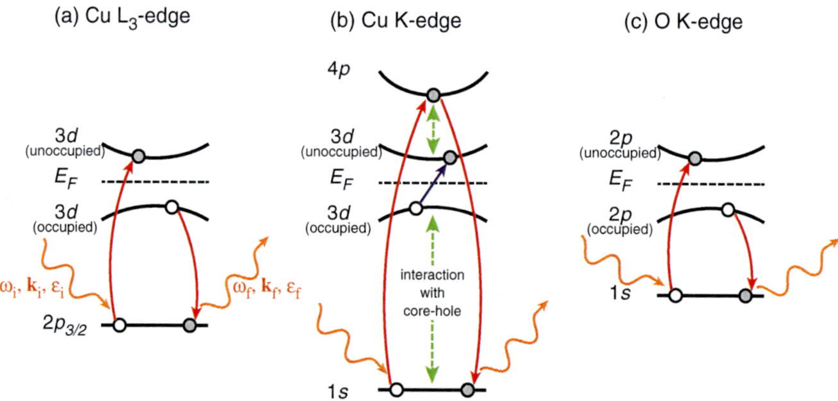

Fig. 1 Schematic figures of RIXS. Cu L_3-edge in **a** and O K-edge in **c**, a sequential process of absorption and emission of a photon between core ($2p_{3/2}$ or $1s$) and valence ($3d$ or $2p$) orbitals directly create an excitation. In the case of Cu K-edge in **b**, an excitation occurs though the $1s$-$3d$ and/or $3d$-$4p$ Coulomb interactions

electrons in the Cu $3d$ orbitals and sometimes ligand atoms interact with the $1s$ core hole via Coulomb interaction so that excitations of the valence electron system are evolved in the intermediate state. After that, the $4p$ electron fills the $1s$ core hole with the emission of a photon. Even though it is less dominant, the Coulomb interaction between the $3d$ electrons and a photoexcited $4p$ electron in the intermediate state can also contribute to the excitation process. We can also use the O K-edge for the study of copper oxides. The $2p$ orbital of oxygen is partially empty through the hybridization to the Cu $3d$ orbitals and is involved in the dipole transition from the $1s$ orbital. In terms of the accessible momentum space, the Cu K-edge in the hard X-ray regime can cover whole Brillouin zone while the Cu L-edge and O K-edge in the soft X-ray regime have limitation.

2.2 Calculation for Isolated Cu^{2+} Ion

As an instructive example, the Cu L_3-edge RIXS cross section of the isolated Cu^{2+} ion under the tetragonal crystal field with the D_{4h} symmetry is calculated below. This is often a good approximation of Cu ions in the high-T_c cuprates and related copper oxides. The RIXS intensity from the second term of Eq. (3) is given by

$$I \propto \sum_f \left| \sum_n \frac{\langle f | \boldsymbol{\varepsilon}_f^* \cdot \mathbf{p} \exp\left(-i\mathbf{k}_f \cdot \mathbf{r}\right) | n \rangle \langle n | \boldsymbol{\varepsilon}_i \cdot \mathbf{p} \exp\left(i\mathbf{k}_i \cdot \mathbf{r}\right) | i \rangle}{E_i - E_n + \hbar\omega_i + i\Gamma_n} \right|^2$$
$$\times \delta\left(E_i - E_f + \hbar\omega_i - \hbar\omega_f\right) \tag{4}$$

Taking the dipole approximation, we can rewrite the scattering intensity as

$$I \propto \sum_f \left| \sum_n \frac{\boldsymbol{\varepsilon}_f^* \cdot \langle f | \mathbf{r} | n \rangle \langle n | \mathbf{r} | i \rangle \cdot \boldsymbol{\varepsilon}_i}{E_i - E_n + \hbar\omega_i + i\Gamma_n} \right|^2 \delta\left(E_i - E_f + \hbar\omega_i - \hbar\omega_f\right). \tag{5}$$

We assume that one hole occupies the $d_{x^2-y^2}$ orbital with down spin in the initial state. The orientation of the up spin is along the arbitrary unit vector $\hat{\mathbf{n}} = (\sin\theta_s \cos\phi_s, \sin\theta_s \sin\phi_s, \cos\theta_s)$ and the eigenvectors of the up spin ($|\uparrow\rangle$) and down spin ($|\downarrow\rangle$) on the basis of eigenstate of s_z are given by

$$|\uparrow\rangle = \begin{pmatrix} e^{i\frac{\phi_s}{2}} \cos\frac{\theta_s}{2} \\ e^{-i\frac{\phi_s}{2}} \sin\frac{\theta_s}{2} \end{pmatrix}$$

$$|\downarrow\rangle = \begin{pmatrix} -e^{-i\frac{\phi_s}{2}} \sin\frac{\theta_s}{2} \\ e^{i\frac{\phi_s}{2}} \cos\frac{\theta_s}{2} \end{pmatrix}. \tag{6}$$

The wavefunctions of the initial state $\left|d_{x^2-y^2}, \downarrow\right\rangle$ and its spin-flipped state $\left|d_{x^2-y^2}, \uparrow\right\rangle$ are

$$\left|d_{x^2-y^2}, \downarrow\right\rangle = \frac{1}{\sqrt{2}}\left[-e^{-i\frac{\phi_s}{2}}\sin\frac{\theta_s}{2}\left(Y^{\uparrow}_{22} + Y^{\uparrow}_{2\bar{2}}\right) + e^{i\frac{\phi_s}{2}}\cos\frac{\theta_s}{2}\left(Y^{\downarrow}_{22} + Y^{\downarrow}_{2\bar{2}}\right)\right]$$

$$\left|d_{x^2-y^2}, \uparrow\right\rangle = \frac{1}{\sqrt{2}}\left[e^{i\frac{\phi_s}{2}}\cos\frac{\theta_s}{2}\left(Y^{\uparrow}_{22} + Y^{\uparrow}_{2\bar{2}}\right) + e^{-i\frac{\phi_s}{2}}\sin\frac{\theta_s}{2}\left(Y^{\downarrow}_{22} + Y^{\downarrow}_{2\bar{2}}\right)\right] \quad (7)$$

where $Y^{\uparrow}_{2l_z} = \left|3d; l=2, l_z, s=\frac{1}{2}, s_z=\frac{1}{2}\right\rangle$ and $Y^{\downarrow}_{2l_z} = \left|3d; l=2, l_z, s=\frac{1}{2}, s_z=-\frac{1}{2}\right\rangle$. The wavefunctions of other $3d$ orbitals can be calculated in a similar manner. Using these wavefunctions, we can calculate the matrix element of the scattering amplitude (F) in Eq. (5) and it is conveniently represented as a 3×3 tensor in the orthogonal coordinate system with two polarization vectors:

$$F \propto \sum_n \boldsymbol{\varepsilon}^*_f \cdot \langle f|\mathbf{r}|n\rangle \langle n|\mathbf{r}|i\rangle \cdot \boldsymbol{\varepsilon}_i$$

$$= \varepsilon^*_{f,\alpha} M_{\alpha\beta} \varepsilon_{i,\beta} \quad (8)$$

The elements of M are given by $M_{\alpha\beta} = \sum_{j_z} \langle f|\alpha\left|j=\frac{3}{2}, j_z\right\rangle \left\langle j=\frac{3}{2}, j_z\right|\beta|i\rangle$ with $\alpha, \beta = x, y, z$. Removing the integral of the radial part which is a common factor for the $2p$-$3d$ transition, the matrix becomes

$$M_{\downarrow} = \begin{pmatrix} \frac{1}{3} & \frac{i}{6}\cos\theta_s & 0 \\ -\frac{i}{6}\cos\theta_s & \frac{1}{3} & 0 \\ 0 & 0 & 0 \end{pmatrix} \quad (9)$$

for the spin-conserving (elastic) scattering ($|i\rangle = |f\rangle = \left|d_{x^2-y^2}, \downarrow\right\rangle$) and

$$M_{\uparrow} = \begin{pmatrix} 0 & \frac{i}{6}\sin\theta_s & 0 \\ -\frac{i}{6}\sin\theta_s & 0 & 0 \\ 0 & 0 & 0 \end{pmatrix} \quad (10)$$

for the spin-flip scattering ($|i\rangle = \left|d_{x^2-y^2}, \downarrow\right\rangle$, $|f\rangle = \left|d_{x^2-y^2}, \uparrow\right\rangle$). The matrix for other d-d transitions can be calculated in the same way.

Here we consider the experimental configuration shown in Fig. 2. The X-rays are irradiated onto the xy-plane and the z-axis of the sample lies in the scattering plane. The angles 2θ, δ, and ϕ are the scattering angle, the rotation angle of the sample from the specular condition around the axis perpendicular to the scattering plane, and the rotation angle of the sample around the z-axis, respectively. The polarization vectors are

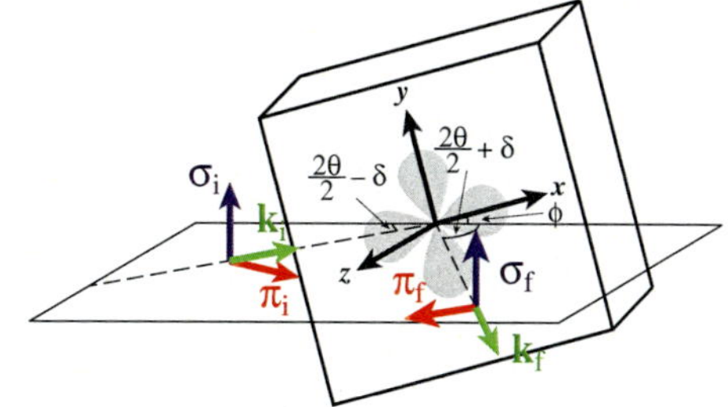

Fig. 2 Schematic figure of experimental configuration. X-rays are irradiated onto the xy-plane of the sample. The shaded four-leaf clover is the $d_{x^2-y^2}$ orbital. The angles 2θ, δ, and ϕ is defined in the text

$$\boldsymbol{\sigma}_i = -\hat{\mathbf{x}}\cos\phi + \hat{\mathbf{y}}\sin\phi$$

$$\boldsymbol{\pi}_i = \left(\hat{\mathbf{x}}\cos\phi + \hat{\mathbf{y}}\sin\phi\right)\sin\left(\frac{2\theta}{2}+\delta\right) + \hat{\mathbf{z}}\cos\left(\frac{2\theta}{2}+\delta\right)$$

$$\boldsymbol{\sigma}_f^* = -\hat{\mathbf{x}}\cos\phi + \hat{\mathbf{y}}\sin\phi$$

$$\boldsymbol{\pi}_i^* = -\left(\hat{\mathbf{x}}\cos\phi + \hat{\mathbf{y}}\sin\phi\right)\sin\left(\frac{2\theta}{2}-\delta\right) + \hat{\mathbf{z}}\cos\left(\frac{2\theta}{2}-\delta\right), \tag{11}$$

where σ and π are the polarization perpendicular and parallel to the scattering plane, respectively. Then, the scattering amplitudes of Eq. (8) become

$$F_{\sigma_i \to \sigma_f} \propto \boldsymbol{\sigma}_f^* M_\downarrow \boldsymbol{\sigma}_i$$

$$= \frac{1}{3}$$

$$F_{\sigma_i \to \pi_f} \propto -\frac{i}{6}\cos\theta_s \sin\left(\frac{2\theta}{2}-\delta\right)$$

$$F_{\pi_i \to \sigma_f} \propto -\frac{i}{6}\cos\theta_s \sin\left(\frac{2\theta}{2}+\delta\right)$$

$$F_{\pi_i \to \pi_f} \propto -\frac{1}{3}\sin\left(\frac{2\theta}{2}-\delta\right)\sin\left(\frac{2\theta}{2}+\delta\right) \tag{12}$$

for spin-conserving scattering and

$$F_{\sigma_i \to \sigma_f} \propto 0$$

$$F_{\sigma_i \to \pi_f} \propto -\frac{i}{6}\sin\theta_s \sin\left(\frac{2\theta}{2}-\delta\right)$$

$$F_{\pi_i \to \sigma_f} \propto -\frac{i}{6}\sin\theta_s \sin\left(\frac{2\theta}{2}+\delta\right)$$

$$F_{\pi_i \to \pi_f} \propto 0 \tag{13}$$

for spin-flip scattering. The scattering amplitude determines the polarization dependence. It is notable that the spin-flip scattering requires a change of the polarization ($\sigma_i \to \pi_f$ or $\pi_i \to \sigma_f$) in the Cu^{2+} ion.

2.3 Calculation for Real Materials

In theoretical calculation for real materials, various approximations are taken. In the case of RIXS at the transition-metal L-edge, a fast-collision approximation, where Γ_n in the second term of Eq. (3) is assumed to be large, is often applied [10]. This approximation neglects the contribution of indirect excitations through the core-hole potential in the intermediate state, and the RIXS intensity is given by dynamical correlation functions multiplied by a prefactor which depends on the experimental geometry and the polarization condition. The prefactor can be evaluated from wavefunctions of the isolated atom (ion) as shown for the Cu^{2+} case in the previous subsection. If only the $d_{x^2-y^2}$ orbital is relevant, the prefactor for spin-conserving scattering and spin-flip scattering is proportional to the absolute squares of Eqs. (12) and (13), respectively.

On the other hand, the core-hole potential in the intermediate state must be taken into account for the RIXS at the transition-metal K-edge. A theoretical technique to calculate the K-edge RIXS cross section is a numerally exact diagonalization of a finite-sized cluster. An advantage of this technique is that one can treat many-body effects and the core-hole potential of the system exactly. Another technique is an analytical approach taking a certain approximation. If the core-hole potential is treated as the first-order perturbation [11, 12], the RIXS intensity is proportional to the dynamical charge correlation function of the $3d$ electrons. Yet another analytical approach is an ultrashort core-hole lifetime approximation [13, 14], where the RIXS intensity is represented by a series expansion assuming Γ_n is the larger than the energy scale of the intermediate states. The lowest order of the approximation without the term of the core-hole interaction corresponds to the fast-collision approximation, and the intensity of the strong limit of the core-hole interaction is directly related to the dynamical charge correlation function of the $3d$ electrons, as with the perturbative approach.

3 RIXS Studies on High-T_c Cuprates

Strongly correlated copper oxides, especially high-T_c superconducting cuprates, are the most intensively studied material with RIXS to date. Various kinds of excitation have been observed. These excitations are discussed in this section and I hope that this will help the reader understand what can be studied using RIXS technique.

3.1 Electronic Structure of Cuprates and Excitations Observed by RIXS

Parent compounds of high-T_c superconducting cuprates and their related materials with divalent copper ions are Mott insulators. Strictly speaking, these copper oxides are classified into charge-transfer type according to the Zaanen–Sawatzky–Allen scheme [15], where the energy of charge transfer (Δ) from the O $2p$ to the Cu $3d_{x^2-y^2}$ levels is smaller than the on-site Coulomb repulsion (U_d) of the $3d$ electrons. The energies Δ and U_d are of the order of several eV, and charge excitations at this energy scale are related to Δ and U_d. Figure 3a shows such an electronic structure. The density of states in the figure represents single-particle excitations; the states below and above the Fermi level correspond to one-electron removal and addition, respectively. When a hole is introduced in the O $2p$ orbital, it couples strongly with a localized hole on the Cu $3d_{x^2-y^2}$ orbital, forming a singlet bound state at the top of the O $2p$ band. This bound state is called the Zhang–Rice singlet [16]. Therefore, the charge excitation at the lowest energy in the insulating copper oxides is a transition of an electron from the Zhang–Rice singlet band (ZRB) to the upper Hubbard band (UHB). I refer to the transition as the Mott gap excitation, taking it in its broad sense.

When carriers (electrons or holes) are doped in the Mott insulators, another charge excitations below the gap arise as intraband excitations, which are related to the dynamics of the doped carriers. The dispersion relation of the intraband excitations follows the band structure where the carriers are doped, and its magnitude is scaled by the nearest neighbor hopping energy between the Cu atoms (t). In high-T_c superconducting cuprates, t is about 0.4 eV.

At an energy scale similar to that of Mott gap excitation, so-called d-d (orbital) excitations sometimes appear in the RIXS spectra. The d-d excitations are associated with the reconfiguration of d electrons, and their the energy is governed by crystal

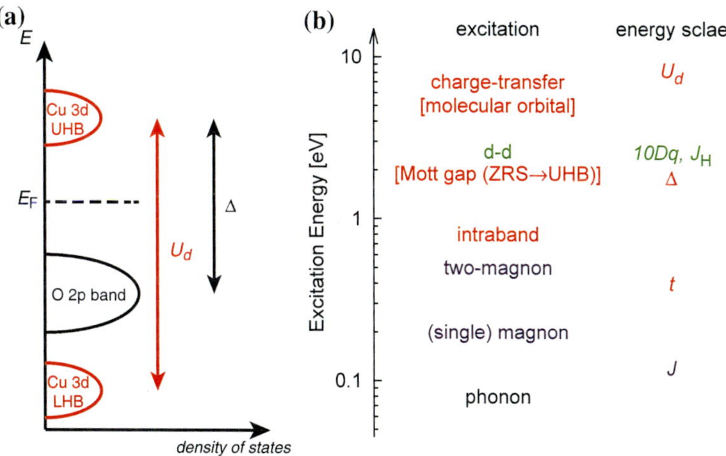

Fig. 3 **a** Schematic electronic structure of insulating copper oxides. The Cu $3d_{x^2-y^2}$ bands are separated into the lower Hubbard band (LHB) and the upper Hubbard band (UHB) by the Coulomb repulsion (U_d). The O $2p$ band is situated in between the Cu $3d_{x^2-y^2}$ bands. Because the charge transfer energy (Δ) is smaller than U_d, the copper oxides are classified as a charge-transfer insulator. **b** Various excitations in copper oxides observed with RIXS. The excitations are scaled to the energy of relevant interactions

field ($10Dq$), Hund coupling (J_H), and so on. The excitations are local in nature in most cases, resulting in small momentum dependence.

Below the charge and d-d excitations, magnetic excitations are observed in undoped copper oxides. The energy of the magnetic excitations is of the order of the magnetic exchange interaction (J) which is about 100 meV in the high-T_c cuprates. Because the energy scale of these magnetic excitations is one order smaller than that of the charge excitations in the insulating cuprates, they are well separated from the charge excitations. Electron dynamics at low energy can be dictated by a Hamiltonian only with the spin degree of freedom, such as the Heisenberg model. In the doped cuprates, magnetic and charge excitations overlap.

RIXS studies on cuprates have been performed at the Cu K-, L-, M- and the O K-edges so far. Generally speaking, charge excitations, which occur via the Coulomb interaction with a core hole in the intermediate state, are prominent at the Cu K-edge, while d-d excitations are dominant at the Cu L and M-edges. At the O K-edge, charge and d-d excitations are almost even. Another difference between these edges is the accessible momentum space; one can measure the entire Brillouin zone at the Cu K-edge while the accessible reciprocal point is restricted at the Cu L_3 and the O K-edges. The momentum transfer at the Cu M-edge is negligible.

The electronic excitations in the copper oxides and their related interactions that scale the energy of the excitations are summarized in Fig. 3b. In the following subsection, the electronic excitations observed by RIXS are described in detail.

3.2 Insulating Cuprates

3.2.1 Charge Excitations

Cu K-edge RIXS has been used in most of the studies of charge excitations in cuprates. Because the intermediate state of RIXS corresponds to the final state of X-ray absorption and observed excitations vary with the selection of the intermediate state, we start with a brief explanation of X-ray absorption spectroscopy (XAS) of the cuprates. In Fig. 4a, b, the XAS spectra of $La_{2-x}Sr_xCuO_4$ and $Nd_{2-x}Ce_xCuO_4$ are shown, respectively. Because of the two-dimensional ($2D$) nature of the materials, the XAS spectra depend on the photon polarization (ε).

Similar to the case of Cu $2p$ photoemission [20], nonlocal screening of the core-hole potential is important in the Cu K-edge XAS. In Fig. 4c, the initial and intermediate states of RIXS are shown schematically for the undoped cuprate. A single band Hubbard model is used for simplicity, and U_d is the Coulomb interaction between $3d$ electrons. The $1s$ and $4p$ levels are also taken into account and U_c is the core-hole potential between the $1s$ core hole and the $3d$ electron. U_d and U_c are taken to be positive, and $U_c > U_d$. There are two intermediate states of RIXS (final states of XAS) in this simple model. One is the well-screened state, where an electron transfers from a neighboring site and two electrons screen the core-hole potential at a cost of U_d. The energy of the well-screened state is $E_{1s-4p} - 2U_c + U_d$, where E_{1s-4p} is

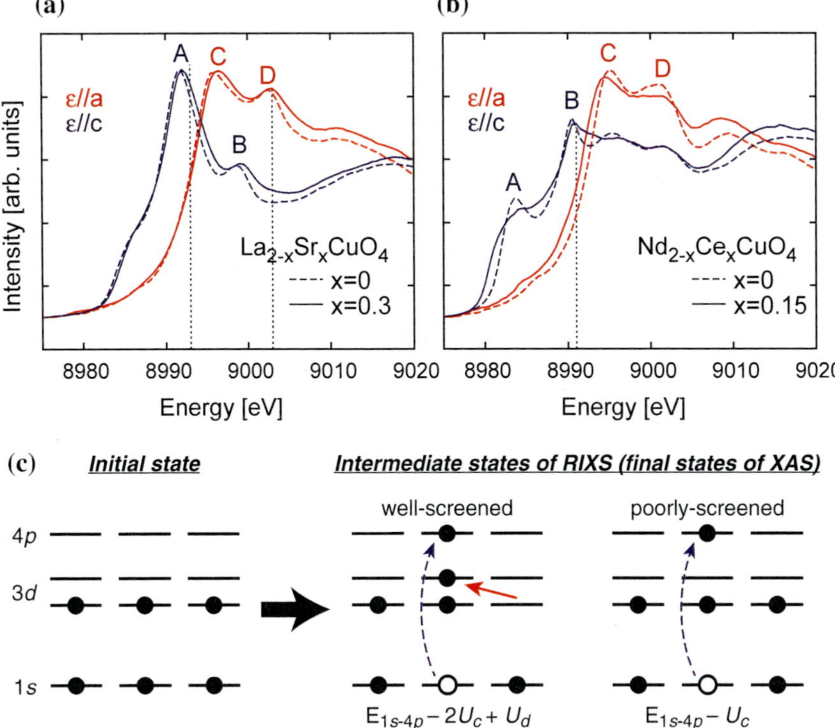

Fig. 4 X-ray absorption spectra of **a** La$_{2-x}$Sr$_x$CuO$_4$ and **b** Nd$_{2-x}$Ce$_x$CuO$_4$ near the Cu K-edge. The figures are based on the data in Refs. [17–19]. *Vertical dotted bars* indicate the incident photon energy used in the study of doping effect. **c** Schematic figures of the initial and intermediate states of RIXS in the Hubbard model. *Filled* and *open circles* represent an electron and a core-hole, respectively

the difference in energy between the 1s and 4p levels. Taking the oxygen orbitals into account, electron transfer occurs from a ligand oxygen, and as a result a hole remains in the O 2p level forming the Zhang–Rice singlet. The other is the poorly-screened state. Only one electron screens the core hole in the poorly-screened state, and the energy is $E_{1s-4p} - U_c$, which is higher than that of the well-screened state. In the configuration-interaction picture, the well-screened and poorly-screened states are expressed by linear combinations of $|\underline{1s}, 3d^9, 4p\rangle$ and $|\underline{1s}, 3d^{10}, \underline{L}, 4p\rangle$, where $\underline{1s}$ and $\underline{L}$ denote a hole in 1s level and ligand oxygen, respectively, and $|\underline{1s}, 3d^{10}, \underline{L}, 4p\rangle$ is dominant in the well-screened state.

The two states form peak features near the absorption edge in XAS spectra. Based on the assignment by Tolentino et al. [21], the peaks A and C labeled in Fig. 4a, b correspond to the well-screened state and the peaks B and D can be ascribed to the poorly-screened state.

After a pioneering RIXS work on NiO by Kao et al. [22], Cu K-edge RIXS studies on cuprates started. At the early stages of this research, local charge transfer excitations were studied in the parent compounds of high-T_c superconducting cuprates. Hill et al. [23] observed a 6-eV excitation in the RIXS spectra of Nd_2CuO_4, and it was identified as a charge-transfer type excitation of an electron from a bonding molecular orbital to an antibonding molecular orbital in a copper oxygen plaquette, which is called molecular orbital excitation. Later on, Kim et al. [24] measured the molecular orbital excitation in various copper oxides systematically and found a relation between the excitation energy and Cu-O bond length, i.e., the energy separation between the bonding and antibonding orbitals decreases with increasing the bond length. On the other hand, Abbamonte et al. [25] observed another charge-transfer excitation at 5–6 eV in La_2CuO_4 and $Sr_2CuO_2Cl_2$ and suggested that the excitation is a transition to an a_{1g} excitonic state. Symmetry argument on the polarization dependence supports the assignment [26].

While finite dispersion was indicated in Ref. [25], Hasan et al. [27] reported the first observation of a dispersive excitation in the entire Brillouin zone of the CuO_2 plane. Figure 5a, b show their measurement of the Mott gap excitation in $Ca_2CuO_2Cl_2$ which is a parent compound of high T_c superconducting cuprates. Peak positions of the Mott gap excitation are indicated by the vertical bars. They found that the energy of the Mott gap excitation is lowest at the zone center of the CuO_2 plane, and the magnitude of the dispersion along the $[\pi, \pi]$ direction is larger than that along the $[\pi, 0]$ direction. Subsequently, Kim et al. [28] reported dependence on incident photon energy (ω_i) for La_2CuO_4 in addition to the momentum dependence. The ω_i-dependence at the zone center $\mathbf{Q} = (3, 0, 0)$ is shown in Fig. 5c. It is clear

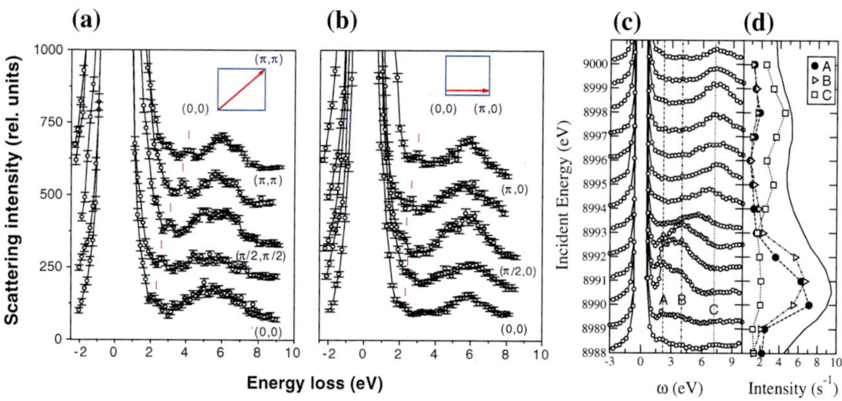

Fig. 5 **a, b** Cu K-edge RIXS spectra of $Ca_2CuO_2Cl_2$ [27]. The *vertical bars* indicate the Mott gap excitations. Reprinted from Ref. [27] with permission from AAAS. **c** Incident photon energy dependence of Cu K-edge RIXS spectra of La_2CuO_4 [28]. The momentum is fixed at the zone center. **d** RIXS intensities at 2.2 (**a**), 4.0 (**b**), and 7.2 eV (**c**) as a function of the incident photon energy. Absorption spectrum is also plotted by the *solid line*. Incident photon polarization is parallel to the c-axis ($\varepsilon_i \parallel \mathbf{c}$). Reprinted figure with permission from Ref. [28]. Copyright 2002 by the American Physical Society

that the spectral shape varies with ω_i. In Fig. 5d, RIXS intensities at 2.2 (A), 4.0 (B), and 7.2 eV (C) are plotted as a function of ω_i together with the XAS spectrum. The 7.2-eV peak in the RIXS spectra is the molecular orbital excitation and its intensity is largest at $\omega_i \simeq 8998$ eV, implying that the excitation resonates at the poorly-screened intermediate state. Prior to their work, a resonant condition of the molecular orbital excitation was found to be the poorly-screened intermediate state in Nd_2CuO_4 [23, 29]. Idé and Kotani theoretically pointed out that the molecular orbital excitation is suppressed at the well-screened intermediate state because the final state of the molecular orbital excitation is a state without forming the Zhang–Rice singlet. In contrast, the Mott gap excitation at 2–4 eV is resonantly enhanced at the well-screened intermediate state [28, 30]. This resonance is reasonable if one takes consideration of the formation of the Zhang–Rice single at the well-screened intermediate state.

Following these works, detailed studies of La_2CuO_4 [31, 32] and other parent compounds, Nd_2CuO_4, [18, 33] and $YBa_2Cu_3O_6$ [18] have been reported. The dispersion relation of the Mott gap excitation of these Mott insulators is qualitatively similar to that of $Ca_2CuO_2Cl_2$, but the resonant condition of the Mott gap excitation seems to be somewhat complicated in Nd_2CuO_4 and $YBa_2Cu_3O_6$ [18]. Apart from the Cu K-edge RIXS studies, Harada et al. [34] observed the Mott gap excitation at the O K-edge. In their study, polarization dependence was crucial for the distinction of the Mott gap excitation from d-d excitations, as predicted in a theoretical study [35].

Theoretically, Tsutsui et al. [36] predicted the dispersion relation of the Mott gap excitation. A half-filled single-band Hubbard model was used to describe the occupied ZRB and unoccupied UHB by mapping ZRB onto the lower Hubbard band (LHB) in the model. They also included Cu $1s$ and $4p$ orbitals in the model to treat the K-edge RIXS process and calculated the spectra by using the numerically exact diagonalization technique on small clusters. Calculated RIXS spectra are shown in

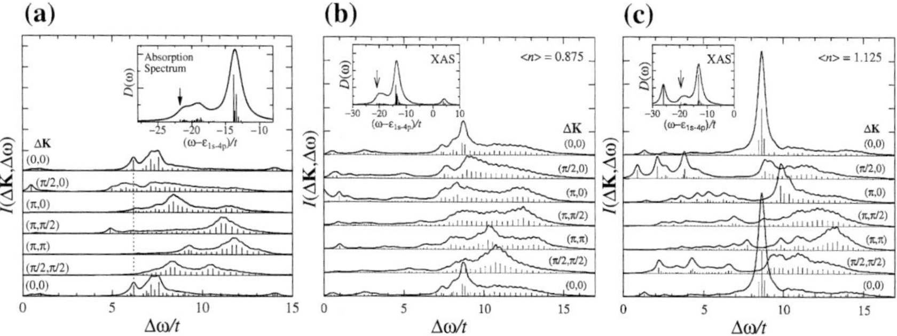

Fig. 6 Theoretical calculation of Cu K-edge RIXS spectra for **a** undoped, **b** hole-doped, and **c** electron-doped cuprates [36–38]. The *insets* of respective figures are the calculated X-ray absorption spectra, and the arrows indicate the incident photon energy for the calculation of the RIXS spectra in the main panel. Reprinted from Ref. [38], with permission from Elsevier

Fig. 6a. A characteristic feature of the Mott gap excitation, i.e., larger dispersion along the $[\pi, \pi]$ direction than the $[\pi, 0]$ direction, is well captured in this model.

Afterwards, another theoretical approach was made to reproduce the RIXS spectra of the parent compounds of high-T_c cuprates. Nomura and Igarashi [12, 39] adopted a three-band Hubbard (d-p) model, where the antiferromagnetic ground state was described using the Hartree–Fock approximation, and the electron correlation effects were taken into account within the random phase approximation. The scattering by the core hole was treated within the Born approximation. The effect of multiple scattering in the intermediate state was also taken into account in their subsequent study [40]. On the other hand, Chen et al. [41] included the Cu e_g ($d_{x^2-y^2}$ and $d_{3z^2-r^2}$) orbitals and bonding O $2p$ orbitals in their model and calculated the spectra using a two-step exact diagonalization algorithm. These calculations of the multiband model have been found to be useful for reproducing the fine features of the experimental spectra in a wide energy range.

3.2.2 d-d Excitations

Around the energy region of the Mott gap excitation, d-d excitations are dominant in the RIXS spectra at the Cu $L_{2,3}$- and $M_{2,3}$-edges. Because the d-d excitations are optically forbidden, the capability to observe d-d excitations with strong intensity is an advantage of RIXS at these edges. Several years before experimental observation, Tanaka and Kotani calculated the d-d excitations of La_2CuO_4 and CuO theoretically [43] and demonstrated that the RIXS spectra of the d-d excitations depend on the photon polarization. Duda et al. [44] found the polarization dependence at the Cu L_3-edge and Kuiper et al. [45] did at the Cu $M_{2,3}$-edges. Ghiringhelli et al. [46] succeeded in measuring a multiplet feature of the d-d excitations by exploiting an improved energy resolution. Recently, energy resolution was improved further, and Moretti Sala et al. [42] could determine the energy and symmetry of Cu-$3d$ states of four parent compounds (La_2CuO_4, $Sr_2CuO_2Cl_2$, $CaCuO_2$, and $NdBa_2Cu_3O_6$) combined with their polarization and scattering-geometry dependence. Figure 7a shows d-d excitations of La_2CuO_4 in their study [42]. Three main features at -1.7, -1.8 and -2.1 eV are identified as the excitations from $d_{3z^2-r^2}$, d_{xy}, and d_{yz}/d_{zx} orbitals to the $d_{x^2-y^2}$ orbital, respectively. In Fig. 7b, Cu L_3-edge RIXS spectra of the four parent compounds are compared. Spectral shape varies with the ligand field. In particular, existence of apical ligands substantially affects the energy of the excitation from the $d_{3z^2-r^2}$. Hozoi et al. [47] recently demonstrated that the energy levels of the d orbitals obtained from an ab initio quantum chemical calculation agree well with the experimental RIXS results.

Cu K pre-edge is another possible edge for accessing the $3d$ states in the intermediate state of RIXS. Kim et al. [48] demonstrated that the d-d excitations can be measured utilizing K pre-edge. They observed a d-d excitation of $Sr_2CuO_2Cl_2$ at 2 eV by tuning the incident energy to the Cu K pre-edge and identified it as the transition from the d_{yz}/d_{zx} orbitals based on the angular dependence of the quadrupole transition-matrix element.

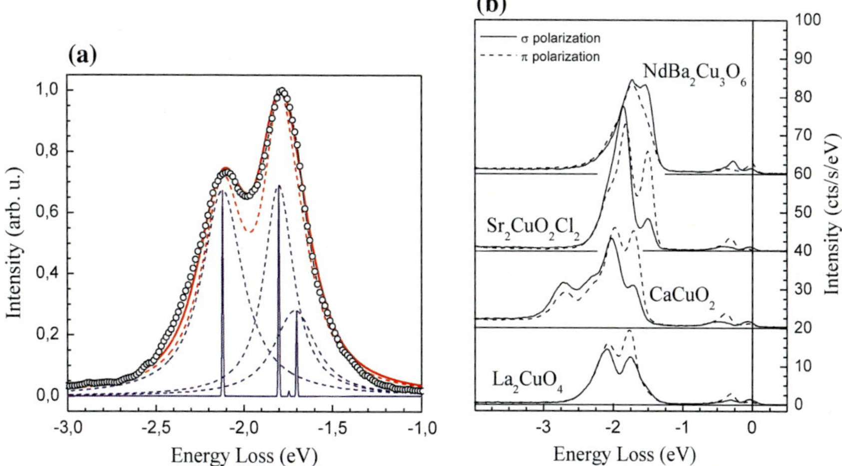

Fig. 7 **a** Cu L_3-edge RIXS spectrum of La_2CuO_4 [42]. *Open circles* show the experimental data taken at $2\theta = 90°$ and $\delta = 0°$ and lines are the theoretical spectra. **b** Cu L_3-edge RIXS spectra of various copper oxides measured at $2\theta = 130°$ and $\delta = 45°$. Reprinted from Ref. [42], © IOP Publishing & Deutsche Physikalische Gesellschaft. CC BY-NC-SA

3.2.3 Spin Excitations

At lower energies than those of the Mott gap and d-d excitations, spin excitations appear in the RIXS spectra. The energy of spin excitations corresponds to the energy scale of J. So far, both single-magnon and two-magnon excitations have been studied in the copper oxides. The selection rule of optical transition allows observation of the former at the Cu $L_{2,3}$-edge, while the latter can be observed at the Cu K-, L-edges, and O K-edges. As the photon has an angular momentum $L = 1$, one can observe the transition with $\Delta L_z = 0$, 1, or 2 as well as two spin-flips ($\Delta S_z = 0$ in total). The two spin-flips are well-known as two-magnon excitations in conventional Raman scattering. For the single spin-flip excitation, it is allowed at the Cu $L_{2,3}$-edge as calculated in Sect. 2.2. Strong spin-orbit interaction at the Cu L-edge split the $2p$ level into $2p_{1/2}$ (L_2-edge) and $2p_{3/2}$ (L_3-edge) states, and it is crucial for the occurrence of the spin-flip excitation such as single-magnon excitation.

The possibility to observe the two-magnon excitations with RIXS was pointed out by theoretical calculations at the Cu K-edge [36] and the O K-edge. [35] Experimentally, Harada et al. [34] found a 0.5-eV feature in the O K-edge RIXS spectra of $Sr_2CuO_2Cl_2$ and it was interpreted as two-magnon excitations based on the polarization dependence. Hill et al. found the 0.5-eV feature in the Cu K-edge RIXS spectra of La_2CuO_4 and Nd_2CuO_4 and measured its momentum dependence [49, 50]. At the Cu K-edge, the intensity of the feature is largest at $\mathbf{q} = (\pi, 0)$. Theoretical calculations [51–54] confirmed that the 0.5-eV feature originates from two-magnon excitations. Later, the two-magnon excitations at the O K-edge were reinvestigated in detail by

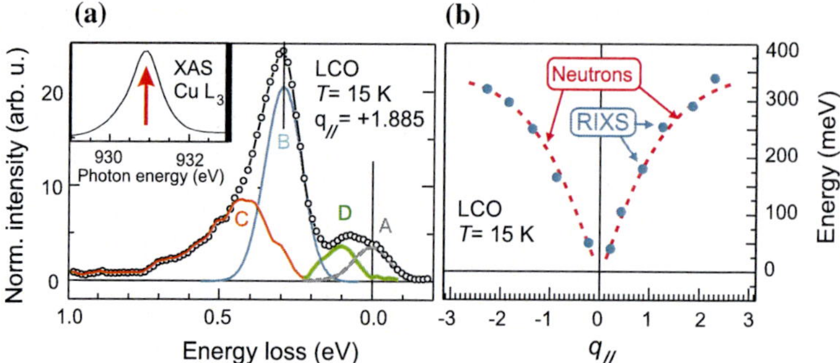

Fig. 8 Low-energy part of Cu L_3-edge RIXS spectrum of La_2CuO_4 [6]. The spectrum is decomposed into **a** elastic scattering, **b** single magnon, **c** multi magnon, and **d** phonons. The *inset* is the X-ray absorption spectrum (XAS) and the *arrow* indicates the incident photon energy for the spectrum in the main panel. **b** Dispersion relation obtained by RIXS and inelastic neutron scattering. Reprinted figure with permission from Ref. [6]. Copyright 2010 by the American Physical Society

Bisogni et al. [55] and compared with the results at the Cu L_3- and Cu K-edges. They found that the O K-edge RIXS is sensitive to the less-dispersive high-energy ridge in the density of states of the two-magnon excitations, while RIXS at the Cu L_3- and K-edges samples the dispersive low-energy ridge. In their paper, they referred to the excitation as bimagnon that are the coherent states two-magnon [56].

A unique capability of the $L_{2,3}$-edge is the investigation of spin-flip excitations. This capability was theoretically proposed more than ten years ago [57], but it was experimentally proved very recently owing to the improvement of energy resolution [4]. In contrast to the local nature of the spin-flip excitations in NiO [4], the excitations in cuprates show a large dispersion of several times J [6]. In Fig. 8a, the low-energy part of the Cu L_3-edge RIXS spectrum of La_2CuO_4 is presented. The spin-flip excitation (single magnon) has the largest spectral weight in the spectrum. The magnetic origin of the excitations was confirmed by their polarization dependence [5, 58] and consistency with the dispersion relation obtained by inelastic neutron scattering, as shown in Fig. 8b. Subsequently, Guarise et al. reported magnetic excitations in another insulating cuprate, $Sr_2CuO_2Cu_2$ [59].

3.2.4 Substitution Effect

Chemical substitution for a specific element is often used in the study of high-T_c superconducting cuprates. Since a core electron participates in the scattering process, RIXS has an advantage of element selectivity. Therefore RIXS can be a powerful tool for probing the electronic states of substituted elements and understanding their interaction with surrounding atoms. RIXS spectra were obtained at the K-edge of each substituted element in Ni-substituted La_2CuO_4 and its counterpart Cu-substituted La_2NiO_4 [60]. Figure 9a shows the RIXS spectra of $La_2Cu_{0.95}Ni_{0.05}O_4$

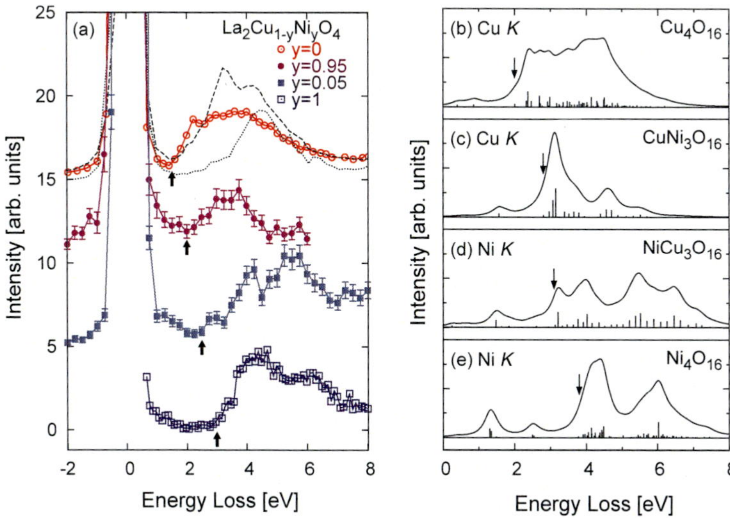

Fig. 9 **a** RIXS spectra of La$_2$Cu$_{1-y}$Ni$_y$O$_4$ measured at the Brillouin zone center [60]. The upper and lower two spectra were measured at the Cu K- and Ni K-edges, respectively. The *dashed* and *dotted lines* in $y = 0$ are the spectra at the momenta of $(\pi, 0)$ and (π, π), respectively. The onset edges of the CT excitations are indicated by the *arrows*. The RIXS spectra calculated on Cu$_{4-z}$Ni$_z$O$_{16}$ clusters with $z = 0$ (**b**), 3 (**c**), 1 (**d**), and 4 (**e**). **b**, **c** shows the RIXS spectra for Cu K-edge, and **d**, **e** for Ni K-edge. *Arrows* indicate the leading edges of CT excitations

and La$_2$Cu$_{0.05}$Ni$_{0.95}$O$_4$. As a reference, the spectra of unsubstituted La$_2$CuO$_4$ and La$_2$NiO$_4$ [28, 31] are also presented. The upper two spectra of $y = 0$ and 0.95 are measured at the Cu K-edge. In this case, the core hole is created at the $1s$ orbital of the Cu site, and the observed excitations centered at $3\,\text{eV}$ are ascribed to the charge-transfer excitations from O $2p$ to Cu $3d$. On the other hand, the lower two spectra of $y = 0.05$ and 1 are measured at the Ni K-edge, and therefore the charge-transfer excitations from O $2p$ to Ni $3d$ are observed. The center of mass of the spectral weight is somewhat similar between the upper two spectra and between the lower two spectra, respectively, which means that the charge-transfer excitation is locally unchanged at the substituted site. Looking at the spectra more precisely, we notice that the onset edge of the CT excitation, which is indicated by the arrows, shifts systematically from the top to the bottom. A theoretical calculation in Fig. 9b–e, which was performed on a Cu$_{4-z}$Ni$_z$O$_{16}$ cluster by numerical exact diagonalization, reproduced the experimentally observed shift quantitatively, and electronic states near the substituted elements were argued from the charge-transfer excitations.

3.3 Doped Cuprates

3.3.1 Charge Excitations

One of the central issues in strongly correlated electron systems is the evolution of their electronic structures on carrier doping, departing from the Mott insulating state. Several groups studied charge excitations of hole-doped $La_{2-x}Sr_xCuO_4$ by Cu K-edge RIXS [19, 31, 61–64]. In addition, superconducting $YBa_2Cu_3O_{7-\delta}$ [65], $HgBa_2CuO_{4+\delta}$ [66], $Ca_{1.8}Na_{0.2}CuO_2Br_2$ [18], $Bi_{1.76}Pb_{0.35}Sr_{1.89}CuO_{6+\delta}$ [19] have been measured.

In Fig. 10a, Cu K-edge RIXS spectra of hole-doped $La_{2-x}(Ba,Sr)_xCuO_4$ are presented. The incident photon energy is 8993 eV, which is close to the peak of the well-screened state indicated by a vertical dotted bar in Fig. 4a. The spectral intensity is normalized to the high-energy tail at 6–8 eV. The main change upon hole doping is that continuum intensity fills the Mott gap associated with the disappearance of the lower part of the Mott gap excitation. Qualitatively, these changes are similar to the doping effect of optical conductivity [67]. The continuum intensity below the Mott gap can be attributed to the dynamics of the doped holes within the Zhang–Rice band, and therefore, it is called the intraband excitation. Excitation across the gap remains in the RIXS spectra, even though its lower part disappears upon hole doping, indicating the persistence of the Mott gap. As a result, the spectral weight of the Mott gap excitation in the doped cuprates is located higher in energy than that in the undoped cuprates. A recent systematic measurement of doping dependence at high-symmetric momenta by Ellis et al. [64] demonstrated that the spectral weight transfer from high energies (2.5–6 eV) to low energies (1–2 eV) is nearly linear with x at least at $\mathbf{q} = (\pi, 0)$. In other words, the spectral weight at low energies is proportional to the hole concentration.

In contrast to the case of hole doping, a clear peak feature with large dispersion was observed as an intraband excitation of the electron-doped cuprates and it has large dispersion [17, 18]. Figure 10b shows electron-doping dependence of the Cu K-edge RIXS spectra in $Nd_{2-x}Ce_xCuO_4$. At $x = 0$, the Mott gap is clearly seen at all the momenta, and the magnitude of the gap at $\mathbf{q} = (0, 0)$ is consistent with an optical study [68]. When electrons are doped in the UHB, the excitation below the Mott gap increases at finite momenta forming a peak. Because the intensity of the excitation is roughly proportional to the electron concentration x, it can be ascribed to the intraband excitation within UHB. With increasing momenta, the spectral weight of the intraband excitation shifts to higher energy accompanied with broadening of the width. At the zone center, the 2-eV peak is independent of the doping, which indicates that the Mott gap persists in the electron-doped cuprates.

Li et al. [33] also measured $Nd_{2-x}Ce_xCuO_4$ but they offered a different interpretation. Along the $[\pi, \pi]$ direction, electron-doping splits the spectral weight of the excitation across the gap of the undoped spectrum, e.g., the peak around 5 eV at $\mathbf{q} = (\pi, \pi)$ separates into two peaks, and the peak at lower energy softens with doping. On the other hand, the spectral weight along the $[\pi, 0]$ direction is found to

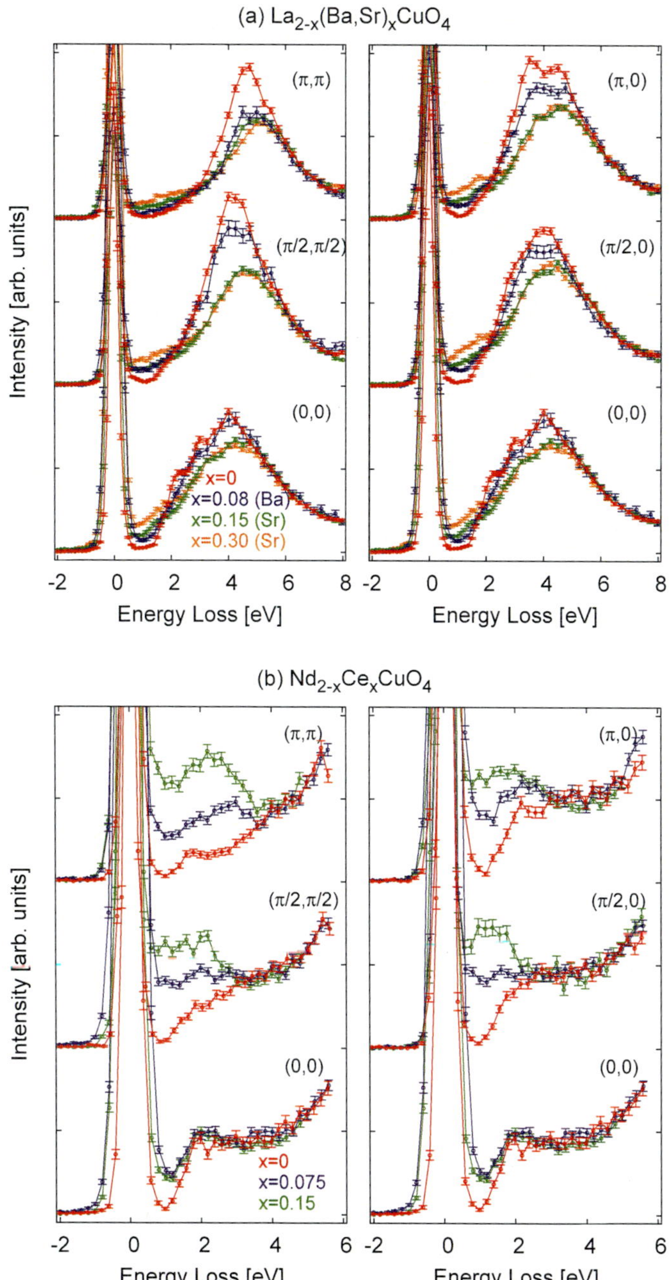

Fig. 10 Cu K-edge RIXS spectra of **a** hole-doped La$_{2-x}$(Ba,Sr)$_x$CuO$_4$ and **b** electron-doped Nd$_{2-x}$Ce$_x$CuO$_4$. The figures are based on the data in Refs. [18, 19]

soften monotonically upon electron doping. Eventually, the charge gap closes in the vicinity of the zone center. Their interpretation relies on the theory that the charge gap in the parent Nd_2CuO_4 is considered to be antiferromagnetic in origin and that it collapses upon electron doping [69, 70]. This dichotomy is reminiscent of the argument whether the parent compound of the electron-doped superconductor, which lacks apical oxygens, is a Mott or Slater insulator [71].

Preceding the experiments, Tsutsui et al. [37] predicted the momentum dependence of the Mott gap excitation in doped cuprates. Similarly to the undoped cuprate [36], the two-dimensional Hubbard model was studied by numerical exact diagonalization. Figure 6b, c show the calculated spectra for hole and electron doping, respectively. In the case of hole doping, the spectral weight above $6t$ is the Mott gap excitation, and its dispersion is found to be smaller than the undoped case. The smaller dispersion was observed experimentally. When the onset energy of the excitation is plotted as a function of momentum transfer, its momentum dependence is rather flat in $x = 0.17$, as shown in Fig. 6 of Ref. [61]. Li et al. [63] achieved the same conclusion in $x = 0.14$ of $La_{2-x}Sr_xCuO_4$. $YBa_2Cu_3O_{7-\delta}$ is another popular system of hole-doped superconductor, but the existence of one-dimensional CuO chains running along the crystalline b-axis next to the double CuO_2 planes complicates matters. Ishii et al. [65] distinguished the Mott gap excitation in the CuO_2 planes from that in the CuO chain by measuring the momentum dependence of a twin-free crystal of optimally doped $YBa_2Cu_3O_{7-\delta}$. The extracted dispersion relation of the Mott gap excitation of the plane is small, which is consistent with the case of $La_{2-x}Sr_xCuO_4$. On the other hand, the Mott gap excitation for electron doping is located above $8t$ in the calculated spectra. The intensity of the Mott gap excitation concentrates at $\sim 8.5t$ at $\mathbf{q} = (0, 0)$ and it corresponds to the 2-eV peak in Fig. 10b. The magnitude of the dispersion is much larger than that of the hole-doped case and is almost comparable to the undoped case. The difference in dispersion between hole- and electron-doped systems is argued with relation to the carrier dependence of short range antiferromagnetic spin correlation; the short-range AF correlation, which is kept in the electron-doped case, influences the momentum dependence of the Mott gap excitation. Experimentally, momentum dependence of the Mott gap excitation for electron doping has not been observed clearly.

In the previous subsection, well-screened and poorly-screened states were considered as intermediate states in undoped cuprates. When carriers are doped, a core hole can be created at the doped site, and this state can be an intermediate state in addition to the other two states. Such an intermediate state is shown in Fig. 11. Energy of the intermediate state is E_{1s-4p} and $E_{1s-4p} - 2U_c$ for hole- and electron doped cases, respectively. The former energy is higher than that of the poorly-screened state ($E_{1s-4p} - U_c$) while the latter is lower than that of the well-screened state ($E_{1s-4p} - 2U_c + U_d$).

In the course of the study of $Nd_{2-x}Ce_xCuO_4$, it was pointed out that the selection of the intermediate state is important in measuring the excitation at low energy [17, 38]. Especially, the intraband excitation is enhanced resonantly when the core-hole is created at a site with a doped electron. The spectra in Fig. 10b were measured at $\omega_i = 8991\,eV$, indicated by the vertical dotted bar in Fig. 4b, and the incident

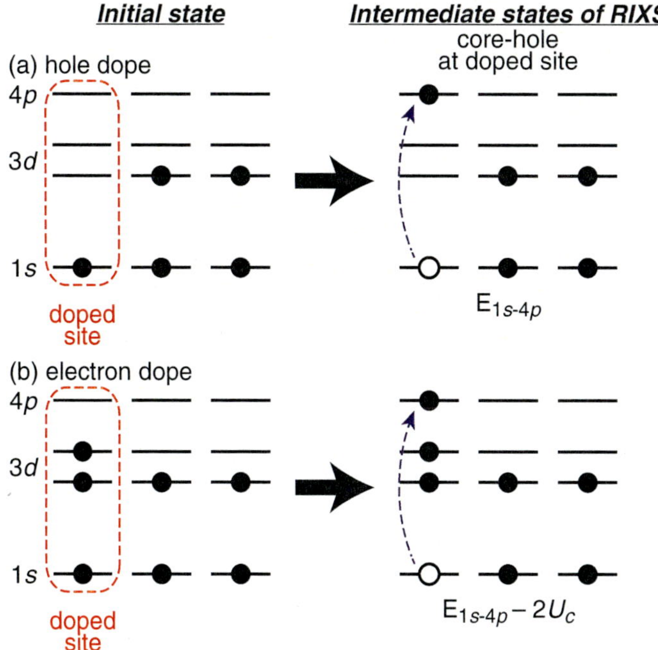

Fig. 11 Schematic figures of the initial and intermediate states of RIXS in **a** hole- and **b** electron-doped systems

photon energy satisfies the condition for $\varepsilon \parallel \mathbf{a}$. Furthermore, it was theoretically demonstrated that the RIXS spectra is qualitatively similar to the dynamical charge correlation function $N(\mathbf{q}, \omega)$ under this condition. Actually, the experimental RIXS spectra capture characteristic features of theoretical calculations of $N(\mathbf{q}, \omega)$. It is also suggested that the same condition for the hole-doped cuprates corresponds to the incident photon energy above the poorly-screened state. Later, Jia et al. [72] performed a systematic calculation of RIXS spectra and $N(\mathbf{q}, \omega)$ versus incident photon energy and reached the same conclusion. Their results are presented in Fig. 12. It is clear in Fig. 12c, f that the intraband excitations at low energy are observed dominantly and that the RIXS spectra track the features of $N(\mathbf{q}, \omega)$. Note that $N(\mathbf{q}, \omega)$ is qualitatively similar between the hole- and electron-doped systems; the spectral weight shifts to higher energy with increasing the momentum. The similarity was also confirmed in a recent work using a t-J model with long range hoppings (t' and t'') [73].

Experimentally, most of the studies of hole-doped cuprates have been measured at the incident photon energy close to the well-screened intermediate state, including the spectra in Fig. 10a and the RIXS spectra below the gap forms a continuum under this condition. Eventually, Wakimoto et al. [19] measured overdoped $La_{2-x}Sr_xCuO_4$ and $Bi_{1.76}Pb_{0.35}Sr_{1.89}CuO_{6+\delta}$ (Bi2201) at higher incident photon energy and succeeded

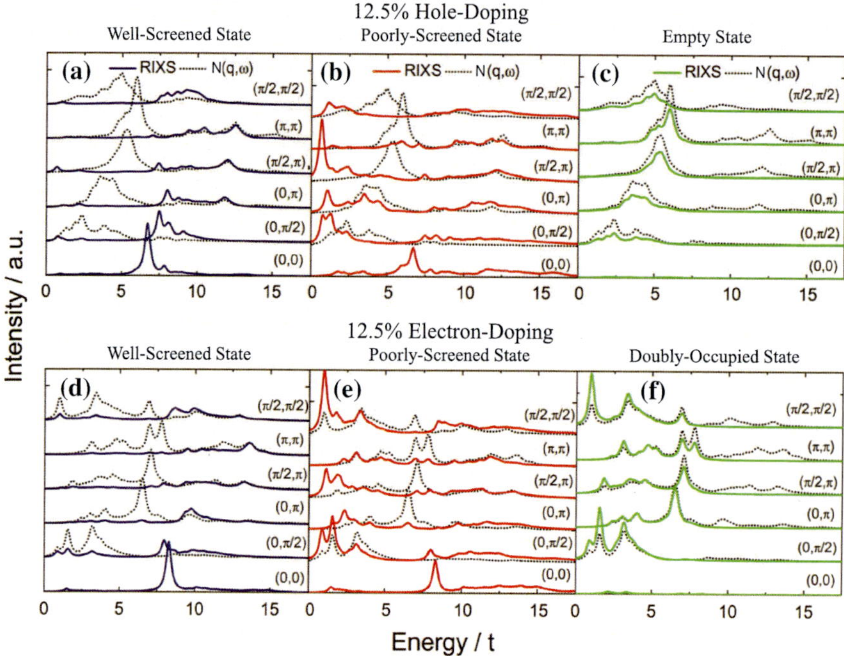

Fig. 12 Calculation of RIXS spectra and dynamical charge correlation functions $N(\mathbf{q}, \omega)$ for (**a**)–(**c**) hole- and (**d**)–(**f**) electron-doped systems of the single-band Hubbard model. *Solid lines* represent the RIXS spectra with the intermediate state having the core-hole well-screened, poorly screened, and created at an empty/doubly occupied site (doped site), respectively. *Dotted lines* are $N(\mathbf{q}, \omega)$. Reprinted from Ref. [72] under a CC-BY license

in observing dispersive excitations below the Mott gap in both compounds. The incident photon energy is 9003 eV, as indicated by the vertical dotted bar in Fig. 4a, which is equal to or slightly higher than the poorly-screened state. The spectra of electron doped electron-doped $Nd_{1.85}Ce_{0.15}CuO_4$ and hole-doped $La_{1.7}Sr_{0.3}CuO_4$, after subtracting the elastic line and the molecular orbital excitations, are compared in Fig. 13. The dispersive intraband excitations are similar between them, suggesting that the incident photon energy selected for the measurement of Fig. 13b fulfills the condition that the core hole is created at the hole site in the intermediate state. Thus, $N(\mathbf{q}, \omega)$ in the 1–2 eV region is proven to be qualitatively similar between hole- and electron-doped cuprates, as expected from the theories mentioned above.

3.3.2 Spin Excitations

It is generally believed that the superconductivity of cuprates is intimately related to the antiferromagnetic spin fluctuation and inelastic neutron scattering (INS) has been widely used for studying spin dynamics in the reciprocal lattice space

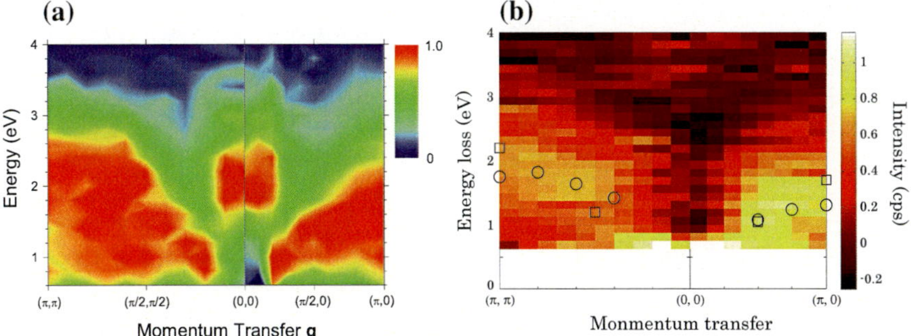

Fig. 13 Low-energy part of Cu K-edge RIXS for **a** electron-doped $Nd_{1.85}Ce_{0.15}CuO_4$ [17] and **b** hole-doped $La_{1.7}Sr_{0.3}CuO_4$ [19]. The incident photon energy is tuned to the condition where dispersive intraband excitations are resonantly enhanced. *Circles* and *squares* in (**b**) indicate the positions of the center of mass of the intraband excitations for $La_{1.7}Sr_{0.3}CuO_4$ and $Nd_{1.85}Ce_{0.15}CuO_4$, respectively

[74, 75]. Recently, high-resolution RIXS at the Cu L_3-edge has become complementary to measuring momentum-resolved spin excitations. RIXS can cover a momentum space away from the magnetic zone center, where high-energy magnetic excitations are hard to measure with INS.

Soon after the observation of the dispersive single-magnon in the parent compound La_2CuO_4 [5, 6], the hole-doping effect on the spin excitations in $YBa_2Cu_3O_{6+x}$ family ranging from an undoped antiferromagnetic insulator to a slightly overdoped superconductor was investigated by the Cu L_3-edge RIXS [76]. So far hole-doped $YBa_2Cu_3O_{6+x}$ [76, 77] and its isostructural materials $(Ca_xLa_{1-x})(Ba_{1.75-x}La_{0.25+x})$ Cu_3O_y [78], $La_{2-x}Sr_xCuO_4$ [6, 79–81], $Tl_2Ba_2CuO_{6+\delta}$ (Tl2201) [82], $Bi_{2+x}Sr_{2-x}$ $CuO_{6+\delta}$, $Bi_{1.5}Pb_{0.55}Sr_{1.6}La_{0.4}CuO_{6+\delta}$ (Bi2201) [83], $Bi_2Sr_2CaCu_2O_{8+\delta}$ (Bi2212) [84, 85], $Bi_2Sr_2Ca_2Cu_3O_{10+\delta}$ (Bi2223) [83], and electron-doped $Nd_{2-x}Ce_xCuO_4$ [86, 87] have been investigated. The spin excitations in relation to the superconductivity are discussed in detail in a review paper [88].

It is shown in the previous subsection that the spectral weight of the Cu L_3-edge RIXS in a sub-eV region is primarily spin excitations in the undoped antiferromagnetic insulator. The peak of the excitations follows the dispersion-relation of the magnon excitations observed by INS, and the width of the excitations is resolution-limited. When carriers are doped, two kinds of excitations can be observed in the energy range potentially. One is the intraband charge excitations and the other is a remnant of the spin excitations. As described later, the spin excitations are dominant when the π-polarized incident beam and the configuration of $\delta > 0$ (see Fig. 2) are chosen. Most of the work on the spin excitations in the doped cuprates has been performed under these conditions.

Cu L_3-edge RIXS spectra of $La_{2-x}Sr_xCuO_4$ are shown in Fig. 14a. Typical hole-doping effects on the spin excitations are seen in the figure; the peak width of the spin excitations broadens upon hole doping, but its position is almost kept unchanged

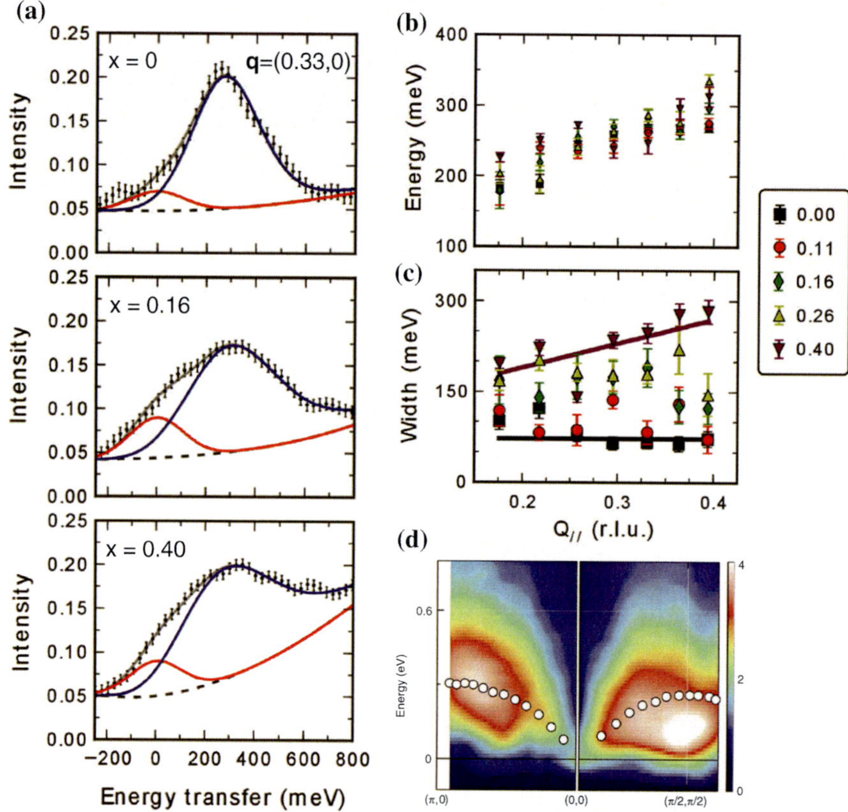

Fig. 14 **a** Cu L_3-edge RIXS spectra of $La_{2-x}Sr_xCuO_4$ [80, 88]. Experimental data (*filled black circles*) are decomposed into elastic scattering (*red lines*), spin excitations (*blue lines*), and the background (*dashed black lines*). **b** Energy dispersion and **c** half-width at half-maximum of spin excitations along the $[\pi, 0]$ direction. Reprinted from [88], with permission from Elsevier. **d** Cu L_3-edge RIXS intensity map of underdoped Bi2212, after subtraction of the elastic scattering [85]. *Open circles* are the magnon dispersion of the antiferromagnetic insulator of Bi2212, $Bi_{1.6}Pb_{0.4}Sr_2YCu_2O_{8+\delta}$. Reprinted by permission from Macmillan Publishers Ltd.: Nature Communications (Ref. [85]), Copyright 2014

(Fig. 14b, c) even in the overdoped $x = 0.40$. Even though a long range antiferromagnetic order disappears at the hole concentration, the dispersive nature of the spin excitations still survives. Therefore, the excitation is often referred to as a paramagnon. The persistence of the paramagnon without change in peak position or intensity is commonly observed in hole-doped cuprates.

The spin excitations exhibit contrasting behavior upon electron doping [86, 87]. Figure 15a shows Cu L_3-edge RIXS spectra of $Nd_{2-x}Ce_xCuO_4$. With increasing electron doping (x), the spectral weight clearly moves to higher energy and the width of the peak broadens. In Fig. 15b, c, the doping dependences of the peak position and the width are summarized.

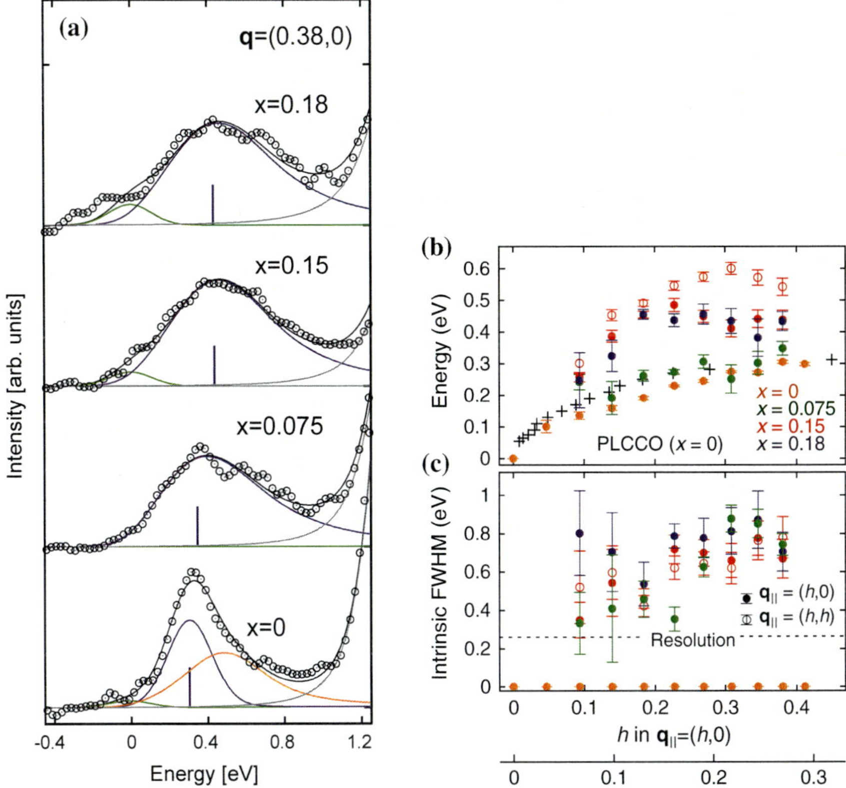

Fig. 15 a Cu L_3-edge RIXS spectra of $Nd_{2-x}Ce_xCuO_4$ [86]. *Open circles* are experimental data and *solid lines* are fitting results of elastic scattering (*green lines*), single-magnon excitations (*blue lines*), multi-magnon excitation (*orange lines*) and the tail of the *d-d* excitations (*grey lines*). In the doped compounds ($x \neq 0$), the multi-magnon component is neglected. **b** Energy dispersion and **c** full-width at half-maximum of spin excitations after deconvolution of the experimental resolution

So far, the assignment of the spin excitations to the dominant spectral weight relies on continuous doping evolution because the experiments were performed under the condition where the spin-flip excitations becomes large; π-polarized incident beam and the configuration of $\delta > 0$ were used. Jia et al. [89] justified this assignment theoretically by demonstrating that the low-energy part of RIXS measured under this condition agrees well with the dynamical spin correlation function $S(\mathbf{q}, \omega)$. Experimentally, the recent achievement of polarization analysis for a scattered beam [90] makes this assignment more rigorous [77]. As described in Sect. 2.3, the RIXS intensity is given by the dynamical correlation functions multiplied by the cross section of the transition for the isolated atom if one takes the fast-collision approximation, which is reasonable for the Cu L_3-edge RIXS. In the case of high-T_c cuprates, the spin moment lies in the xy-plane ($\theta_s = 90°$), and the cross section of the spin-conserving and spin-flip transitions in this spin orientation are shown as a function

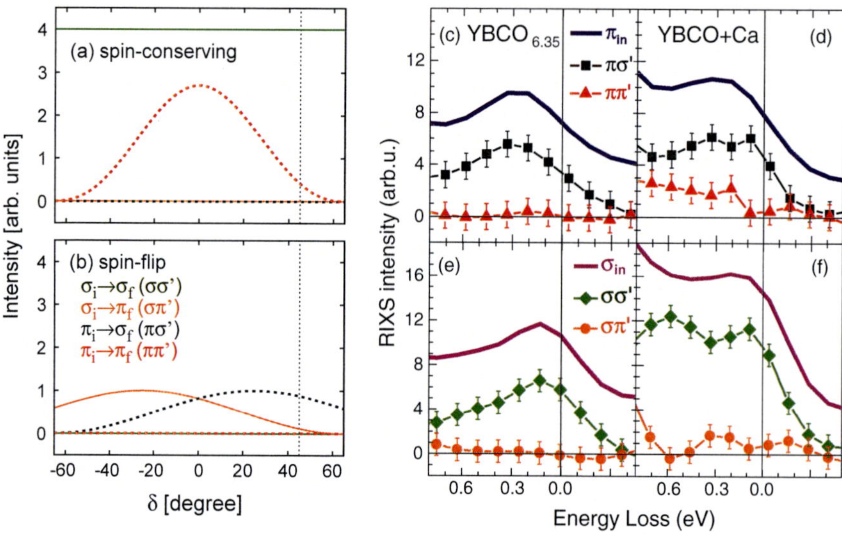

Fig. 16 **a**, **b** Polarization dependence of Cu L_3-edge RIXS cross section for a single Cu^{2+} ion within the $d_{x^2-y^2}$ orbital. Spin-conserving and spin-flip transitions for $2\theta = 130°$ and $\theta_s = 90°$ are calculated from the absolute square of Eqs. 12 and 13. Maximum of the spin-flip transitions is normalized to unity. The *vertical dotted lines* indicate the experimental conditions under which spectra in **c–f** are measured. **c–f** Polarization-resolved RIXS spectra of $YBa_2Cu_3O_{6.35}$ (YBCO$_{6.35}$) and $Y_{0.85}Ca_{0.15}Ba_2Cu_3O_{6+x}$ (YBCO+Ca) [77]. The spectra of π_{in} and σ_{in} are the spectra without analyzing the polarization of the scattered beam. Reprinted figure with permission from Ref. [77]. Copyright 2015 by the American Physical Society

of δ in Fig. 16a, b, respectively. The cross section of the spin-conserving transition appears in the polarization of $\sigma_i \to \sigma_f$ or $\pi_i \to \pi_f$ while spin-flip transition does in $\sigma_i \to \pi_f$ or $\pi_i \to \sigma_f$. Consequently, the RIXS intensity is proportional the dynamical charge correlation function $N(\mathbf{q}, \omega)$ for the $\sigma_i \to \sigma_f$ and $\pi_i \to \pi_f$ conditions, while it is given by the dynamical spin correlation function $S(\mathbf{q}, \omega)$ for the $\sigma_i \to \pi_f$ and $\pi_i \to \sigma_f$ conditions [73].

Polarization-resolved RIXS spectra of $YBa_2Cu_3O_{6.35}$ (YBCO$_{6.35}$, hole concentration $p \sim 0.062$) and $Y_{0.85}Ca_{0.15}Ba_2Cu_3O_{6+x}$ (YBCO+Ca, $p \sim 0.21$) are presented in Fig. 16. They are measured at $2\theta = 130°$ and $\delta = 45°$. If one uses the π-polarized incident beam in the underdoped YBCO$_{6.35}$ (Fig. 16c), all the spectral weight appears in the $\pi\sigma'$ channel, namely, it is an experimental proof that the excitations come from the spin-flip transition. On the other hand, the spectral weight for the σ-polarized incident beam arieses from charge excitation (plus phonons) observed in the $\sigma\sigma'$ channel (Fig. 16d). In the overdoped YBCO + Ca (Fig. 16e, f), the same argument is still effective, even though the intensity of charge excitation increases and appears in the $\pi\pi'$ channel.

To describe low-energy electronic properties of doped cuprates, t-J or t-t'-t''-J model, which is derived from the Hubbard model by eliminating doubly occupied sites, is often used. The models predict that the spin excitations soften with hole doping [91]. Contrary to this prediction, the energy of spin excitations of hole doped cuprates observed by RIXS is almost unchanged from those of undoped antiferromagnetic insulators. Jia et al. [89] pointed out that this discrepancy is resolved by introducing a three-site exchange interaction including a doped site. The three-site exchange interaction is omitted in the t-J model but is held in the Hubbard model. The simulated $S(\mathbf{q}, \omega)$ of the Hubbard model reproduces the doping-independent dispersion relation for hole doping as well as the high-energy shift for electron doping, as shown in Fig. 17.

Doping-independent dispersion of the paramagnon was commonly observed in hole-doped cuprates in the early stage of research, but all the spectra were measured along the $[\pi, 0]$ direction. Recently, the momentum dependence of underdoped and optimally doped Bi2212 were compared between the $[\pi, 0]$ and $[\pi, \pi]$ directions; the spin excitations of the hole-doped compounds softens along the $[\pi, \pi]$ direction compared to those of the parent insulating compound [85]. Figure 14d shows the Cu L_3-edge RIXS intensity map of underdoped Bi2212, after subtraction of the elastic scattering. The open circles present the magnon dispersion of the antiferromagnetic insulator of Bi2212. It is clear that the spectral weight of the $[\pi, \pi]$ direction is located at lower energy than the magnon dispersion of the antiferromagnetic insulator shown by the open circles. Such softening along the $[\pi, \pi]$ direction is also indicated in the measurements of Bi2201, Bi2223 [83], and $La_{2-x}Sr_xCuO_4$ ($x = 0.25$ and 0.30) [81]. These observations are argued in relation to the itinerant nature of electrons in the cuprates. Electrons in a doped Mott insulator have both localized and itinerant characters, and compromising this duality is a central issue in strongly correlated electron systems. RIXS on the doped cuprates can give important clues to examine the issue.

While the paramagnon excitations in the doped cuprates have recently been a topic of interest, two-magnon excitations were also studied in $La_{2-x}Sr_xCuO_4$. Cu K-edge RIXS shows that the spectral weight of the two-magnon excitations is rapidly suppressed upon hole doping; the weight survives up to at least $x = 0.07$ but is absent

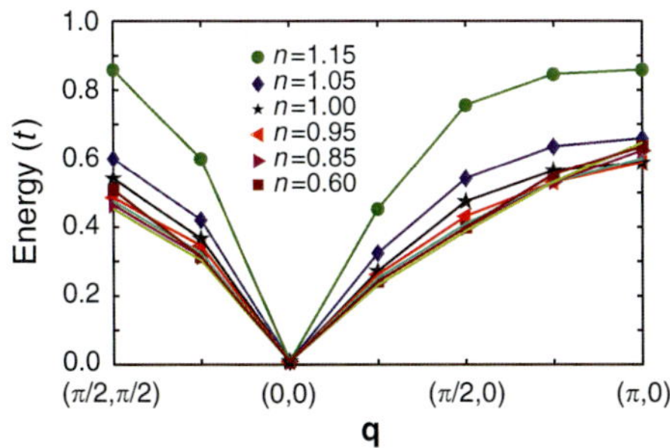

Fig. 17 Dispersion relation of the peak position in dynamical spin correlation function $S(\mathbf{q}, \omega)$ of the Hubbard model using determinant quantum Monte Carlo simulations. n is the electron concentration per site. Reprinted by permission from Macmillan Publishers Ltd.: Nature Communications (Ref. [89]), Copyright 2014

at $x = 0.17$ [49, 50]. This doping dependence of the Cu K-edge RIXS is quite similar to that observed by a Raman study [92]. In contrast, the two-magnon excitations are observed by the O K-edge RIXS up to an overdoped concentration of $x = 0.22$ with almost constant weight [93], which is compatible with the persistence of paramagnon in Cu L_3-edge RIXS and a recent resonant Raman scattering study on $HgBa_2CuO_{4+\delta}$ [94]. Further studies including other hole-doped compounds are necessary to resolve the discrepancy of the doping dependence of the two-magnon between the Cu K- and O K-edges.

3.3.3 Observation of Charge Order

In the last part of this subsection, charge order in doped cuprates is presented. Charge order in cuprates was originally discovered in $(La,Nd)_{2-x}Sr_xCuO_4$ as a charge stripe order in 1995 by Tranquada et al. [95]. Exactly speaking, lattice distortion induced by the charge order was detected by neutron scattering. Later, Abbamonte et al. [96] observed charge modulation directly in a resonant X-ray scattering experiment on $La_{1.875}Ba_{0.125}CuO_4$. In the ordered state, doped holes form one-dimensional charge stripes in the CuO_2 plane which separate antiferromagnetic domains. Because the superconducting transition temperature is suppressed around the hole concentration where the charge order is observed, the charge order is considered to be a competing state of the superconductivity. It had been a long-standing issue whether the charge order is universal in the doped cuprates or specific in the so-called La-214 family.

Eventually, in 2012, Ghiringhelli et al. [97] succeeded in observing the charge order in another cuprate; it was found in Cu L_3-edge RIXS spectra in $Nd_{1.2}Ba_{1.8}Cu_3O_7$ (hole concentration $p = 0.11$) as shown in Fig. 18a. The (quasi-)elastic scattering has a maximum intensity at $\mathbf{q} = (-0.31, 0)$ while the intensity of the spin and d-d excitations changes monotonically as a function of momentum. Polarization dependence confirms that the peak of the elastic scattering has charge origin. The energy resolution for the scattered photons in RIXS is a great advantage in detecting broad and weak modulations of the (quasi-)elastic signal; if one measures the signal in energy-integrated mode as does in an usual resonant elastic scattering experiment, huge d-d excitations and spin excitations appear as a background. This is clearly seen in Fig. 18b, c, where the charge-order peaks in $YBa_2Cu_3O_{6.6}$ ($p = 0.11$) are compared between energy-resolved and energy-integrated measurements. Similarly, Hashimoto et al. [98] performed a RIXS experiment and found a charge order in optimally-doped $Bi_{1.5}Pb_{0.6}Sr_{1.54}CaCu_2O_{8+\delta}$ (Pb-Bi2212). RIXS was also used in the study of the charge order in $HgBa_2CuO_{4+\delta}$ (Hg-1201) [99]. After all, the charge order has been observed by energy-integrated resonant X-ray scattering in almost all cuprate families, including $Bi_2Sr_{2-x}La_xCuO_{6+\delta}$ (Bi2201) [100], $Bi_2Sr_2CaCu_2O_{8+x}$ (Bi2212) [101], and even in electron-doped $Nd_{2-x}Ce_xCuO_4$ [102]. In this sense, RIXS plays a crucial role in the study of the charge order in cuprates. The experimental facts that the charge order is universally observed in the cuprates produce a renewed interest of the relation between the charge order and superconductivity.

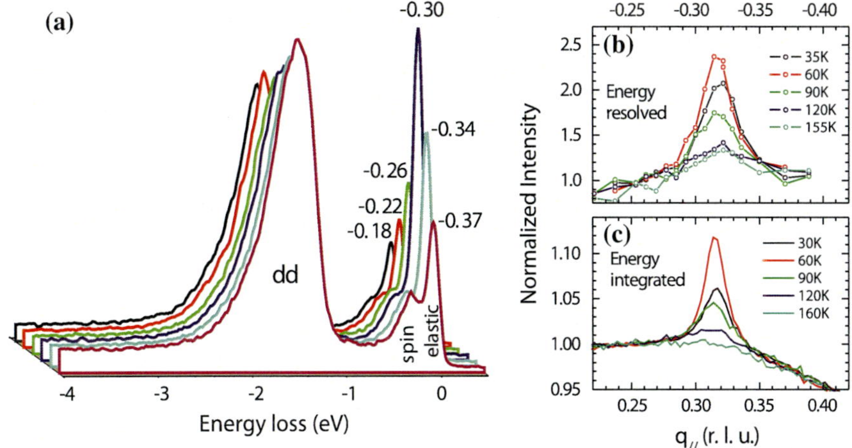

Fig. 18 **a** Cu L_3-edge RIXS spectra of $Nd_{1.2}Ba_{1.8}Cu_3O_7$ [97]. The numbers close to the elastic peak indicate the momentum h in $\mathbf{q} = (h, 0)$. **b, c** Comparison of the charge-order peak in $YBa_2Cu_3O_{6.6}$ between energy-resolved and energy-integrated measurements. Reprinted from Ref. [97] with permission from AAAS

As for the excitations related to the charge order, Wakimoto et al. [103] reported enhancement of the intraband excitations at the propagation vector of the charge order $\mathbf{q}_{CO} = (1/4, 0)$ in a Cu K-edge RIXS study of the charge-stripe-ordered $La_{2-x}(Ba,Sr)_xCuO_4$ ($x \sim 1/8$). Collective stripe excitations and anomalous softening of the charge excitonic modes of the in-gap states were proposed as a possible origin of this enhancement. However, their subsequent study [19] showed that the enhancement near $\mathbf{q}_{CO}$ extends to the overdoped region, indicating that direct relation of the enhancement to the charge-stripe order is unlikely. On the other hand, the magnetic excitations observed in the Cu L_3-edge RIXS by Dean et al. [79] might change slightly across the $\mathbf{q}_{CO}$. Examination of the excitations related to the charge order still remains to be done.

4 RIXS Studies on Other Copper Compounds

4.1 Zero-Dimensional Cuprates

In addition to the high-T_c superconductors, RIXS has been also applied to the study of other copper oxides. In zero-dimensional (0D) cuprates, such as Bi_2CuO_4 [104] and CuB_2O_4 [105], CuO_4 plaquettes are isolated electronically, and the Mott gap excitation is found to be suppressed in the Cu K-edge RIXS because the Zhang–Rice singlet cannot be formed in the isolated plaquette. Instead, d-d excitations are observed around 2 eV in CuB_2O_4 [105]. Observation of the d-d excitations by the Cu

K-edge RIXS was also achieved in Li_2CuO_2 [106], $CuGeO_3$ [107]and $KCuF_3$ [108]. In these compounds, the Mott (charge-transfer) gap is large (>3 eV), and excitations across the Mott gap do not overlap with the d-d excitations.

In CuB_2O_4, a molecular orbital excitation between the bonding and antibonding orbitals in a CuO_4 plaquette is observed at the energy region around 6 eV. The Gaussian Line shape of the molecular orbital excitation as a function of energy and incident-energy dependence of the peak position of the excitation indicate the coupling to the lattice degree of freedom and are discussed in terms of a Franck–Condon effect [109].

4.2 One-Dimensional Cuprates

One-dimensional (1D) electronic properties are realized in some copper oxide compounds. Some of these are ideal materials of the 1D $S = 1/2$ antiferromagnetic Heisenberg model and attract great interest because of their fascinating properties such as spin-charge separation [110]. The 1D Cu-O chains are classified into two types. One is the edge-sharing chain, where neighboring CuO_4 plaquettes share their edge and the other is the corner-sharing chain whose CuO_4 plaquettes share oxygen atoms at their corner and the angle of the Cu-O-Cu bond is 180°. These are shown in Fig. 19a, b, respectively. The most important difference between the two types of chain is the magnitude of the transfer energy between the neighboring Cu sites via O $2p$ orbitals. In the edge-sharing chain, an O $2p$ orbital hybridizing with a $3d_{x^2-y^2}$ orbital of Cu ion is almost orthogonal to that of the next Cu ion, and as result, the transfer energy is very small. Li_2CuO_2, $CuGeO_3$, and $Ca_{2+5x}Y_{2-5x}Cu_5O_{10}$ are categorized in the edge-sharing chain. In $Ca_{2+5x}Y_{2-5x}Cu_5O_{10}$, holes can be doped in the chain by substituting Ca for Y. On the other hand, in the corner-sharing chain, a O $2p$ orbital hybridizes with the two neighboring Cu $3d_{x^2-y^2}$ orbitals and the transfer

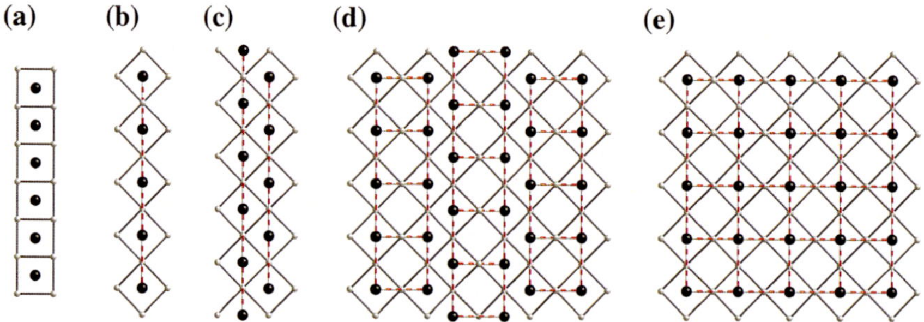

Fig. 19 Various Cu-O networks formed by planar CuO_4 plaquettes, **a** edge-sharing chain, **b** corner-sharing chain, **c** zigzag chain, **d** two-leg ladder, and **e** square lattice. *Black* and *white circles* represent Cu and O atoms, respectively. The 180° Cu-O-Cu bonds, along which the electron transfer and magnetic interaction are large, are connected by the *dashed lines*

energy becomes large as does in the $2D$ high-T_c cuprates. The magnitude of the transfer energy governs the charge dynamics of the cuprates. Furthermore, because the superexchange interaction between the nearest-neighbor Cu spins is determined by the transfer energy, the magnitude of the magnetic interaction differs between the two types of chain; it is more than one order of magnitude larger in the corner-sharing chain than in the edge-sharing chain. The corner-sharing chain appears in Sr_2CuO_3 and $SrCuO_2$. In the latter compound, there are two chains forming a zigzag chain, which is shown in Fig. 19c. Within the energy scale studied by RIXS, the two chains can be regarded as independent because the interactions along the 180° Cu-O-Cu bond (connected by dashed lines in Fig. 19) are much larger than those along the 90° bond.

4.2.1 Charge Excitations

Charge excitations in the edge-sharing chain in Li_2CuO_2, $CuGeO_3$, and $Ca_2Y_2Cu_5O_{10}$ were studied using Cu K-edge [106, 107, 111–114] and O K-edge RIXS [115–118]. The momentum dependence of the RIXS spectra of these compounds was found to be small, which is consistent with the small transfer energy.

In the edge-sharing chain, the angle of the Cu-O-Cu is close to 90° and the superexchange interaction between the nearest-neighbor Cu spins are strongly suppressed compared to in the corner-sharing chain. In reality, the magnetic interaction is sensitive to the angle and even the sign depends on the material; it is ferromagnetic and antiferromagnetic in Li_2CuO_2 and $CuGeO_3$, respectively [119]. The magnitude of the interaction is $\sim 10\,meV$ in both cases. This difference of sign affects the formation of the Zhang–Rice singlet because this singlet state can be excited when the Cu spin configuration next to the oxygen ion at the core-hole site is antiferromagnetic [120]. Figure 20a, b show the temperature dependence of O K-edge RIXS spectra for Li_2CuO_2 and $CuGeO_3$, respectively [118]. The charge-transfer excitation from ZRB to UHB, which is referred to as the Mott gap excitation (taking this term in its broad sense throughout this chapter), in Li_2CuO_2 and $CuGeO_3$ is observed at 3.2 and 3.8 eV, respectively. In Li_2CuO_2, ferromagnetic correlation develops at low temperature and, as a result, the formation of the Zhang–Rice singlet is suppressed and the intensity of the peak at 3.2 eV decreases with decreasing temperature. In contrast, the peak at 3.8 eV in $CuGeO_3$ shows opposite behavior because of the antiferromagnetic interaction between the nearest-neighbor Cu spins.

Charge excitations in the corner-sharing chain were measured using Cu K-edge [107, 111, 113, 121–125] RIXS and argued mostly in terms of spin-charge separation. Figure 21a shows a simplified picture of the propagation of a hole created by a photon in a half-filled $1D$ antiferromagnetic insulator. Hopping of the hole to the left site breaks an antiferromagnetic coupling indicated by a vertical wavy line. However, additional hopping to the left does not create further broken antiferromagnetic coupling. The broken antiferromagnetic coupling can move to the right. Thus, the motion of the hole (holon) and the broken antiferromagnetic coupling (spinon) decouples from each other, and it is the so-called spin-charge separation. In the Mott

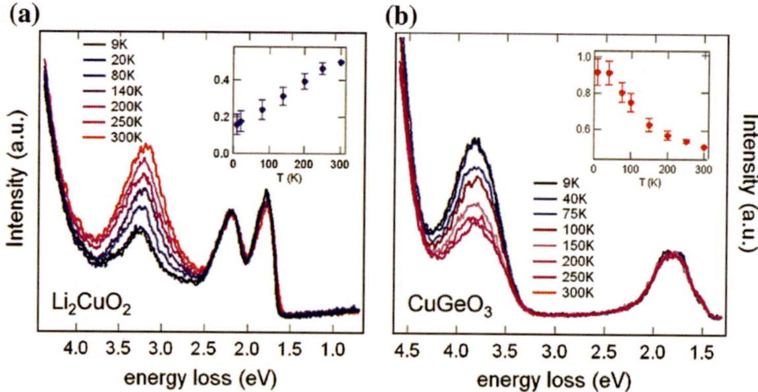

Fig. 20 Temperature dependence of O K-edge RIXS spectra for **a** Li_2CuO_2 and **b** $CuGeO_3$ [118]. The *insets* show the integrated intensities of Mott gap excitation. Reprinted figure with permission from Ref. [118]. Copyright 2013 by the American Physical Society

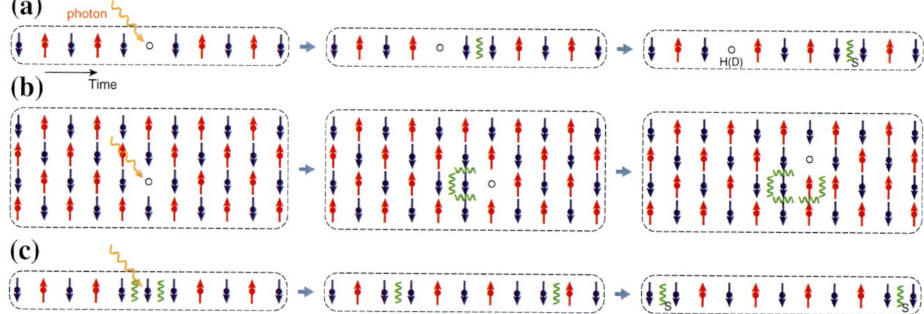

Fig. 21 **a**, **b** Propagation of a hole in a half-filled **a** $1D$ and **b** $2D$ $S = 1/2$ antiferromagnetic insulator. **c** Propagation of a single spin-flip in the $1D$ antiferromagnetic insulator. Photon indicated by a wavy arrow in the *left* of each figure creates a hole or spin-flip and they propagate in the lattice. Wavy lines represent broken antiferromagnetic couplings. H, D, and S denote holon, doublon and spinon, respectively

gap excitation in RIXS, a doubly-occupied site (doublon) is also created, and the motion of the doublon is the same as the holon. The dispersion of the doublon and the holon follows the upper Hubbard band and the lower Hubbard band (more precisely, the Zhang–Rice singlet band), respectively. In contrast, as shown in Fig. 21b, the motion of a hole in the $2D$ antiferromagnetic insulator breaks the antiferromagnetic couplings one after another, and it cannot be independent of the motion of the spin.

Hasan et al. [121] measured the Cu K-edge RIXS spectra of Sr_2CuO_3 and $SrCuO_2$, which are shown in Fig. 22a. The Mott gap excitation is observed around 2–4 eV and was found to be dispersive along the chain direction, in contrast to the edge-sharing chain. Such dispersion is consistent with theoretical calculations using exact diagonalization on a finite-size cluster [126, 127] and Hartree–Fock theory with a

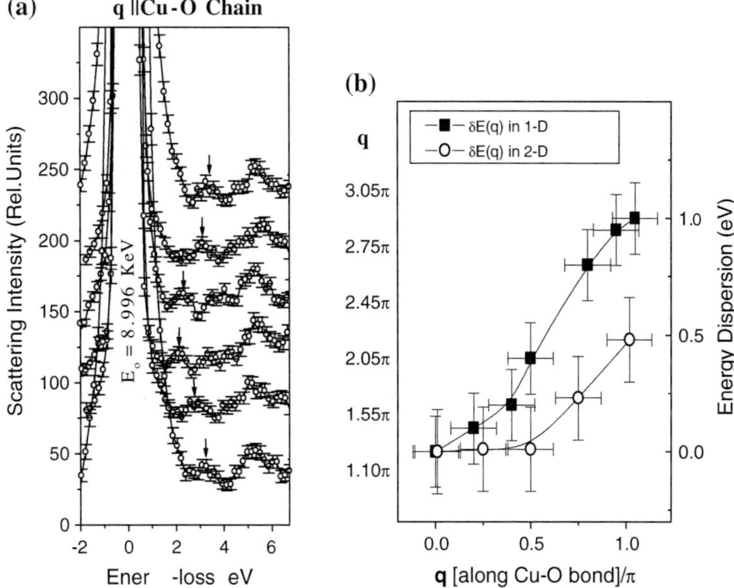

Fig. 22 **a** Cu K-edge RIXS spectra of the corner-sharing chain cuprate [121]. **b** Comparison of the dispersion relation between $1D$ and $2D$ cuprates. Reprinted figure with permission from Ref. [121]. Copyright 2002 by the American Physical Society

random phase approximation [128]. Hasan et al. [121] pointed out that the dispersion of the $1D$ chain is larger that of the $2D$ square lattice, as shown in Fig. 22b, and it is manifestation that the charge motion is easier in $1D$ than in $2D$ due to the decoupling from the spin degree of freedom.

4.2.2 Spin and d-d Excitations

Schlappa et al. [129] studied spin and orbital (d-d) excitations of the corner-sharing $1D$ cuprate using Cu L_3-edge RIXS of Sr_2CuO_3. The intensity map of the RIXS spectra is shown in Fig. 23a. Spin excitations are observed below 0.8 eV. In the $1D$ $S = 1/2$ spin chain, a single spin-flip created by a photon (or a neutron) fractionalizes into two spinons, as shown in Fig. 21c. Motion of the two spinons forms a continuum with a lower boundary with period π and an upper boundary with period 2π in the spin excitation spectra and this momentum dependence is clearly seen in Fig. 23a.

The spectral weight between 1.5 and 3.5 eV comes from d-d (orbital) excitations. It shows a characteristic dispersion with period π. To interpret the experimental observations, Schlappa et al. proposed the concept of "spin-orbital separation", where an orbital excitation (orbiton) propagates separately from spin excitations (spinons) in the $1D$ lattice, on the analogy of spin-charge separation [129, 130]. Figure 23b

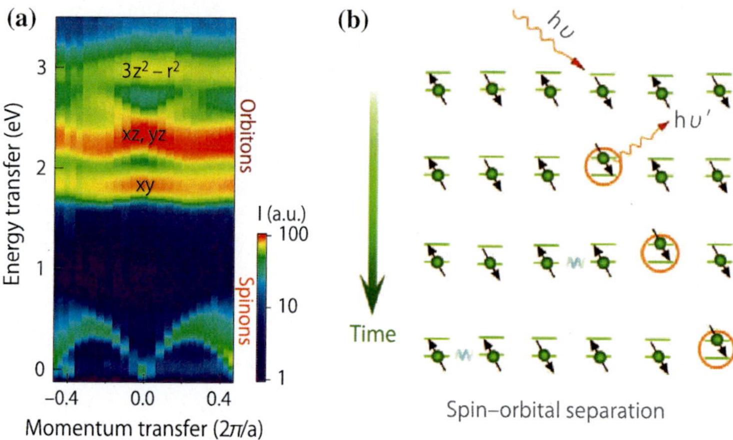

Fig. 23 **a** Cu L_3-edge RIXS intensity map of spectra of Sr_2CuO_3 [129]. **b** Schematic figure of spin-orbital separation. The lower and upper lines represent ground-state and excited-state orbitals, respectively. Reprinted by permission from Macmillan Publishers Ltd.: Nature (Ref. [129]), Copyright 2012

shows the concept schematically. A photon creates an orbital excitation (orbiton) in a $1D$ antiferromagnetic chain. Hopping of the orbiton to the right site breaks one antiferromagnetic coupling and forms a spinon, which is indicated by a wavy line between two spins. After that, the orbiton and spinon can move independently without creating any additional broken antiferromagnetic coupling. The Kugel–Khomskii Hamiltonian is used to describe the orbiton and spinon, and the calculation of the excitation spectra agrees well with experimental observations. The magnitude of dispersion depends on the orbital, and it is ascribed to the difference of the hopping amplitude along the chain. In Sr_2CuO_3, dispersion is large for the transition to d_{xz} and d_{xy} orbitals (in hole representation), where the chain runs along the x direction and the z direction is perpendicular to the Cu-O plaquette.

Subsequently, spin-orbital separation was also found in $CaCu_2O_3$ [131]. The Cu-O plaquettes in $CaCu_2O_3$ form an anisotropic two-leg ladder; the Cu-O-Cu bond angle along the rung is 123°, which results in anisotropic exchange interactions $J_{rung} \simeq \frac{1}{10}J_{leg}$. Because hopping of the d_{xz} orbital is confined along the leg (x) direction, the separation occurs in the d_{xz} orbital, as in the case of Sr_2CuO_3. In contrast, hopping of the d_{xy} orbital becomes two-dimensional in this material and the separation is not observed in the d_{xy} orbital. In the same compound, Bisogni et al. [132] studied femtosecond dynamics of spin excitations in Cu L_3-edge RIXS. They demonstrated by comparing experiment and theory that double spin flips require a finite time of the order of femtoseconds ($\sim$ core-hole lifetime) to be generated, while excitations with a single spin flip is instantaneous.

In the edge-sharing chain, d-d excitations were identified in the RIXS spectra at the O-K [115–118], Cu L_3- [133], and Cu K-edges [106, 107]. Recent measurement of the Cu L_3-edge RIXS spectra of $Ca_2Y_2Cu_5O_{10}$ with high energy resolution by Lee et al. [133] showed abrupt shifts in the d-d peak positions by 50–80 meV as a function of incident photon energy and Gaussian line shape of excitations. Similarly to $0D$ CuB_2O_4, they analyzed the observed features based on a Franck–Condon treatment for the electron-lattice coupling. It should be noted that the electron-lattice coupling also affects the phonon excitations observed by the O K-edge RIXS on the same compound [134].

4.3 Ladder Cuprates

The two-leg ladder lies between the $1D$ chain and the $2D$ square lattice from a structural point of view. However the physical properties of the two-leg ladder are different from those of one- and two-dimensions [135]. Long-range antiferromagnetic order occurs in an antiferromagnetically coupled $S = 1/2$ spins on the square lattice, while strong quantum fluctuations prevent a purely $1D$ $S = 1/2$ spin chain from long-range order. In contrast to these two cases, a pair of spins forms a spin singlet state with a finite energy gap in the two-leg ladder. Furthermore, it is predicted that holes introduced into the two-leg ladder tend to form binding pairs through the rung, which might condense into superconductivity [136].

Electronic excitations of copper oxides with two-leg ladder structure have been studied in $(La,Sr,Ca)_{14}Cu_{24}O_{41}$. It is a composite crystal in which a two-leg ladder and an edge-sharing chain coexist with different periodicity. In the parent $Sr_{14}Cu_{24}O_{41}$, the nominal valence of Cu is $+2.25$ and holes are predominantly located in the chain sites. The substitution of Ca for Sr brings about a transfer of the holes from the chain to the ladder [137, 138]. On the other hand, holes decrease in both the chain and ladder sites when the concentration of trivalent La increases. Because the transfer energy of carriers between the neighboring Cu sites is small in the edge-sharing chain, the ladder part with a large transfer energy is responsible for most physical properties. As predicted in a theory [136], superconductivity is realized at high Ca concentration under high pressure [139, 140].

RIXS experiments at the Cu K-edge on $Sr_{14}Cu_{24}O_{41}$ were performed by three groups independently [114, 125, 141]. A dispersive excitation observed around 2–4 eV is attributed to the Mott gap excitation of the ladder part, as is proven by the periodicity of the momentum dependence. The Mott gap excitation shifts to higher energy with the momentum along the ladder. When the spectra are compared along the rung direction, the spectral weights of the Mott gap excitation of $\mathbf{q} = (\pi, \pi)$ are located at a slightly higher energy region than those of $\mathbf{q} = (0, \pi)$. Here the reduced momentum transfer $\mathbf{q}$ is represented as $\mathbf{q} = (q_{\mathrm{rung}}, q_{\mathrm{leg}})$. After that, it was found that the RIXS spectra at 1–5 eV are found to be dependent on temperature in $Sr_{14}Cu_{24}O_{41}$, which is possibly related to the charge order of this material [142].

Ishii et al. [114] also measured $La_5Sr_9Cu_{24}O_{41}$ and $Sr_{2.5}Ca_{11.5}Cu_{24}O_{41}$ in addition to the parent $Sr_{14}Cu_{24}O_{41}$ and studied doping dependence on the ladder. Hole concentration in the ladder is smallest in $La_5Sr_9Cu_{24}O_{41}$, while it is largest in $Sr_{2.5}Ca_{11.5}Cu_{24}O_{41}$. Figure 24a shows RIXS spectra of the three compounds. Except

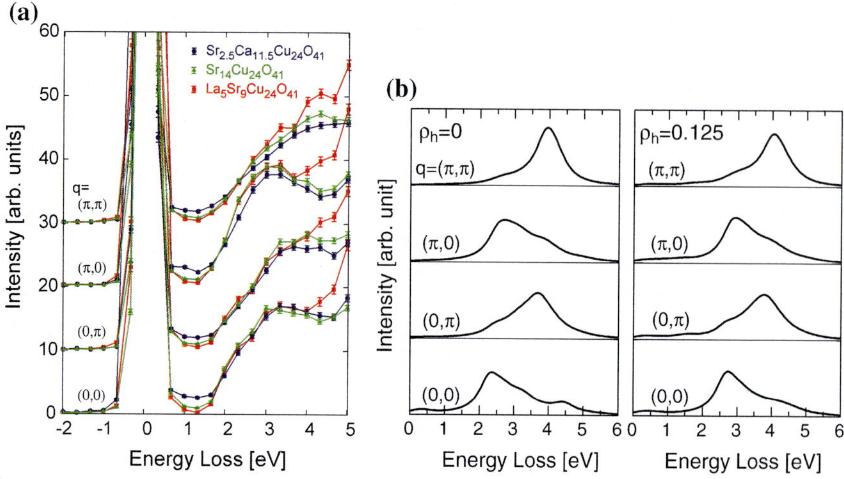

Fig. 24 **a** Cu K-edge RIXS spectra of $(La,Sr,Ca)_{14}Cu_{24}O_{41}$ [114]. The reduced momentum transfer **q** is represented as $\mathbf{q} = (q_{rung}, q_{leg})$. **b** Theoretical calculation of RIXS spectra for undoped (*left*) and hole-doped (*right*) two-leg ladder

for the slight shift to higher energy, which is probably caused by the shift of Fermi energy, the dispersion relation of the Mott gap is almost unchanged upon doping. These features of doping dependence are reproduced by a numerical diagonalization calculation for a single-band Hubbard model, as shown in Fig. 24b. They argued from the insensitivity of the Mott gap excitation to hole doping that holes introduced into the ladder can be paired so as not to destroy the local singlet states along the rungs in the undoped Cu sites. In contrast, as described above, hole-doping to the $2D$ square lattice induces a smaller dispersion of the Mott gap excitation, which is related to the reduction of the antiferromagnetic spin correlation. On the other hand, intraband excitation in the ladder appears as a continuum spectral weight below the gap and the intensity is roughly proportional to the hole concentration in the ladder. These features for the intraband excitation are similar to the CuO_2 square lattice shown in Fig. 10a.

Because superconductivity is observed, it is important to study the electronic structure under high pressure. Cu K-edge RIXS can be a powerful tool for the purpose taking advantage of the large penetration length of the hard X-rays. Yoshida et al. [143] succeeded in measuring momentum-resolved Cu K-edge RIXS spectra of $Sr_{2.5}Ca_{11.5}Cu_{24}O_{41}$ at 3 GPa using a diamond anvil cell. The result indicates that the number of holes in the ladders increases by applying pressure, and this might be a necessary condition for the occurrence of superconductivity.

RIXS at the Cu L_3-edge was also performed in $Sr_{14}Cu_{24}O_{41}$, and spin excitations were investigated. Because spin singlets are formed at the ground state in the ladder, spin triplet excitations (triplons) are observed with a finite energy gap. Schlappa et al. [144] reported dispersive two-triplon excitations and determined the value of the energy gap.

4.4 Polarization-Analyzed RIXS in $KCuF_3$

In general, the RIXS spectrum is a function of energy, momentum, and polarization of both the incident and scattered photons. Most RIXS studies discussed in this chapter so far have focused on energy and momentum. Even though the polarization is an inherent and important characteristic of the photon, it is sometimes overlooked. In this subsection, the role of photon polarization is discussed. While polarization of the incident photon polarization is relatively easy to control, analysis of the scattered photon polarization has become available only recently in both soft [90] and hard X-ray regimes [145, 146].

Like conventional Raman spectroscopy, polarization in RIXS is connected to the symmetry of excitations and it is useful for the assignment of the excitations in RIXS. In Cu L_3 edge RIXS, polarization appears explicitly in the $2p$-$3d$ transition matrix elements and its role is clear. An instructive example in shown in Sect. 2.2 and used for distinction between spin and charge excitations in the spectra in Fig. 16. In contrast, the situation is more complicated in Cu K-edge RIXS. One can choose the symmetry of the $4p$ orbitals by analyzing the polarization, but the $4p$ electron is often treated as a spectator in the RIXS process. This is because the interaction between $3d$ and $4p$ electrons in the intermediate state is weaker than that between $1s$ core-hole and $3d$ electrons. If the $3d$-$4p$ interactions are taken into account, polarization dependence certainly appears in the RIXS spectra [147], and in particular, the interaction becomes important in the d-d excitations observed at the main K-absorption edge of the transition metal [148].

An experimental study of polarization-analyzed RIXS at the Cu K-edge was performed on $KCuF_3$ [108]. While $KCuF_3$ has long been known to display quantum $1D$ antiferromagnetic properties along the c axis, it is also an archetypal compound of the e_g orbital order. In real tetragonal structure, the degeneracy of the e_g orbitals is lifted accompanied with the occurrence of the strong Jahn–Teller distortion. The e_g hole orbitals are ordered according to the pattern shown in the inset of Fig. 25a. There are two types of d-d excitations in $KCuF_3$: One excitation is a transition of an electron from the t_{2g} orbital to the e_g orbital (t_{2g} excitation) and the other one is between the e_g orbitals (e_g excitation).

A typical RIXS spectrum of $KCuF_3$ is shown in Fig. 25b. The d-d excitations are observed around $1.2\,\mathrm{eV}$ while the spectral weight above $6\,\mathrm{eV}$ is the excitations across the charge-transfer gap. The polarization-analyzed RIXS spectra of the 1.2-eV feature are presented in Fig. 25c–f for four different crystal orientations schematized in the corresponding insets. The spectra show a clear dependence on the scattered photon polarization, that is, additional intensity is found around $1\,\mathrm{eV}$ in the spectrum of $\pi \to \pi'$ compared with $\pi \to \sigma'$ in the configurations of (c) and (d), whereas the spectra in (e) and (f) are almost identical for the two polarization conditions.

The observed excitations in the polarization-analyzed RIXS spectra were identified by a phenomenological consideration of the symmetry of the RIXS process, which is also described in Sect. 3.3 of Chap. 1 from a theoretical point of view. When the electronic excitations occur at the local site where the X-ray is absorbed, there

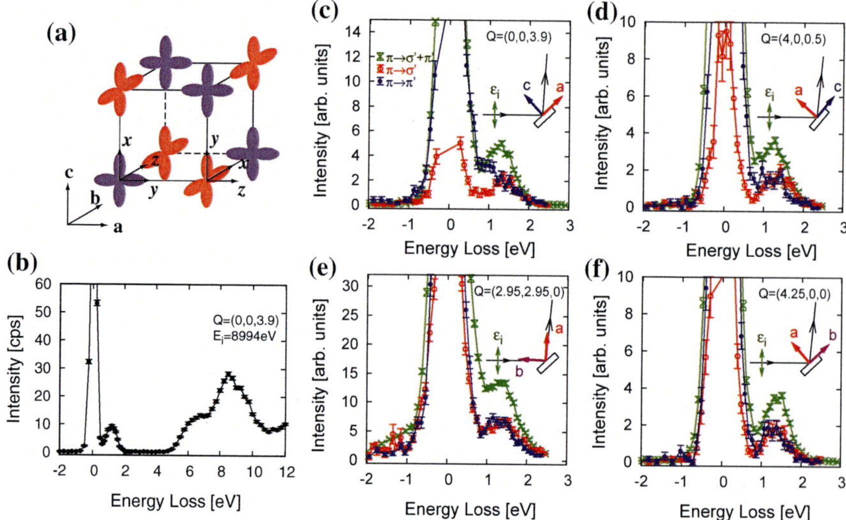

Fig. 25 **a** A schematic representation of the orbital order of KCuF$_3$ in the hole picture. **b** Typical Cu K-edge RIXS spectrum of KCuF$_3$ [108]. **c–f** Polarization analyzed RIXS spectra of KCuF$_3$. Spectra with *filled* and *open circles* are measured in the $\pi \to \sigma'$ and $\pi \to \pi'$ polarization conditions, respectively. Spectra without polarization analysis are also shown by *crosses*. Corresponding experimental geometries are shown in the *insets*

is a selection rule for RIXS that at least one common symmetry should be shared in reduced product-representations $\Gamma_i \times \Gamma_f$ and $P_i \times P_f$. Here $\Gamma_{i(f)}$ and $P_{i(f)}$ are the irreducible representations for the initial (final) electronic orbital and those of the incident (scattered) photon polarization, respectively. The unit vectors **x**, **y**, and **z** are taken so that the hole orbital is represented as $x^2 - y^2$ and there are two polarization conditions in the xyz coordinates because of the two orbital sublattices, as illustrated in Fig. 25a. The local symmetry of the Cu atom is D_{4h} instead of the exact site symmetry of D_{2h} because the Cu-F distances along the x and y directions are almost equal. If only considering onsite excitations, the possible excitation modes in each polarization configuration of the measurements are given by $P_i \times P_f$. These are listed in Table 1. The symmetry of $P_i \times P_f$ has to be either A_{2g} ($\Gamma_{xy} \times \Gamma_{x^2-y^2}$) or E_g ($\Gamma_{yz}, \Gamma_{zx} \times \Gamma_{x^2-y^2}$) for the observation of the t_{2g} excitation and either symmetry is always present in all four experimental configurations of Fig. 25c–f. Accordingly, the t_{2g} excitations are ascribed to the spectral weight observed around 1.3 eV for all configurations and polarization conditions. This is seen to form a high-energy tail feature of the $\pi \to \pi'$ spectra, as observed in Fig. 25c, d. On the other hand, e_g excitation requires B_{1g} ($\Gamma_{3z^2-r^2} \times \Gamma_{x^2-y^2}$) polarization symmetry, which exists in the $\pi \to \pi'$ condition for the configurations shown in Fig. 25c, d, f. Assuming that the symmetry argument gives a necessary condition for observation by RIXS, the additional intensity observed around 1 eV in the $\pi \to \pi'$ polarization of the Fig. 25c, d is ascribed to the e_g excitation. To explain the absence of e_g excitation in the spec-

Table 1 Summary of the polarization conditions of Fig. 25c–f [108]

Configuration	Polarization	ε_i	ε_f	Symmetry of $P_i \times P_f$
(c)	$\pi \to \sigma'$	$\mathbf{y} + \mathbf{z}$	$\mathbf{x}$	$A_{2g} + B_{2g} + E_g$
	$(\mathbf{a} + \mathbf{c}) \to \mathbf{b}$	$\mathbf{x} + \mathbf{y}$	$\mathbf{z}$	E_g
	$\pi \to \pi'$	$\mathbf{y} + \mathbf{z}$	$\mathbf{y} - \mathbf{z}$	$A_{1g} + B_{1g} + E_g$
	$(\mathbf{a} + \mathbf{c}) \to (\mathbf{a} - \mathbf{c})$	$\mathbf{x} + \mathbf{y}$	$\mathbf{x} - \mathbf{y}$	$B_{1g} + A_{2g}$
(d)	$\pi \to \sigma'$	$\mathbf{y} + \mathbf{z}$	$\mathbf{x}$	$A_{2g} + B_{2g} + E_g$
	$(\mathbf{a} + \mathbf{c}) \to \mathbf{b}$	$\mathbf{x} + \mathbf{y}$	$\mathbf{z}$	E_g
	$\pi \to \pi'$	$\mathbf{y} + \mathbf{z}$	$\mathbf{y} - \mathbf{z}$	$A_{1g} + B_{1g} + E_g$
	$(\mathbf{a} + \mathbf{c}) \to (\mathbf{a} - \mathbf{c})$	$\mathbf{x} + \mathbf{y}$	$\mathbf{x} - \mathbf{y}$	$B_{1g} + A_{2g}$
(e)	$\pi \to \sigma'$	$\mathbf{z}$	$\mathbf{y}$	E_g
	$\mathbf{a} \to \mathbf{c}$	$\mathbf{y}$	$\mathbf{x}$	$A_{2g} + B_{2g}$
	$\pi \to \pi'$	$\mathbf{z}$	$\mathbf{x}$	E_g
	$\mathbf{a} \to \mathbf{b}$	$\mathbf{y}$	$\mathbf{z}$	E_g
(f)	$\pi \to \sigma'$	$\mathbf{x} + \mathbf{z}$	$\mathbf{y}$	$A_{2g} + B_{2g} + E_g$
	$(\mathbf{a} + \mathbf{b}) \to \mathbf{c}$	$\mathbf{y} + \mathbf{z}$	$\mathbf{x}$	$A_{2g} + B_{2g} + E_g$
	$\pi \to \pi'$	$\mathbf{x} + \mathbf{z}$	$\mathbf{x} - \mathbf{z}$	$A_{1g} + B_{1g} + E_g$
	$(\mathbf{a} + \mathbf{b}) \to (\mathbf{a} - \mathbf{b})$	$\mathbf{y} + \mathbf{z}$	$\mathbf{y} - \mathbf{z}$	$A_{1g} + B_{1g} + E_g$

trum obtained in the $\pi \to \pi'$ condition of Fig. 25f, it may be necessary to carry out a theoretical evaluation of the intensity including a microscopic description of the RIXS process. These assignments of the d-d excitations are consistent with an optical absorption study [149], where optically-forbidden d-d excitations become allowed through the assistance of phonons. Later on, Nomura analyzed the RIXS spectra of KCuF$_3$ using a microscopic theoretical model and obtained qualitative agreement by including the $3d$-$4p$ interactions in the model [150].

5 Summary

During the last two decades, RIXS has proven to be a powerful technique for investigating momentum-resolved electronic excitations. Development of X-ray sources and related experimental techniques has enabled the measurement of various kinds of excitation by RIXS. In this chapter, charge, spin, and orbital (d-d) excitations in the copper oxides were overviewed. Experimentally and theoretical studies on strongly correlated copper oxides has led to progress in this technique. It is partly because of the relatively simple electronic structure of copper oxides, where only a few (Cu $3d_{x^2-y^2}$ and O $2p_{x,y}$) orbitals are relevant for most electronic properties and simple theoretical models can be applied.

In the next decade, the development of RIXS will continue. An energy resolution of 25 meV has been achieved in a recent Cu K-edge RIXS study [151] and it will be further improved upon, hopefully down to 10 meV in both hard and soft X-ray regimes. This will accelerate research on copper oxides using RIXS. For example, limitation of the sample volume in INS is overcome and low-energy spin excitations can be measured systematically in all high-T_c cuprate families. Although I have made little mention of it, observation of phonon excitations will become possible. The resonant effect on phonon excitations enables the evaluation of the strength of electron-phonon coupling, as already demonstrated in an edge-shared 1D cuprate [134] and a titanate [152]. A more challenging theme is observation of the excitation across the superconducting gap. It was theoretically proposed that the dynamical structure factor of a superconductor observed in RIXS is strongly affected by the superconducting gap and coherence factors and therefore is very sensitive to the symmetry of the order parameter [153].

Finally, I mention some capabilities of next-generation X-ray sources in relation to the study of electron dynamics. Facilities of X-ray free-electron lasers (XFELs) have recently started user operation around the world. The duration of pulsed X-rays has improved from several tens of picoseconds for the storage rings to several tens of femtoseconds for XFELs, and the duration of XFELs corresponds to an energy of 100 meV. This means that electron dynamics can be studied in the time domain ($\mathbf{q}$-t space) using XFELs, which must be complementary to the energy domain ($\mathbf{q}$-ω space) of RIXS.

In forthcoming years, RIXS will continue to develop as an important tool for the study of elementary excitations not only in copper oxides but also in other correlated electron systems.

References

1. A. Kotani, S. Shin, Rev. Mod. Phys. **73**, 203 (2001)
2. L.J.P. Ament, M. van Veenendaal, T.P. Devereaux, J.P. Hill, J. van den Brink, Rev. Mod. Phys. **83**, 705 (2011)
3. J.P. Rueff, A. Shukla, Rev. Mod. Phys. **82**, 847 (2010)
4. G. Ghiringhelli, A. Piazzalunga, C. Dallera, T. Schmitt, V.N. Strocov, J. Schlappa, L. Patthey, X. Wang, H. Berger, M. Grioni, Phys. Rev. Lett. **102**, 027401 (2009)
5. L.J.P. Ament, G. Ghiringhelli, M.M. Sala, L. Braicovich, J. van den Brink, Phys. Rev. Lett. **103**, 117003 (2009)
6. L. Braicovich, J. van den Brink, V. Bisogni, M. Moretti Sala, L.J.P. Ament, N.B. Brookes, G.M. De Luca, M. Salluzzo, T. Schmitt, V.N. Strocov, G. Ghiringhelli, Phys. Rev. Lett. **104**, 077002 (2010)
7. K. Ishii, T. Tohyama, J. Mizuki, J. Phys. Soc. Jpn. **82**, 021015 (2013)
8. M. Blume, J. Appl. Phys. **57**, 3615 (1985)
9. W. Schülke, *Electron Dynamics by Inelastic X-Ray Scattering* (Oxford University Press, Oxford, 2007)
10. M. van Veenendaal, Phys. Rev. Lett. **96**, 117404 (2006)
11. P.M. Platzman, E.D. Isaacs, Phys. Rev. B **57**, 11107 (1998)
12. T. Nomura, J. Igarashi, Phys. Rev. B **71**, 035110 (2005)

13. J. van den Brink, M. van Veenendaal, Europhys. Lett. **73**, 121 (2006)
14. L.J.P. Ament, F. Forte, J. van den Brink, Phys. Rev. B **75**, 115118 (2007)
15. J. Zaanen, G.A. Sawatzky, J.W. Allen, Phys. Rev. Lett. **55**, 418 (1985)
16. F.C. Zhang, T.M. Rice, Phys. Rev. B **37**, 3759 (1988)
17. K. Ishii, K. Tsutsui, Y. Endoh, T. Tohyama, S. Maekawa, M. Hoesch, K. Kuzushita, M. Tsubota, T. Inami, J. Mizuki, Y. Murakami, K. Yamada, Phys. Rev. Lett. **94**, 207003 (2005)
18. K. Ishii, M. Hoesch, T. Inami, K. Kuzushita, K. Ohwada, M. Tsubota, Y. Murakami, J. Mizuki, Y. Endoh, K. Tsutsui, T. Tohyama, S. Maekawa, K. Yamada, T. Masui, S. Tajima, H. Kawashima, J. Akimitsu, J. Phys. Chem. Solids **69**, 3118 (2008)
19. S. Wakimoto, K. Ishii, H. Kimura, K. Ikeuchi, M. Yoshida, T. Adachi, D. Casa, M. Fujita, Y. Fukunaga, T. Gog, Y. Koike, J. Mizuki, K. Yamada, Phys. Rev. B **87**, 104511 (2013)
20. M.A. van Veenendaal, H. Eskes, G.A. Sawatzky, Phys. Rev. B **47**, 11462 (1993)
21. H. Tolentino, M. Medarde, A. Fontaine, F. Baudelet, E. Dartyge, D. Guay, G. Tourillon, Phys. Rev. B **45**, 8091 (1992)
22. C.C. Kao, W.A.L. Caliebe, J.B. Hastings, J.M. Gillet, Phys. Rev. B **54**, 16361 (1996)
23. J.P. Hill, C.C. Kao, W.A.L. Caliebe, M. Matsubara, A. Kotani, J.L. Peng, R.L. Greene, Phys. Rev. Lett. **80**, 4967 (1998)
24. Y.J. Kim, J.P. Hill, G.D. Gu, F.C. Chou, S. Wakimoto, R.J. Birgeneau, S. Komiya, Y. Ando, N. Motoyama, K.M. Kojima, S. Uchida, D. Casa, T. Gog, Phys. Rev. B **70**, 205128 (2004)
25. P. Abbamonte, C.A. Burns, E.D. Isaacs, P.M. Platzman, L.L. Miller, S.W. Cheong, M.V. Klein, Phys. Rev. Lett. **83**, 860 (1999)
26. P. Abbamonte, C.A. Burns, E.D. Isaacs, P.M. Platzman, L.L. Miller, M.V. Klein. arXiv:cond-mat/9911215
27. M.Z. Hasan, E.D. Isaacs, Z.X. Shen, L.L. Miller, K. Tsutsui, T. Tohyama, S. Maekawa, Science **288**, 1811 (2000)
28. Y.J. Kim, J.P. Hill, C.A. Burns, S. Wakimoto, R.J. Birgeneau, D. Casa, T. Gog, C.T. Venkataraman, Phys. Rev. Lett. **89**, 177003 (2002)
29. K. Hämäläinen, J.P. Hill, S. Huotari, C.C. Kao, L.E. Berman, A. Kotani, T. Idé, J.L. Peng, R.L. Greene, Phys. Rev. B **61**, 1836 (2000)
30. L. Lu, J.N. Hancock, G. Chabot-Couture, K. Ishii, O.P. Vajk, G. Yu, J. Mizuki, D. Casa, T. Gog, M. Greven, Phys. Rev. B **74**, 224509 (2006)
31. E. Collart, A. Shukla, J.P. Rueff, P. Leininger, H. Ishii, I. Jarrige, Y.Q. Cai, S.W. Cheong, G. Dhalenne, Phys. Rev. Lett. **96**, 157004 (2006)
32. D.S. Ellis, J.P. Hill, S. Wakimoto, R.J. Birgeneau, D. Casa, T. Gog, Y.J. Kim, Phys. Rev. B **77**, 060501 (2008)
33. Y.W. Li, D. Qian, L. Wray, D. Hsieh, Y. Xia, Y. Kaga, T. Sasagawa, H. Takagi, R.S. Markiewicz, A. Bansil, H. Eisaki, S. Uchida, M.Z. Hasan, Phys. Rev. B **78**, 073104 (2008)
34. Y. Harada, K. Okada, R. Eguchi, A. Kotani, H. Takagi, T. Takeuchi, S. Shin, Phys. Rev. B **66**, 165104 (2002)
35. K. Okada, A. Kotani, Phys. Rev. B **65**, 144530 (2002)
36. K. Tsutsui, T. Tohyama, S. Maekawa, Phys. Rev. Lett. **83**, 3705 (1999)
37. K. Tsutsui, T. Tohyama, S. Maekawa, Phys. Rev. Lett. **91**, 117001 (2003)
38. T. Tohyama, K. Tsutsui, S. Maekawa, J. Phys. Chem. Solids **66**, 2139 (2005)
39. T. Nomura, J. Igarashi, J. Phys. Chem. Solids **67**, 262 (2006)
40. J. Igarashi, T. Nomura, M. Takahashi, Phys. Rev. B **74**, 245122 (2006)
41. C.C. Chen, B. Moritz, F. Vernay, J.N. Hancock, S. Johnston, C.J. Jia, G. Chabot-Couture, M. Greven, I. Elfimov, G.A. Sawatzky, T.P. Devereaux, Phys. Rev. Lett. **105**, 177401 (2010)
42. M.M. Sala, V. Bisogni, C. Aruta, G. Balestrino, H. Berger, N.B. Brookes, G.M. de Luca, D.D. Castro, M. Grioni, M. Guarise, P.G. Medaglia, F.M. Granozio, M. Minola, P. Perna, M. Radovic, M. Salluzzo, T. Schmitt, K.J. Zhou, L. Braicovich, G. Ghiringhelli, New J. Phys. **13**, 043026 (2011)
43. S. Tanaka, A. Kotani, J. Phys. Soc. Jpn. **62**, 464 (1993)
44. L.C. Duda, G. Dräger, S. Tanaka, A. Kotani, J. Guo, D. Heumann, S. Bocharov, N. Wassdahl, J. Nordgren, J. Phys. Soc. Jpn. **67**, 416 (1998)

45. P. Kuiper, J.H. Guo, C. Såthe, L.C. Duda, J. Nordgren, J.J.M. Pothuizen, F.M.F. de Groot, G.A. Sawatzky, Phys. Rev. Lett. **80**, 5204 (1998)
46. G. Ghiringhelli, N.B. Brookes, E. Annese, H. Berger, C. Dallera, M. Grioni, L. Perfetti, A. Tagliaferri, L. Braicovich, Phys. Rev. Lett. **92**, 117406 (2004)
47. L. Hozoi, L. Siurakshina, P. Fulde, J. van den Brink, Sci. Rep. **1**, 65 (2011)
48. J. Kim, D.S. Ellis, T. Gog, D. Casa, Y.J. Kim, Phys. Rev. B **81**, 073109 (2010)
49. J.P. Hill, G. Blumberg, Y.J. Kim, D.S. Ellis, S. Wakimoto, R.J. Birgeneau, S. Komiya, Y. Ando, B. Liang, R.L. Greene, D. Casa, T. Gog, Phys. Rev. Lett. **100**, 097001 (2008)
50. D.S. Ellis, J. Kim, J.P. Hill, S. Wakimoto, R.J. Birgeneau, Y. Shvyd'ko, D. Casa, T. Gog, K. Ishii, K. Ikeuchi, A. Paramekanti, Y.J. Kim, Phys. Rev. B **81**, 085124 (2010)
51. F.H. Vernay, M.J.P. Gingras, T.P. Devereaux, Phys. Rev. B **75**, 020403 (2007)
52. J. van den Brink, Europhys. Lett. **80**, 47003 (5pp) (2007)
53. T. Nagao, J. Ichi Igarashi, Phys. Rev. B **75**, 214414 (2007)
54. F. Forte, L.J.P. Ament, J. van den Brink, Phys. Rev. B **77**, 134428 (2008)
55. V. Bisogni, L. Simonelli, L.J.P. Ament, F. Forte, M. Moretti Sala, M. Minola, S. Huotari, J. van den Brink, G. Ghiringhelli, N.B. Brookes, L. Braicovich, Phys. Rev. B **85**, 214527 (2012)
56. J. Lorenzana, G.A. Sawatzky, Phys. Rev. Lett. **74**, 1867 (1995)
57. F.M.F. de Groot, P. Kuiper, G.A. Sawatzky, Phys. Rev. B **57**, 14584 (1998)
58. L. Braicovich, M. Moretti Sala, L.J.P. Ament, V. Bisogni, M. Minola, G. Balestrino, D. Di Castro, G.M. De Luca, M. Salluzzo, G. Ghiringhelli, J. van den Brink, Phys. Rev. B **81**, 174533 (2010)
59. M. Guarise, B. Dalla Piazza, M. Moretti Sala, G. Ghiringhelli, L. Braicovich, H. Berger, J.N. Hancock, D. van der Marel, T. Schmitt, V.N. Strocov, L.J.P. Ament, J. van den Brink, P.H. Lin, P. Xu, H.M. Rønnow, M. Grioni, Phys. Rev. Lett. **105**, 157006 (2010)
60. K. Ishii, K. Tsutsui, K. Ikeuchi, I. Jarrige, J. Mizuki, H. Hiraka, K. Yamada, T. Tohyama, S. Maekawa, Y. Endoh, H. Ishii, Y.Q. Cai, Phys. Rev. B **85**, 104509 (2012)
61. Y.J. Kim, J.P. Hill, S. Komiya, Y. Ando, D. Casa, T. Gog, C.T. Venkataraman, Phys. Rev. B **70**, 094524 (2004)
62. S. Wakimoto, Y.J. Kim, H. Kim, H. Zhang, T. Gog, R.J. Birgeneau, Phys. Rev. B **72**, 224508 (2005)
63. Y. Li, M. Hasan, H. Eisaki, S. Uchida, J. Phys. Chem. Solids **66**, 2207 (2005)
64. D.S. Ellis, J. Kim, H. Zhang, J.P. Hill, G. Gu, S. Komiya, Y. Ando, D. Casa, T. Gog, Y.J. Kim, Phys. Rev. B **83**, 075120 (2011)
65. K. Ishii, K. Tsutsui, Y. Endoh, T. Tohyama, K. Kuzushita, T. Inami, K. Ohwada, S. Maekawa, T. Masui, S. Tajima, Y. Murakami, J. Mizuki, Phys. Rev. Lett. **94**, 187002 (2005)
66. L. Lu, G. Chabot-Couture, X. Zhao, J.N. Hancock, N. Kaneko, O.P. Vajk, G. Yu, S. Grenier, Y.J. Kim, D. Casa, T. Gog, M. Greven, Phys. Rev. Lett. **95**, 217003 (2005)
67. S. Uchida, T. Ido, H. Takagi, T. Arima, Y. Tokura, S. Tajima, Phys. Rev. B **43**, 7942 (1991)
68. Y. Onose, Y. Taguchi, K. Ishizaka, Y. Tokura, Phys. Rev. B **69**, 024504 (2004)
69. R.S. Markiewicz, A. Bansil, Phys. Rev. Lett. **96**, 107005 (2006)
70. S. Basak, T. Das, H. Lin, M.Z. Hasan, R.S. Markiewicz, A. Bansil, Phys. Rev. B **85**, 075104 (2012)
71. C. Weber, K. Haule, G. Kotliar, Nat. Phys. **6**, 574 (2010)
72. C.J. Jia, C.C. Chen, A.P. Sorini, B. Moritz, T.P. Devereaux, New J. Phys. **14**, 113038 (2012)
73. T. Tohyama, J. Electron Spectrosc. Relat. Phenom. **200**, 209 (2015). Special Anniversary Issue: Volume 200
74. R.J. Birgeneau, C. Stock, J.M. Tranquada, K. Yamada, J. Phys. Soc. Jpn. **75**, 111003 (2006)
75. M. Fujita, H. Hiraka, M. Matsuda, M. Matsuura, J.M. Tranquada, S. Wakimoto, G. Xu, K. Yamada, J. Phys. Soc. Jpn. **81**, 011007 (2012)
76. M. Le Tacon, G. Ghiringhelli, J. Chaloupka, M.M. Sala, V. Hinkov, M.W. Haverkort, M. Minola, M. Bakr, K.J. Zhou, S. Blanco-Canosa, C. Monney, Y.T. Song, G.L. Sun, C.T. Lin, G.M. De Luca, M. Salluzzo, G. Khaliullin, T. Schmitt, L. Braicovich, B. Keimer, Nat. Phys. **7**, 725 (2011)

77. M. Minola, G. Dellea, H. Gretarsson, Y.Y. Peng, Y. Lu, J. Porras, T. Loew, F. Yakhou, N.B. Brookes, Y.B. Huang, J. Pelliciari, T. Schmitt, G. Ghiringhelli, B. Keimer, L. Braicovich, M. Le Tacon, Phys. Rev. Lett. **114**, 217003 (2015)
78. D.S. Ellis, Y.B. Huang, P. Olalde-Velasco, M. Dantz, J. Pelliciari, G. Drachuck, R. Ofer, G. Bazalitsky, J. Berger, T. Schmitt, A. Keren, Phys. Rev. B **92**, 104507 (2015)
79. M.P.M. Dean, G. Dellea, M. Minola, S.B. Wilkins, R.M. Konik, G.D. Gu, M. Le Tacon, N.B. Brookes, F. Yakhou-Harris, K. Kummer, J.P. Hill, L. Braicovich, G. Ghiringhelli, Phys. Rev. B **88**, 020403 (2013)
80. M.P.M. Dean, G. Dellea, R.S. Springell, F. Yakhou-Harris, K. Kummer, N.B. Brookes, X. Liu, Y.J. Sun, J. Strle, T. Schmitt, L. Braicovich, G. Ghiringhelli, I. Boović, J.P. Hill, Nat. Mater. **12**, 1019 (2013)
81. S. Wakimoto, K. Ishii, H. Kimura, M. Fujita, G. Dellea, K. Kummer, L. Braicovich, G. Ghiringhelli, L.M. Debeer-Schmitt, G.E. Granroth, Phys. Rev. B **91**, 184513 (2015)
82. M. Le Tacon, M. Minola, D.C. Peets, M. Moretti Sala, S. Blanco-Canosa, V. Hinkov, R. Liang, D.A. Bonn, W.N. Hardy, C.T. Lin, T. Schmitt, L. Braicovich, G. Ghiringhelli, B. Keimer, Phys. Rev. B **88**, 020501 (2013)
83. M.P.M. Dean, A.J.A. James, A.C. Walters, V. Bisogni, I. Jarrige, M. Hücker, E. Giannini, M. Fujita, J. Pelliciari, Y.B. Huang, R.M. Konik, T. Schmitt, J.P. Hill, Phys. Rev. B **90**, 220506 (2014)
84. M.P.M. Dean, A.J.A. James, R.S. Springell, X. Liu, C. Monney, K.J. Zhou, R.M. Konik, J.S. Wen, Z.J. Xu, G.D. Gu, V.N. Strocov, T. Schmitt, J.P. Hill, Phys. Rev. Lett. **110**, 147001 (2013)
85. M. Guarise, B.D. Piazza, H. Berger, E. Giannini, T. Schmitt, H.M. Rnnow, G.A. Sawatzky, J. van den Brink, D. Altenfeld, I. Eremin, M. Grioni, Nat. Commun. **5**, 5760 (2014)
86. K. Ishii, M. Fujita, T. Sasaki, M. Minola, G. Dellea, C. Mazzoli, K. Kummer, G. Ghiringhelli, L. Braicovich, T. Tohyama, K. Tsutsumi, K. Sato, R. Kajimoto, K. Ikeuchi, K. Yamada, M. Yoshida, M. Kurooka, J. Mizuki, Nat. Commun. **5**, 3714 (2014)
87. W.S. Lee, J.J. Lee, E.A. Nowadnick, S. Gerber, W. Tabis, S.W. Huang, V.N. Strocov, E.M. Motoyama, G. Yu, B. Moritz, H.Y. Huang, R.P. Wang, Y.B. Huang, W.B. Wu, C.T. Chen, D.J. Huang, M. Greven, T. Schmitt, Z.X. Shen, T.P. Devereaux, Nat. Phys. **10**, 883 (2014)
88. M.P.M. Dean, J. Magn. Mag. Mater. **376**, 3 (2015)
89. C.J. Jia, E.A. Nowadnick, K. Wohlfeld, Y.F. Kung, C.C. Chen, S. Johnston, T. Tohyama, B. Moritz, T.P. Devereaux, Nat. Commun. **5**, 3314 (2014)
90. L. Braicovich, M. Minola, G. Dellea, M. Le Tacon, M. Moretti Sala, C. Morawe, J.C. Peffen, R. Supruangnet, F. Yakhou, G. Ghiringhelli, N.B. Brookes, Rev. Sci. Instrum. **85**, 115104 (2014)
91. W. Chen, O.P. Sushkov, Phys. Rev. B **88**, 184501 (2013)
92. S. Sugai, S.I. Shamoto, M. Sato, Phys. Rev. B **38**, 6436 (1988)
93. V. Bisogni, M. Moretti-Sala, A. Bendounan, N.B. Brookes, G. Ghiringhelli, L. Braicovich, Phys. Rev. B **85**, 214528 (2012)
94. Y. Li, M. Le Tacon, Y. Matiks, A.V. Boris, T. Loew, C.T. Lin, L. Chen, M.K. Chan, C. Dorow, L. Ji, N. Barišić, X. Zhao, M. Greven, B. Keimer, Phys. Rev. Lett. **111**, 187001 (2013)
95. J.M. Tranquada, B.J. Sternlieb, J.D. Axe, Y. Nakamura, S. Uchida, Nature **375**, 561 (1995)
96. P. Abbamonte, A. Rusydi, S. Smadici, G.D. Gu, G.A. Sawatzky, D.L. Feng, Nat. Phys. **1**, 155 (2005)
97. G. Ghiringhelli, M. Le Tacon, M. Minola, S. Blanco-Canosa, C. Mazzoli, N.B. Brookes, G.M. De Luca, A. Frano, D.G. Hawthorn, F. He, T. Loew, M.M. Sala, D.C. Peets, M. Salluzzo, E. Schierle, R. Sutarto, G.A. Sawatzky, E. Weschke, B. Keimer, L. Braicovich, Science **337**, 821 (2012)
98. M. Hashimoto, G. Ghiringhelli, W.S. Lee, G. Dellea, A. Amorese, C. Mazzoli, K. Kummer, N.B. Brookes, B. Moritz, Y. Yoshida, H. Eisaki, Z. Hussain, T.P. Devereaux, Z.X. Shen, L. Braicovich, Phys. Rev. B **89**, 220511 (2014)
99. W. Tabis, Y. Li, M.L. Tacon, L. Braicovich, A. Kreyssig, M. Minola, G. Dellea, E. Weschke, M.J. Veit, M. Ramazanoglu, A.I. Goldman, T. Schmitt, G. Ghiringhelli, N. Barišić, M.K. Chan, C.J. Dorow, G. Yu, X. Zhao, B. Keimer, M. Greven, Nat. Commun. **5**, 5875 (2014)

100. R. Comin, A. Frano, M.M. Yee, Y. Yoshida, H. Eisaki, E. Schierle, E. Weschke, R. Sutarto, F. He, A. Soumyanarayanan, Y. He, M. Le Tacon, I.S. Elfimov, J.E. Hoffman, G.A. Sawatzky, B. Keimer, A. Damascelli, Science **343**, 390 (2014)
101. E.H. da Silva Neto, P. Aynajian, A. Frano, R. Comin, E. Schierle, E. Weschke, A. Gyenis, J. Wen, J. Schneeloch, Z. Xu, S. Ono, G. Gu, M. Le Tacon, A. Yazdani, Science **343**, 393 (2014)
102. E.H. da Silva Neto, R. Comin, F. He, R. Sutarto, Y. Jiang, R.L. Greene, G.A. Sawatzky, A. Damascelli, Science **347**, 282 (2015)
103. S. Wakimoto, H. Kimura, K. Ishii, K. Ikeuchi, T. Adachi, M. Fujita, K. Kakurai, Y. Koike, J. Mizuki, Y. Noda, K. Yamada, A.H. Said, Y. Shvyd'ko, Phys. Rev. Lett. **102**, 157001 (2009)
104. J. Kim, D.S. Ellis, H. Zhang, Y.J. Kim, J.P. Hill, F.C. Chou, T. Gog, D. Casa, Phys. Rev. B **79**, 094525 (2009)
105. J.N. Hancock, G. Chabot-Couture, Y. Li, G.A. Petrakovskii, K. Ishii, I. Jarrige, J. Mizuki, T.P. Devereaux, M. Greven, Phys. Rev. B **80**, 092509 (2009)
106. Y.J. Kim, J.P. Hill, F.C. Chou, D. Casa, T. Gog, C.T. Venkataraman, Phys. Rev. B **69**, 155105 (2004)
107. S. Suga, S. Imada, A. Higashiya, A. Shigemoto, S. Kasai, M. Sing, H. Fujiwara, A. Sekiyama, A. Yamasaki, C. Kim, T. Nomura, J. Igarashi, M. Yabashi, T. Ishikawa, Phys. Rev. B **72**, 081101 (2005)
108. K. Ishii, S. Ishihara, Y. Murakami, K. Ikeuchi, K. Kuzushita, T. Inami, K. Ohwada, M. Yoshida, I. Jarrige, N. Tatami, S. Niioka, D. Bizen, Y. Ando, J. Mizuki, S. Maekawa, Y. Endoh, Phys. Rev. B **83**, 241101 (2011)
109. J.N. Hancock, G. Chabot-Couture, M. Greven, New J. Phys. **12**, 033001 (2010)
110. C. Kim, A.Y. Matsuura, Z.X. Shen, N. Motoyama, H. Eisaki, S. Uchida, T. Tohyama, S. Maekawa, Phys. Rev. Lett. **77**, 4054 (1996)
111. M.Z. Hasan, Y.D. Chuang, Y. Li, P.A. Montano, Z. Hussain, G. Dhalenne, A. Revcolevschi, H. Eisaki, N. Motoyama, S. Uchida, Int. J. Mod. Phys. B **17**, 3519 (2003)
112. M.Z. Hasan, Y. Li, Y.D. Chuang, P.A. Montano, Z. Hussain, H. Eisaki, N. Motoyama, S. Uchida, Int. J. Mod. Phys. B **17**, 3513 (2003)
113. D. Qian, Y. Li, M. Hasan, D. Casa, T. Gog, Y.D. Chuang, K. Tsutsui, T. Tohyama, S. Maekawa, H. Eisaki, S. Uchida, J. Phys. Chem. Solids **66**, 2212 (2005)
114. K. Ishii, K. Tsutsui, T. Tohyama, T. Inami, J. Mizuki, Y. Murakami, Y. Endoh, S. Maekawa, K. Kudo, Y. Koike, K. Kumagai, Phys. Rev. B **76**, 045124 (2007)
115. L.C. Duda, J. Downes, C. McGuinness, T. Schmitt, A. Augustsson, K.E. Smith, G. Dhalenne, A. Revcolevschi, Phys. Rev. B **61**, 4186 (2000)
116. F. Bondino, M. Zangrando, M. Zacchigna, G. Dhalenne, A. Revcolevschi, F. Parmigiani, Phys. Rev. B **75**, 195106 (2007)
117. T. Learmonth, C. McGuinness, P.A. Glans, J.E. Downes, T. Schmitt, L.C. Duda, J.H. Guo, F.C. Chou, K.E. Smith, Europhys. Lett. **79**, 47012 (5pp) (2007)
118. C. Monney, V. Bisogni, K.J. Zhou, R. Kraus, V.N. Strocov, G. Behr, J.C.V. Málek, R. Kuzian, S.L. Drechsler, S. Johnston, A. Revcolevschi, B. Büchner, H.M. Rønnow, J. van den Brink, J. Geck, T. Schmitt, Phys. Rev. Lett. **110**, 087403 (2013)
119. Y. Mizuno, T. Tohyama, S. Maekawa, T. Osafune, N. Motoyama, H. Eisaki, S. Uchida, Phys. Rev. B **57**, 5326 (1998)
120. K. Okada, A. Kotani, Phys. Rev. B **63**, 045103 (2001)
121. M.Z. Hasan, P.A. Montano, E.D. Isaacs, Z.X. Shen, H. Eisaki, S.K. Sinha, Z. Islam, N. Motoyama, S. Uchida, Phys. Rev. Lett. **88**, 177403 (2002)
122. M.Z. Hasan, Y.D. Chuang, Y. Li, P. Montano, M. Beno, Z. Hussain, H. Eisaki, S. Uchida, T. Gog, D.M. Casa, Int. J. Mod. Phys. B **17**, 3479 (2003)
123. Y.J. Kim, J.P. Hill, H. Benthien, F.H.L. Essler, E. Jeckelmann, H.S. Choi, T.W. Noh, N. Motoyama, K.M. Kojima, S. Uchida, D. Casa, T. Gog, Phys. Rev. Lett. **92**, 137402 (2004)
124. J.W. Seo, K. Yang, D.W. Lee, Y.S. Roh, J.H. Kim, H. Eisaki, H. Ishii, I. Jarrige, Y.Q. Cai, D.L. Feng, C. Kim, Phys. Rev. B **73**, 161104 (2006)
125. L. Wray, D. Qian, D. Hsieh, Y. Xia, H. Eisaki, M.Z. Hasan, Phys. Rev. B **76**, 100507 (2007)

126. K. Tsutsui, T. Tohyama, S. Maekawa, Phys. Rev. B **61**, 7180 (2000)
127. K. Okada, A. Kotani, J. Phys. Soc. Jpn. **75**, 044702 (2006)
128. T. Nomura, J. Ichi Igarashi, J. Phys. Soc. Jpn. **73**, 1677 (2004)
129. J. Schlappa, K. Wohlfeld, K.J. Zhou, M. Mourigal, M.W. Haverkort, V.N. Strocov, L. Hozoi, C. Monney, S. Nishimoto, S. Singh, A. Revcolevschi, J.S. Caux, L. Patthey, H.M. Ronnow, J. van den Brink, T. Schmitt, Nature **485**, 82 (2012)
130. K. Wohlfeld, S. Nishimoto, M.W. Haverkort, J. van den Brink, Phys. Rev. B **88**, 195138 (2013)
131. V. Bisogni, K. Wohlfeld, S. Nishimoto, C. Monney, J. Trinckauf, K. Zhou, R. Kraus, K. Koepernik, C. Sekar, V. Strocov, B. Büchner, T. Schmitt, J. van den Brink, J. Geck, Phys. Rev. Lett. **114**, 096402 (2015)
132. V. Bisogni, S. Kourtis, C. Monney, K. Zhou, R. Kraus, C. Sekar, V. Strocov, B. Büchner, J. van den Brink, L. Braicovich, T. Schmitt, M. Daghofer, J. Geck, Phys. Rev. Lett. **112**, 147401 (2014)
133. J.J. Lee, B. Moritz, W.S. Lee, M. Yi, C.J. Jia, A.P. Sorini, K. Kudo, Y. Koike, K.J. Zhou, C. Monney, V. Strocov, L. Patthey, T. Schmitt, T.P. Devereaux, Z.X. Shen, Phys. Rev. B **89**, 041104 (2014)
134. W.S. Lee, S. Johnston, B. Moritz, J. Lee, M. Yi, K.J. Zhou, T. Schmitt, L. Patthey, V. Strocov, K. Kudo, Y. Koike, J. van den Brink, T.P. Devereaux, Z.X. Shen, Phys. Rev. Lett. **110**, 265502 (2013)
135. E. Dagotto, T.M. Rice, Science **271**, 618 (1996)
136. E. Dagotto, J. Riera, D. Scalapino, Phys. Rev. B **45**, 5744 (1992)
137. M. Kato, K. Shiota, Y. Koike, Phys. C **258**, 284 (1996)
138. T. Osafune, N. Motoyama, H. Eisaki, S. Uchida, Phys. Rev. Lett. **78**, 1980 (1997)
139. M. Uehara, T. Nagata, J. Akimitsu, H. Takahashi, N. Môri, K. Kinoshita, J. Phys. Soc. Jpn. **65**, 2764 (1996)
140. K.M. Kojima, N. Motoyama, H. Eisaki, S. Uchida, J. Electron Spectrosc. Relat. Phenom. **117–118**, 237 (2001)
141. A. Higashiya, S. Imada, T. Murakawa, H. Fujiwara, S. Kasai, A. Sekiyama, S. Suga, K. Okada, M. Yabashi, K. Tamasaku, T. Ishikawa, H. Eisaki, New J. Phys. **10**, 053033 (2008)
142. M. Yoshida, K. Ishii, K. Ikeuchi, I. Jarrige, Y. Murakami, J. Mizuki, K. Tsutsui, T. Tohyama, S. Maekawa, K. Kudo, Y. Koike, Y. Endoh, Phys. C **470**, S145 (2010)
143. M. Yoshida, K. Ishii, I. Jarrige, T. Watanuki, K. Kudo, Y. Koike, K. Kumagai, N. Hiraoka, H. Ishii, K.D. Tsuei, J. Mizuki, J. Synchrotron Radiat. **21**, 131 (2014)
144. J. Schlappa, T. Schmitt, F. Vernay, V.N. Strocov, V. Ilakovac, B. Thielemann, H.M. Ronnow, S. Vanishri, A. Piazzalunga, X. Wang, L. Braicovich, G. Ghiringhelli, C. Marin, J. Mesot, B. Delley, L. Patthey, Phys. Rev. Lett. **103**, 047401 (2009)
145. X. Gao, C. Burns, D. Casa, M. Upton, T. Gog, J. Kim, C. Li, Rev. Sci. Instrum. **82**, 113108 (2011)
146. K. Ishii, I. Jarrige, M. Yoshida, K. Ikeuchi, T. Inami, Y. Murakami, J. Mizuki, J. Electron Spectrosc. Relat. Phenom. **188**, 127 (2013)
147. S. Ishihara, S. Ihara, J. Phys. Chem. Solids **69**, 3184 (2008)
148. M. van Veenendaal, X. Liu, M.H. Carpenter, S.P. Cramer, Phys. Rev. B **83**, 045101 (2011)
149. J. Deisenhofer, I. Leonov, M.V. Eremin, C. Kant, P. Ghigna, F. Mayr, V.V. Iglamov, V.I. Anisimov, D. van der Marel, Phys. Rev. Lett. **101**, 157406 (2008)
150. T. Nomura, J. Phys. Soc. Jpn. **83**, 064707 (2014)
151. D. Ketenoglu, M. Harder, K. Klementiev, M. Upton, M. Taherkhani, M. Spiwek, F.U. Dill, H.C. Wille, H. Yavaş, J. Synchrotron Radiat. **22**, 961 (2015)
152. S. Moser, S. Fatale, P. Krüger, H. Berger, P. Bugnon, A. Magrez, H. Niwa, J. Miyawaki, Y. Harada, M. Grioni, Phys. Rev. Lett. **115**, 096404 (2015)
153. P. Marra, S. Sykora, K. Wohlfeld, J. van den Brink, Phys. Rev. Lett. **110**, 117005 (2013)